SOLDERING HANDBOOK

FOR

PRINTED CIRCUITS

AND

SURFACE MOUNTING

SOLDERING HANDBOOK

FOR

PRINTED CIRCUITS

AND

SURFACE MOUNTING

Design, Materials, Processes,
Equipment, Trouble-Shooting,
Quality, Economy, and
Line Management

HOWARD H. MANKO
Manko Associates Inc.
Teaneck, New Jersey

VNR VAN NOSTRAND REINHOLD
_____ New York

Library of Congress Catalog Card Number 85-31449
ISBN 0442-26423-2

Manufactured in the United States of America

Published by Van Nostrand Reinhold
115 Fifth Avenue
New York, New York 10003

Van Nostrand Reinhold International Company Limited
11 New Fetter Lane
London EC4P 4EE, England

Van Nostrand Reinhold
480 La Trobe Street
Melbourne, Victoria 3000, Australia

Nelson Canada
1120 Birchmount Road
Scarborough, Ontario M1K 5G4, Canada

15   14   13   12   11   10   9   8   7   6   5   4

Library of Congress Cataloging-in-Publication Data

Manko, Howard H.
  Soldering handbook for printed circuits & surface
mounting.

  Bibliography: p.
  Includes index.
  1. Printed circuits — Handbooks, manuals, etc.
2. Solder and soldering — Handbooks, manuals, etc.
I. Title.
TK7868.P7M36   1986   621.381'74   85-31449
ISBN 0-442-26423-2

As always to Mira

PREFACE

The printed circuit industry has achieved maturity and universal acceptance. No known interconnection technology threatens to render it obsolete in the foreseeable future. It offers two unique advantages that are important for any assembly technology: quality (reliability) and economy.

The mode of component attachment to printed circuit boards, however, is undergoing a radical change. Technical and economic pressures are forcing the industry to convert some or all of its assembly to surface-mounting techniques. We are moving away from the traditional large through-the-hole connection with its mechanical security. It is being replaced by a small surface butt and/or lap joint, sometimes with no added mechanical support to the solder. This change requires a complete reassessment of design, production, and inspection techniques. A major portion of this book is devoted to the changes imposed by surface mounting. This recent development is an extension of the established hybrid (thick- and thin-film) industry. Yet when it is applied to conventional printed circuits, there are major differences.

One must view the printed circuit board as a planar surface designed to provide interconnections between electronic devices. The electronic industry is using them for mass-production techniques to join discrete, integrated, and special components (leaded and leadless). This book applies to all board variations including single-sided, double-sided, multi-layer, and flexible circuits.

This work concentrates on assembly, soldering, and cleaning, with emphasis on manufacturing techniques. Unfortunately, space limitations prevent us from dealing with the full intricacies of printed circuit board fabrication. We will, however, touch on those aspects that have a direct impact on the final quality of the solder joint, as well as its inspection and repair: board design and material considerations for reliable soldering, assembly techniques and their impact on joining, and management techniques. For obvious reasons, soldering is closely related to cleaning. Flux removal is only part of the true need for clean assemblies.

Space does not permit an in-depth discussion of solder technology. For a detailed study of this science, the reader is referred to a comprehensive soldering text by the author.* We will, however, include enough description of soldering practices to make this book a self-contained unit. Our main emphasis will be on methods of design and material selection, mass production processing, touchup, and repair techniques.

The quality of the final printed circuit assembly is not neglected. There is a chapter on quality standards and inspection techniques. It takes the reader from incoming inspection through process control to final inspection.

Finally, the book addresses important management issues of cost and control. In light of aggressive international competition, these are crucial problems. The health and future of the electronics industry in general and printed circuit manufacturing in particular depends on our ability to produce high-reliability products at the lowest possible cost.

Howard H. Manko

---

* *Solders and Soldering,* 2nd ed., McGraw-Hill Book Co., New York, 1979.

ACKNOWLEDGMENTS

No one has an internal source of information; we all gain our knowledge from a series of other people: parents, teachers, friends, co-workers, clients. We are enriched by reading and learning from a chain of many previous intellects, each one adding a little of themselves to the synthesis of the human experience.

Thus, it is truly impossible to acknowledge the numerous contributors to the present book, especially since many are not obvious even to the conscious mind. However, the author wishes to thank his colleagues at Alpha Metals, Hollis Automation, and Hexacon Electric for the specific help they have given in the preparation of this book. Special mention should go to Ronald A. Bulwith, who has prepared many of the cross-sections presented herein.

An attempt has been made to represent many diverse pieces of equipment. The author wishes to emphasize that no endorsement of any models or makes is implied in their presentation. He has selected an extensive cross-section of those pictures vendors were kind enough to provide and which seemed appropriate to back up the text.

CONTENTS

SOLDERING HANDBOOK

FOR

PRINTED CIRCUITS

AND

SURFACE MOUNTING

1
DESIGN FOR GOOD SOLDERING AND CLEANING

1-1 PRINTED CIRCUIT PROCESSING—A LOOK INTO THE FUTURE

There are a number of basic methods used to produce the printed circuit board. The processes used to generate the image and the impact on soldering are important. Let us review them from the historical vantage point.

The oldest technique used is called *subtractive* and can be outlined as follows. The printed circuit board is made from a copper- clad laminate that is either single- or double-sided. A protective pattern is deposited on the copper surface, and the unwanted copper is etched away hence the name subtractive, which fits this method well. This process can be further divided according to the type of *etch resist* used.

On single sided boards and single layers of a multilayer board, the etch resist may be an organic coat (applied by screening or photographic techniques). After the exposed unwanted copper is etched away, the resist coating is removed. This leaves a bare copper pattern on the boards. This simple technique is also referred to as *print and etch*.

When a plated-through hole is generated, however, a tin-lead plating is used as the etch resist. This is the most common process for all double sided and multilayer boards. These boards end up with a tin-lead plating on the surface of the lands and pads. This solder coat is usually reflowed for technical reasons (see Section 1-24).

A more recent technique develops the metallic pattern on a laminate which is not copper clad. This method is referred to as the *additive* process. All boards made this way end up as bare copper. Coating them with solder is an additional unrelated step to the pattern development.

Today environmental impact is a major issue. The subtractive etching process requires a larger amount of effluent treatment relative to the

1

additive process, which makes it more expensive. If it were not for the substantial investment made in older subtractive equipment and technology, more additive boards would be available. This situation incidentally, gave rise to an intermediate process called the *ultra-thin* or *semi additive* method. Here an extremely thin layer of copper is clad to the surface, protected by an easily removable coating like aluminum. After the top aluminum layer has been stripped, the pattern is build up with more copper. The whole structure is then minimally etched, which removes the thin layer of unwanted copper. A totally bare copper board is thus generated, using the conventional subtractive process.

Several board properties are important for soldering. One is the type of metallic trace exposed to the solder mask (see Section 3-16). The ideal combination is a nonfusible coating like copper under the mask, often referred to as *solder mask over bare copper* (SMOBC). This guarantees that the solder resist will not shift during soldering, preventing flux entrapment, flaking and other problems. We are also interested in the quality of the adhesion of the copper to the laminate, termed the *copper peel strength*. This can be directly related to the joint strength and to thermal damage during soldering like lifted pads and measling (see Section 7-26). Finally, we are concerned with the quality of the laminate surface after etching. This affects the cleaning process and the board's ability to absorb chemicals which tend to lower the surface insulation properties.

It is always difficult to predict future developments, but there are obvious trends. Cheaper laminates, thinner boards, and polymer thick films will help reduce cost. Technical improvements like higher temperature materials, higher insulation resistance, and modified coefficients of expansion are on the horizon. The major foreseeable changes, however, will be in two areas: *fine line* dense boards and *surface mounted devices* (SMD). Both strive to increase package density and reduce the number of total interconnections. This will raise the price of the finished board per square inch while lowering total assembly and device costs.

Fine line circuitry suits low current circuitry because surface insulation resistance becomes the limiting factor. We have the technology to make 2 mil lines on 2 mil centers, but today's consensus regards this as the lower limit. Surface mounting is discussed throughout the book, especially in Chapter 5.

Surface mounting has necessitated additional board materials and construction, with a controlled thermal expansion in the X and Y directions. These materials must match or at least come close to the *thermal expansion coefficient* (TCE) of the surface mounted devices. Otherwise, the solder joints with their diminished size and strength, cannot withstand large temperature excursions. Increased and modified resin coatings of the laminate offer one solution. Different re- inforcing fibers in the weave or internal layers of stabilizing metals can also modify board properties in this direction. Finally, changes in the metal layers laminated to the board also promise good TCE matching.

1-2 THE IMPACT OF BOARD MATERIAL

The grade of plastic used in the laminate has a great effect on the soldering process. To date, boards made of all the available resin grades can be readily soldered. The grade selected for a particular application usually represents a trade-off between price (economy) and scientific or engineering requirements (quality). Once a material has been selected, it is easy to set soldering parameters compatible with the laminate.

Here are the most important properties to consider during material selection. They affect both soldering and final quality, and are listed in order of importance:

1. *Warpage.* Unfortunately, the flatness of the printed circuit board in all stages of production is not easily maintained. It is normally due to residual internal stresses and/or the effect of additional mechanical operations. The lower cost laminates are more susceptible to warpage that starts in the drilling and etching stage. With multilayer, boards special stabilizing baking operations are needed to relieve the lamination stresses. The soldering process also induces warpage for obvious reasons.

   Since warpage always results in higher production costs, one must carefully weigh the material savings against this unnecessary labor increase. A true cost analysis will often reveal that a better grade laminate is actually less expensive and improves product performance.

   Warpage is also a problem in postsolder processing. Automatic lead cutting and testing are operations that depend on board flatness. The ability to install or plug the board into the equipment is another factor to consider.

   In surface mounting, the mechanical stability of the board becomes more important as compared to through-the-hole bonding. The accuracy of SMD device placement, the reproducibility of joint formation, and stresses under use conditions dictate stable geometries.

   The width of the board, which is a design function, also has an effect on warpage. Board thickness can be related to its ability to support its own weight. If the inherent board properties are exceeded, the printed circuit will also flex while in use. Boards 0.062 in. (1.6 mm) thick stay stable up to 7-9 in. (17.5-22.8 cm). Thick laminates can be wider, while thin laminates are always subject to warpage. When the boards are too wide, they tend to warp even without the weight of their components. The structure is further weakened by nonuniform etch patterns and hole location. It is sometimes possible to add permanent or temporary stability through the use of stiffeners (see Section 3-18). In some cases, a stabilizing bake above the glass transition temperature will increase board flatness.

2. *Surface resistance as a function of absorption and contamination.* In the design stage, it is easy to choose a material from a handbook for its surface insulation properties. These values, however, are meaningful only for ideal conditions, with clean surfaces, and no adsorbed or absorbed chemicals. Such virgin surfaces are seldom achievable in industry because of the contamination that is collected after assembly and in use. Remember that:

- Processing solutions (mostly organic solvents) may be absorbed into the surface, lowering insulation resistance with increased humidity.
- Conductive (mostly ionic) contamination collects on the surface due to handling, storage and in-use circumstances. Again, under humid conditions surface resistivity is impaired.
- Underetching, or solder smears from mechanical handling, will effectively shorten the gap between adjacent conductors.

  Thus the designer or materials engineer should consider the level of contamination his boards will encounter. High-impedance boards, as well as high-voltage designs, may require the use of conformal coatings. The use of a solder resist may offer another partial solution. In both cases, the coatings must not be applied over contaminated surfaces.

3. *Hole-generating ability.* When a material is selected, its ability to be punched or drilled becomes paramount. This is particularly important for plated-through hole quality. A rough hole exterior can cause difficulties in the plating process. This leads to a variety of problems in the eventual use of the plated-through hole during assembly and soldering. If crevices are formed during punching or drilling, they become saturated with process solutions. The plated wall covers these faults, making them unwashable. Later the heat of the soldering operation will cause the trapped liquid to rupture the plated wall and blowholes to form. This could also cause the simultaneous release of the corrosive chemicals (see Section 7-17).

Table 1-1. Cost Comparison
for Printed Circuits.
(Cents per Hole)

BOARD TYPE	PROCESS	COST f
Single side	Punched	1.0
Double sided	Punched	2.5
	Drilled	5.2
Multi-layer		
4 Layers	Drilled	22.0
10 Layers	Drilled	120.0
>10 Layers	Drilled	380.0

All calculations are for glass epoxy, without
fine lines or small holes, and based on operations and scrap cost (1984).

   Obviously, cost plays a major role in this design decision. For a
general comparison, see Table 1-1. The cheaper punched holes are
not recommended for plated-through hole use.
   The drilling production parameters are also critical. Drill speed
(RPM), number of hits per minute, number of boards per stack,
cooling metal entry and backup layers, and other factors all increase
the cost but improve the quality. A good designer will never leave
such decisions to the discretion of the board producer or the purchasing department, but will specify them. Although there is a way
of cleaning a rough hole, called *etch back*, this too increases cost.
4. *Thermal coefficient of expansion (TCE).* This vital property varies
   from one board material to another and with the laminate manufacturer. The situation is further complicated by the differences in value
   for the three dimensional directions of the same material. The planar
   directions (X and Y axes) are generally smaller than the perpendicular direction (Z axis) through the material. This is due to the construction of the laminate, which consists of a structural weave and a
   resin. The X and Y directions are influenced by the properties of the
   weave and the resin. The Z direction is affected only by the resins
   used. Thus all assemblies which might undergo many thermal cycles
   should be designed with a full understanding of this problem.
   Table 1-2 gives the average TCE values for some standard printed
circuit materials, in the useful temperature range, all below their
glass transition temperature (T_g). Keep in mind that the values of
the TCE increase dramatically above their T_g. The table also lists the
TCE of typical metallic conductors from which component leads are
made. A quick review of the values will help dramatize the impor-

Table 1-2. Linear Coefficients of Thermal Expansion. (Values Reported In 10^{-6} inch/inch/°C.)

PRINTED CIRCUIT BOARD MATERIALS
(Average values[a]—20°C–190°C)
$\times 10^{-6}$ inch/inch/°C.

TYPE	$T_g{}^b$	X	Y	Z^c
FR- 2	173°C	17.5	30.9	225.4
FR- 3	125°C	18.1	18.6	132.0
FR- 4	140°C	9.5	11.1	146.5
FR-45	132°C	15.0	13.0	71.0
PC-45	120°C	19.0	11.0	208.0
G -10	110°C	Neg.	Neg.	164.0
XXXP	130°C	6.0	16.0	219.0

[a] All values may deviate 20%.
[b] Above glass transition temperature double these values.
[c] Thickness of material.

METALS AND ALLOYS
(Temperature range—0°C–200°C)
$\times 10^{-6}$ inch/inch/°C.

NAME	SYMBOL	COEFFICIENT
Eutectic solder	SN63	24.56
Aluminum	Al	23.80
Copper	Cu	16.80
Gold	Au	14.20
Iron	Fe	12.10
Kovar[a]	n/a	5.00
Iron-Nickel[a]	Alloy 42	4.30

[a] Matched for glass to metal sealing.

CONFORMAL COATINGS
(Temperature range from −55°C to +71°C)
$\times 10^{-5}$ inch/inch/°C.

NAME	COEFFICIENT
Polyurethane	10–20
Epoxy	4,5–6.5

tance of proper joint design. Also, note the TCE of the resins normally used for SMD adhesives and conformal coatings.

The TCE is a parameter of vital importance to assemblies with surface mounted components (see Section 5-3). The laminate values for the X and Y planes must be linked to the device expansion. The full effect of any mismatch is obviously directly proportional to the component dimensions. Furthermore, as the component size increases, solder joint dimensions decrease as a function of geometry. Thus the smaller the joint, the bigger the stress, and that limits the device size for every mismatch.

1-3 THE EFFECT OF SURFACE MOUNTING

The diminutive size of the surface mounted component is not the only reason for board design changes. As we will see later, the number of *input/output* (I/O) interconnects per device also increases. As a result we need what is referred to as a *fine line board*. It is hard to define such a constantly changing technology as fine line circuitry. We have been moving towards more closely spaced boards for years, using a variety of concepts. In that respect we have already achieved 5 mil lines on 5 mil centers, with plated-through hole diameters as small as 6 mil. The consensus today puts 2 to 3 mil lines, with 2 to 3 mil centers as a theoretical limit to the print, plate, and etch technology. With rotating drills we look at a hole limit in the 5 to 6 mil range. Laser technology may some day permit even smaller holes, but there are still some uniformity limitations in laser drilling today.

Let us analyze the impact of the nature of surface mounted components and the high densities on the printed circuit board.

I. First, there is an effect on the laminate materials, for processing and the end use. Here we need:

- Closer spacing
- Narrower conductors
- Planar stability
- Controlled TCE
- Smaller holes

These are examples of the dramatic changes needed. They affect board material, design and processing. In many cases they push the state of the art.

The effects of these changes on the board can be further divided into:

1. The materials of construction
2. The metallurgy of the conductors
3. The mechanical board construction

Let us look at each of these elements briefly:

1. *The materials of construction* have to be different then the historical laminates we have been using. Such common materials as FR4 (glass epoxy) may suffice for the simpler low density single- and double-sided board surface mounting applications. For more complex and multilayer boards, materials with a controlled TCE are needed, and many are in the development stages.

The reduced spacing between conductors also requires higher surface insulation resistance. In addition, we need closer dimensional stability under temperature and humidity cyclic conditions. This problem may be solved by the use of different resin systems like Bismaleimide-Triazine or Polyimide. The reeenforcing materials may also change from glass to Kevlar, quartz, or graphite etc. (Ref. 1-2).

2. *The metallurgy of the board conductors* is another important issue. The tin-lead deposit over copper that is the common finish in the subtractive process is not really suitable for high density surface mounting. One major reason is the solder mask needed on fine line circuitry (for increased surface insulation resistance and ease of soldering). Placing a solder mask over a fusible surface causes shifting during soldering, with possible shorts and chemical entrapment (see sections 1-21, 3-16, and 7-30). This applies to plated and fused finishes, though leveled boards fare somewhat better because of the limited amount of solder left on the surface.

Stripping the solder plating from subtractive boards or switching to additive technology solves the mask problem. It enables us to use the solder mask over bare copper (SMOBC), but requires additional steps to protect the solderability of the exposed pads. This protection can be achieved by air-leveling the masked board, and thus reintroducing solder. Studies of nonfusible and solderable finishes are presently underway.

3. *Mechanical board construction* refers to the anticipated shift to more complex types of construction. As boards become smaller and denser, we will see a shift from double-sided to multilayer boards. We will also see a trend away from surface conductors, leaving pads only on the outside layers.

At first glance, all the above steps would appear to increase the cost of the end product. Table 1-1 shows a typical cost factor escalation with board complexity. But there are several other factors that counter the price increase, such as reduced board size, lower cost components, simplified assembly, and so on.

II. There are also some very important consequences in the production of fine line boards. These can be roughly divided into:

1. Small hole drilling
2. Large aspect ratio plating
3. Precision etching

These changes not only require new and improved production equipment but also push the state of the art. Not every shop will be in a position to provide reproducible, high reliability boards. Let us discuss each item in detail:

1. *Small hole drilling* has been made possible by new and improved machines and drills. Holes of approximately 0.006 in. (0.15 mm) can be drilled under careful conditions. They are obviously used only as via holes to connect surfaces and/or inner layers, as well as for engineering change and repair applications. Their price is high, and they should be used only if needed.

The larger holes needed for pinned or leaded components are still smaller than those on conventional boards. They require precision drilling since the quality of the hole greatly affects plating, soldering, and long range reliability. Accuracy of hole location is also critical for high-pin-count components. This is usually beyond the capability of the older equipment now in use.

2. *Large aspect ratio plating* refers to the ability to produce a uniform adherent coating inside a long, narrow hole. As boards get thicker and the diameter becomes smaller, the aspect ratio (hole height to diameter) increases. This poses serious problems in electro-plating because of current distribution, solution depletion in the hole, and circulation difficulties. Similar problems exist in electroless plating processes. The average plater has much to learn in this area.

Plating quality in the hole is as important to reliability as drilling. In addition, it provides the vital interconnects for multi- layer boards. The plated barrel lies in the vulnerable Z direction, where the TCE is largest, and it has to withstand the rigors of assembly, soldering, shipping, and in-use stresses.

3. *Precision etching* is needed in fine line circuitry to avoid shorts, insufficient spacing, reduced conductor width, and undercutting. The narrow conductors and small spacing leave little room for errors. Here the common use of 1 oz copper [0.0014 in. (0.035 mm) without plating buildup] is no longer practical.

As a rule of thumb the average undercutting ratio is 1 : 1 per side (undercut section = the height of the total copper). Thus a 1 oz copper board with a 0.001 in. (0.025 mm) copper plating will have a potential undercut of 0.0024 in. (0.06 mm). This can be tolerated on conventional boards but is not acceptable for fine-line circuitry for obvious reasons.

Similar problems exist in underetching, where surface insulation is affected. No residual copper specs between closely spaced lines can be allowed.

Thinner copper laminates can be used in the semi-additive process, but a complete change to additive processing may be more economical. Either process reduces etching difficulties and, at the same time solves many precision imaging problems.

III. The design of this new generation of printed circuits poses a number of new problems. Material compatibility with the devices and the process depends on:

- Dimensional stability (to maintain tolerances)
- Electrical parameters (impedance, capacitance, Rf..)
- Assembly techniques (mounting, soldering, cleaning, testing and inspection)
- Cost justification

We will cover some of the solder joint and cleaning considerations later in this book.

In summary, fine-line boards are different from today's printed circuits. They require a fresh approach in terms of material selection, production processes, and equipment. Because only a small number of board manufacturing facilities are in a position to generate them today, the full utilization of surface mounting will be hampered for a while. Once the industry has adjusted to the new conditions and has had time to control the process, we can move forward.

For this and other practical reasons, in the near term SMD devices will be mounted on conventional boards with lower densities. We will gain an increase in total component density by adding surface mounted components to the unused bottom of the board. This trend will be driven by the cost pressures and production capabilities.

With time, however, more truly fine-line boards will be used for a better utilization of surface mounting. It now appears that only a percentage of future boards will be populated with surface mounted devices only. There is a range of components that may never be suitable for surface mounting. There are devices that do not lend themselves to simple surface mounting techniques. Size and weight, for example, make surface mounting without mechanical anchorage impossible (transformers, coils, relays, etc.). In addition, manually operated devices like switches must be secure to withstand manipulation.

1-4 A WORD ABOUT SOLDER JOINT DESIGN

The detailed design of a solder joint is a straight forward proposition. It is covered in great detail in Chapter 4, pp 124-164 in "SOLDERS AND SOLDERING" (Ref. 1-1). This design outline covers material selection and compatibility checks, electrical and mechanical calculation for the 14 joint configurations, and much more. We will not duplicate any of that material, but concentrate on the plated-through hole joint and four typical SMD joints.

In general, each solder joint must meet six major design criteria (see list below). These can be divided into three material-oriented factors, which can be modified by selecting a different material system, and three design-oriented parameters, which affect reliability and cost. The designer must pay close attention to all these details to achieve fast, efficient soldering.

Unfortunately, solder joints are seldom designed, they just happen. Once it is decided to use a combination of boards, through holes, and/or surface mounting, the joint configuration is fixed. The size and shape of the pad may be the only items a designer may consider for the solder joint. The same holds true for the materials selected, and the tendency is to use familiar items. The problems are left to production and quality control personnel to resolve. This chapter, therefore, can serve as an excellent tool to help an organization to rationalize problems on existing designs. For example, it can help them decide if production imperfections are tolerable, and determine which ones need rework. It also points to simple design changes that would solve difficult process problems.

The solder joint design requirements are:

1. MATERIAL ORIENTED
   a. Electrical conductivity
   b. Mechanical stability
   c. Heat dissipation

2. DESIGN ORIENTED
   a. Ease of manufacturing
   b. Simplicity of repair
   c. Fully inspectable

We will use this design list in our discussion of the specific joints throughout this chapter.

1-5 ANALYSIS OF THROUGH-THE-HOLE JOINT CONFIGURATION

We shall start with the traditional leaded component joint to the board through a hole. The solder fillet is a function of joint structure and varies from single sided boards to plated-through holes. It also depends on lead wire arrangement and differs from straight through to bent-over leads. The selection of a single-sided versus a double-sided plated-through board is seldom based on the quality of solder joints. The decision to use a bent-over versus a straight lead must be made with great care. The analysis that follows will help the reader determine whether his design meets his own assembly's physical needs.

This analysis is also a very powerful tool in the decision making stage during the process of setting up workmanship standards (see Section 8-2). Acceptance criteria can be directly related to a calculation of how much imperfection can be tolerated without affecting the integrity of the solder joint. This, in turn can be translated into meaningful touch-up and repair procedures on the production line (see Chapter 9).

The material presented here is basically analytical and is concerned only with pure tensile forces for strength and electrical conductivity. Following the same rationale, the designer can calculate other properties with which he is concerned (shear strength, resistance to vibration, electrical impedance, etc.).

To simplify the analysis, five separate zones have been identified on solder fillets; they are listed in Fig. 1-1. Zones I to IV are shown on a bent lead placed in a plated-through hole. Zone V represents the base for a single-sided printed circuit board with a straight through lead. A combination of these five zones can thus represent any known configuration of solder joints as follows:

- Plated-through hole with bent over lead;
  Zones I, II, III and IV
- Plated-through hole with straight-through lead;
  Zones V, III, and IV
- Single sided board with bent over lead;
  Zones I and II
- Single sided board with straight through lead;
  Zone V

Each zone is analyzed separately. Thus the reader can adapt those zones which apply to him without any difficulties. For example, the analysis can be used to calculate if a plated-through hole fillet has to exceed 75% of board thickness, an imperfection allowed by the IPC (Ref. 1-3),

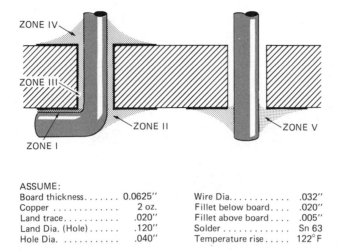

ASSUME:

Board thickness	0.0625″	Wire Dia.	.032″
Copper	2 oz.	Fillet below board	.020″
Land trace	.020″	Fillet above board	.005″
Land Dia. (Hole)	.120″	Solder	Sn 63
Hole Dia.	.040″	Temperature rise	122°F

Fig. 1-1. A schematic of the average fillet used for this calculation.

and many Government documents. In this case, one merely calculates the properties required (strength, conductivity, etc.) for the appropriate fillet sections. In our example, just omit zone IV completely, take 75% of zone III, and either zones I and II (for bent-over leads) or zone V (for a straight connection). In the total fillet evaluation, all zones are arithmetically added.

All assumptions made and specified in this book were conservative and based on minimum inspection criteria. A set of dimensions was arbitrarily selected which represents an average structure of a good solder fillet. Since any minor changes in dimensions may affect the conductivity and strength of the solder joint, the reader is warned not to use these figures without referring to the dimensions. Figure 1-1 defines explicitly the conditions used in our example.

Let us review each of the zones and analyze their unique characteristics. Their benefits must be weighed against their shortcomings. This will help shape our thinking about quality and reliability.

1-6 THE ADVANTAGES OF A CLINCHED LEAD

Zone I represents the important component of a clinched lead (Fig. 1-2). This part of the fillet is in reality a lap joint between the bent wire and the printed circuit board. By itself, a lap solder joint is a preferred configuration for mechanical strength only, because the solder in a lap joint is stronger than in any other stress mode. However, in the total stress

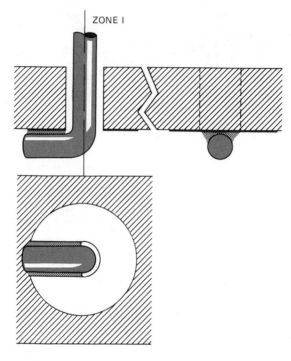

Fig. 1-2. A schematic of Zone I.

analysis of a plated-through hole, this does not hold true. The bent-over lead wire does not allow any degree of freedom in the upper direction, causing problems under slow cycle fatigue conditions. Note that zone I is the only joint that is entirely independent of the hole-to-wire ratio.

Zone I as defined in this analysis does not include the filled portion of the hole, which is labeled as zone II. If zone I alone is used, without zone II, we would have half or a partially open hole. This is sufficient for conductivity and strength purposes. We must only make sure that there is good wetting of the pad and the wire. Furthermore, the solder must extend at least half of the circumference of the hole, in order to give the copper foil added peel strength.

On the average printed circuit board, the lands and pads are less conductive than the component lead. The copper peel strength is also weaker than the lead wire. Based on these facts, let us analyze the conductivity and strength of zone I. Obviously we will compare it to the board, not to the component termination. In those cases where the reverse is true, it is easy to use the same process to adjust the interpretation.

The calculations given here are based on design guidelines and a mathematical methodology developed in "SOLDERS AND SOLDERING" Chapter 4. (Ref. 1-1). The following material stands on it's own, but gives no elaboration to avoid duplication. Reference is given to specific data in the other book.

1. **Electrical conductivity of zone I**; In order to calculate the electrical conductivity of zone I, we have made the following assumptions (see also Fig. 1-1):

- Land width—0.020 in. (0.5 mm) leading away from the hole
- Wire diameter—0.032 in. (0.8 mm)
- Hole diameter—0.040 in. (1 mm)
- Outside diameter of land—0.120 in. (3 mm)
- Copper/Sn63 resistivity ratio $\mu = 8.25$

Using the above data and the formula for a round to flat joint from Table 4-8 in Ref. 1-1, we get;

$$L_j = \pi/4 \; \mu \; D \qquad (1\text{-}1)$$

Since L_j is fixed in zone I, we add a multiplier X to establish the ratio between the lead conductivity and the joint. (REMEMBER—$1/X \times 100$ = percentage difference). Thus:

$$X = \pi/4 \; \mu \; D/L_j \qquad (1\text{-}2)$$
$$X = 0.785 \times 8.25 \times 0.032/0.040 = 5.18 \qquad (1\text{-}3)$$
$$1/X = 0.193 \text{ or } 19.30\% \qquad (1\text{-}4)$$

From Table 4-7 in Ref. 1-1, we get that an AWG 20 wire can conduct 9.8 amps for a temperature rise of 122°F (50°C). Then:

$$9.8 \times 19.30/100 = 1.90 \text{ amps} \qquad (1\text{-}5)$$

Therefore zone I can conduct 1.9 amps for a temperature rise of 122°F (50°C), or 2.5 amps for a temperature rise of 158°F (70°C.). This can also be translated into 25.33% of the conductivity of a 2 oz copper conductor [2 oz copper with a land width of 0.020 in. (0.5 mm) has a reported conductivity of 7.5 amps for the temperature quoted].

An analysis of printed circuits indicates that solder joints are seldom expected to carry such currents. This configuration (zone—I) has thus always been considered to be entirely safe.

2. Mechanical (Tensile) Strength of zone I; We will now calculate the mechanical strength of this solder joint under tensile conditions. Remember, however, that in use it is seldom under any pure stress.

A design review indicates that the limiting factor is the copper peel strength rather than the solder. Any load (pressure) from the component side will be translated into a peeling mode in zone I. Our calculations are intended to simulate the normal behavior of components on printed circuit boards under vibration, temperature cycling, and handing conditions. They are also guidelines for the assessment of conditions in abnormal abuse.

Let us calculate the strength by using Tables 4-5, 4-6, and 4-7 of Ref. 1-1. We must make the following assumptions:

- The lead is made of OFHC copper
- The solder used is SN63 (Eutectic)
- The Strength ratio of OFHC/SN63 is $\beta = 5.85$
- AWG 20 wire (0.032 in.) has a tensile strength of 31 lb.
- Lap joint length $L_j = 0.040$ in. (1 mm)

Taking the formula from Table 4-5 of Ref. 1-1 we get;

$$L_j = \pi/4 \ \beta \ D \qquad (1-6)$$

Therefore we can calculate the multiplier (remember, L_j here is fixed) as follows:

$$X = \pi \ \beta \ D \ / \ \pi \ L_j \qquad (1-7)$$
$$X = 0.785 \times 5.85 \times 032/0.040 = 3.675 \qquad (1-8)$$
$$1/X = 0.272 \text{ or } 27.2\% \qquad (1-9)$$

zone I therefore is equal in strength to 27.2% of the 0.032 in. copper tensile strength. (31 lb for AWG 20 OFHC wire)

$$31.0 \times 27.2/100 = 8.43 \text{ lb} \qquad (1-10)$$

This means that zone I can support 8.43 lb under pure tensile forces.

We must now equate the joint strength to the peel strength of the copper laminate on the printed circuit board. The peel strength of 1 oz copper laminate falls somewhere between 7 and 10 lb for an entire inch of width (2 oz copper falls between 8 and 14 lb). It is easy to see that the solder joint is much stronger than the copper inside the hole circumference where peel failure would start. This is best understood if the force is applied from the component side.

If we assume then that:

- Lead diameter = 0.032 in. (0.8 mm)
- Minimal solder cross section = 0.040 in. (hole diameter)
- Hole is wetted 180° around
- Peel strength—1 oz Copper—FR4 = 10 lb/in.

then we can calculate the length of the critical section for the copper as follows:

$$L = \pi/2\ D = 1.57 \times 0.040 = 6.28 \times 10^{-2} \qquad (1\text{-}11)$$

This can be translated into the actual joint strength under peel:

Joint strength (S) = peel strength $\times$ conductor width

$$S = 10 \times 6.28 \times 10^{-2} = 0.628 \text{ lb} \qquad (1\text{-}12)$$

Thus we see that the solder joint is several orders of magnitude larger than the peel strength of the joint for 1 oz of copper. The superiority of the solder joint has been confirmed by numerous tests throughout the industry. Tests of solder joints in zones I and II show that they can support between 2 and 6 lb of upward pull.

1-7 THE EFFECT OF THE COMPLETED FILLET

Clinched-over leads (zone I) seldom exist by themselves where the back of the hole is not filled with solder. They usually have an additional solder fillet to the rest of the pad, as described in zone II (Fig. 1-3). Let us calculate the additional conductivity and strength which are added by this second part of the solder fillet. All the assumptions made are identical to those in zone I.

1. **Electrical conductivity of zone II:** In order to calculate the electrical conductivity of zone II, we must first calculate the critical area in the solder. For this we need to determine the critical circumference of the solder C_s. This, in turn, is the difference between the width of the lead which creates the lap joint for zone I and the total circumference in equation 1-13:

Where C_s is the critical solder area, D_h is the diameter of the hole and D_w is the diameter of the wire;

ZONE II

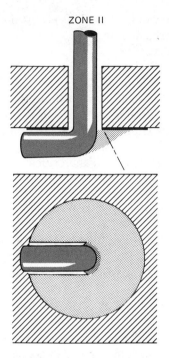

Fig. 1-3. A schematic of Zone II.

$$C_s = \pi(D_h - D_w) = 3.14 \times (0.040 - 0.032) = 9.37 \times 10^{-2} \quad (1\text{-}13)$$

We can then calculate the critical solder area (A_s) using the factor f from Table 1-3, and 0.020 for 2 oz copper height;

$$A_s = C_s f H = 0.0937 \times 0.778 \times 0.020 = 1.46 \times 10^{-3} \quad (1\text{-}14)$$

If we add the multiplication factor to the standard equation, we get:

$$X \times A_s = \mu \times A_c \quad (1\text{-}15)$$

Let us take $100/X$, which is the percent conductivity of the AWG 20 wire:

$$1/X = 0.00146 \times 100/(8.25 \times 8.04 \times 10^{-4}) = 22.01\% \quad (1\text{-}16)$$

If we take the current carrying capacity of the wire as 9.8 amps for an allowable temperature rise of 122°F. (50°C) we derive the conductivity of zone II as follows:

$$9.8 \times 22.01/100 = 2.16 \text{ amps} \qquad (1\text{-}17)$$

If the conductivity of the land in the printed circuit board is 7.5 amps for the same temperature rise, we get the percent conductivity of the solder joint as follows:

$$2.16 \times 100/7.5 = 28.76\% \qquad (1\text{-}18)$$

which means that the zone II solder joint can carry 28.76% of the conductivity of the land.

2. **Strength (tensile) of zone II:** The calculations for the solder strength of zone II are based on the same critical circumference as the conductivity. A quick calculation will reveal that this circumference is 74.50% of the full circumference, and if we equate the solder joint strength to that of zone V, we get a solder strength of 9.62 lb (Table 1-3). This is much higher than the peel strength of the laminated copper. In reality the strength of zone II is identical to that of zone I under these load conditions, since we have made the assumption that zone I encompasses one half of the circumference of the hole. Zone II thus has a theoretical strength of 8.43 Lbs.

In order to approximate the combined strength of a bent-over lead with a complete fillet, one merely adds zone I and zone II arithmetically for electrical conductivity and mechanical strength. Remember that the clinched lead adds mechanical strength and conductivity to the combined joint strength. In practice, however, stress distributions and voltage gradients do not utilize the full potential of the solder in the fillet.

1-8 THE ANATOMY OF A STRAIGHT THROUGH LEAD

For economic reasons, many leads are not clinched and are fitted straight through the hole. This configuration has many desirable advantages in design (less space), stress distribution, assembly, wave soldering, and inspection. Swaged edges are sometimes used to retain the leads in the board and prevent them from falling out. In our analysis, we will first treat zone V alone. This is a pure case of a straight through lead on a single-sided printed circuit. We will then extend this to zones III and IV, which apply to a plated-through hole.

This joint may be loaded under tensile or sheer, and the affected zone would be the same (see Fig. 1-4). The configuration, however, is very sensitive to the wire to hole ratio (see Section 1-11).

In this analysis, we will consider a contour joint which can be approximated by a straight plane starting at the edge of the land and terminating

ZONE V

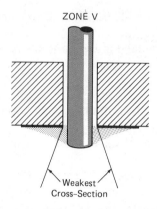

Weakest
Cross-Section

Fig. 1-4. A schematic of Zone V.

at 0.020 in. (0.5 mm) on the protruding lead. With the standard techniques of fillet shape and size control (see Section 1-18), different fillet contours can be achieved. We must further assume that the fillet is perfect and continuous 360° around the wire, and that good wetting has occurred. The other assumptions are the same as in those for the clinched lead.

1. **Electrical conductivity of zone V:** In order to calculate the electrical conductivity of this solder joint section, we have made the following assumptions (see also Fig. 1-1):

- Land width—0.020 in. (0.5 mm) leading away from the hole.
- Diameter of wire—0.032 in. (0.8 mm)
- Diameter of hole—0.040 in. (1 mm)
- Outside diameter of land—0.120 in. (3 mm)
- Copper to Sn63 resistibility ratio $\mu = 8.25$

Using the above data and the geometrical formula describing the weakest cross section we determine the critical area of solder (see Eq. 1-14) as follows:

$$A_s = \pi \times 0.778 \times 0.020/0.040 = 1.96 \times 10^{-3} \qquad (1\text{-}19)$$

Here the factor represents the lead height and is taken from Table 1-5 (see also Section 1-11). For a tolerance of 0.008 in. (0.2 mm), as in our case, f is equal to 77.8% or 0.778.

For the conductivity formula:

$$A_s = \mu \, A_c; \qquad A_c = A_s/8.25 = 2.37 \times 10^{-4} \qquad (1\text{-}20)$$

This means that the solder joint can carry as much power as the equivalent area of the conductor designated by A_c. Compare this first to the current-carrying capacity of an AWG 20 copper conductor (see table 5-7, Ref. 1-1) for a temperature rise of 122°F (50°C) and an area A of:

$$A_{(AWG20)} = \pi/4 \times D^2 = 0.785 \times 0.032^2 = 8.04 \times 10^{-4} \qquad (1\text{-}21)$$

By equating these two areas, we can see that zone V can conduct 29.48% of wire current or:

$$C_v = 2.37 \times 10^{-4} \times 9.8/8.04 \times 10^{-4} = 2.88 \text{ amp} \qquad (1\text{-}22)$$

Next, let us calculate the value for the printed circuit board. If we take the conductivity of an 0.020 in. conductor made of 2 oz copper as 7.5 amps, we get the conductivity of the zone V solder joint as:

$$C_v = 2.88/7.5 = 0.3914 \text{ or } 38.14\% \qquad (1\text{-}23)$$

This means that the zone V solder joint can conduct 38.14% of the current that the land can conduct by itself for the same temperature rise.

The current-carrying capacity of the solder in a straight-through connection, as in zone V, is larger than that of the crimped joint in zone I. The reason is obviously the larger mass of solder alloy. In both cases, the conductivity is still far greater than the currents, which are normally carried on printed circuit boards. Therefore, this joint too is considered to be safe for most applications. The use of straight-through design, however, requires a good wire-to-hole ratio, and all surfaces must have good solderability.

2. **Mechanical (tensile) strength of zone V;** As we calculate the mechanical strength of zone V, we must remember the critical cross section (see Fig. 1-4). It always extends from the edge of the hole to the curvature of the solder by the shortest route. For an analysis of the size of this parameter as a function of the hole-to-wire ratio, see Section 1-11.

This joint is truly under tensile stress when the forces are exerted from the component side. We must, therefore, equate the strength of the critical solder area to the peel strength of the copper on the hole circumference. All other considerations are similar to those stated for zone I.

Let us calculate the strength by making the following assumptions:

- The lead is made of OFHC Copper
- The solder used is SN63 (eutectic tin-lead)
- Lead diameter = 0.032 in. (0.8 mm)

- Strength ratio is $\beta = 5.85$
- Strength of AWG 20 copper lead is 31 lb

To calculate the strength of this joint, disregarding the strength of the printed circuit board, we use the same formula as in Eq. 1-14:

$$A_s = \pi \, D \, f \, H = 3.14 \times 0.040 \times 0.778 \times 0.020 = 1.96 \times 10^{-3} \quad \text{(1-24)}$$

If we introduce the multiplier X we can equate the area of the solder A_s to the cross sectional area of the lead and the strength ratio β.

$$X \, A_s = \beta \, A_c; \quad \text{(1-25)}$$

$$1/X = 1.96 \times 10^{-3}/(5.85 \times 8.04 \times 10^{-4}) = 0.4167 \text{ or } 41.67\% \quad \text{(1-26)}$$

Thus 41.67% of the strength of the lead is the strength of the critical area in the solder joint. Since this area has 31 lb UTS (Ultimate Tensile Strength) we get:

$$T_s = 41.67 \times 31.0/100 = 12.91 \text{ lb} \quad \text{(1-27)}$$

Therefore, the strength of the critical area of the solder (T_s) is 12.91 lb. This is much larger than the copper peel strength in the solder joint area. Thus the printed circuit board is again the limiting factor in direct tensile stress.

1-9 THE CONTRIBUTION OF THE PLATED-THROUGH HOLE

The plated-through hole fulfills a double function. First, it provides electrical interconnection for the circuitry on both surfaces of the printed circuit board and the inside layers in multi-layer board (MLB) designs. Second, it serves as an anchorage area for the solder fillet. Since the soldered connection enhances the electrical conductivity, the two properties define the joint in zone III. The quality and reliability of this design depends on the characteristics of the plated barrel in the printed circuit board (see Fig. 1-5).

The quality of the solder joint is directly related to the integrity of the plated-through hole process. Unfortunately, little uniformity can be expected in the wall finish because of the variety of processes used in board manufacturing. We depend on drilling and hole finish, plating adhesion and thickness, cleanliness and solderability, and so on. As a result, there is a tendency to mistrust the plated-through hole completely and to de-

ZONE III

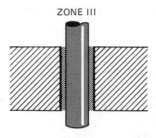

Fig. 1-5. A schematic of Zone III.

pend entirely on the solder for strength and conductivity. This will be shown to be a poor practice because the solder strength in zone III depends on plating adherence. Nor is the solder capable of inter- connecting internal conductors of a multilayer construction when there are barrel problems. Finally, the conductivity of the solder from top to bottom is normally less than that of the lead wire in the joint.

In this analysis, we have assumed a peel strength of 3 lb per linear inch inside the hole. This value is a conservatives average based on a series of in house tests. It represents only about 35% of the peel strength of the external laminated layer. The true strength contribution from the plated portion of the hole is impractical to assess, since empirical measurements vary not only from process to process, but also from lot to lot and across the board.

Table 1-3 presents the strength of the solder in the plated-through hole in zone III, with the contribution of the plated portion of PTH. It is suggested that the reader adjust these values according to the quality of his particular process.

The electrical conductivity of the plated-through hole exceeds by far that of the 0.020 in. (0.5 mm) widths land used in our analysis. Therefore, the plated-through hole is considered to be superior to the land, and the electrical conductivity of the hole is always adequate. This also explains why we do not have to fill the hole to the brim, but accept a depression of 75% of board thickness (provided, of course, that we have good wetting on the top). It also explains why every hole does not have to be filled.

Zone III is not sensitive to the hole-to-wire ratio for both electrical conductivity and strength. In automatic wave soldering, however, the hole-to-wire ratio is critical because we desire to fill the plated-through hole each time (see Tables 1-4 and 1-5 and Section 1-11).

This joint may be loaded in tensile or shear, and the effect on the critical cross section—the plated portion of the barrel—would be the same. For this analysis, we will assume complete filleting of zone III, with no imperfections such as blow holes, entrapments, and discontinuities (See Fig. 1-5). The other assumptions are the same as those in Section 1-4 on clinched leads.

1. **Electrical conductivity of Zone III.** In order to calculate the electrical conductivity of zone III, we must make the following assumptions. (see also Fig. 1-1):

- Width of lead—0.020 in. (0.5 mm)
- Diameter of wire—0.032 in. (0.8 mm)
- Diameter of hole—0.040 in. (1 mm)
- Copper to SN63 resistivity ratio $\mu = 8.25$

Using the above data and the geometric formula describing the wire-to-hole resistivity (Table 4-8, item 6, Ref. 1-1):

$$L_j = 1/4\mu D \qquad (1\text{-}28)$$

If we introduce the multiplier X, we get:

$$X = 1/4 \times 8.25 \times 0.032/0.0625 = 1.056 \qquad (1\text{-}29)$$

If we desire to calculate the percent conductivity of zone III, we must take $1/X$, which equals:

$$1/X = 0.95 \text{ or } 95\% \qquad (1\text{-}30)$$

If we take the conductivity of the wire to be 9.8 amp (for a 122°F or 50°C temperature rise), we get:

$$9.8 \times 0.95 = 9.3 \text{ amp} \qquad (1\text{-}31)$$

Thus the critical area of the solder can conduct 9.3 amp in our joint, in comparison to the land conductivity of 7.5 amp. Or:

$$9.3 \times 100/7.5 = 124\% \qquad (1\text{-}32)$$

This means that the plated-through hole can conduct more current than the copper land for the same temperature rise. This joint, therefore, is safe for all conventional applications.

2. **Mechanical (Tensile) Strength of Zone III;** In order to calculate the mechanical strength of zone III, we must again make some assumptions.

We will consider the strength of the solder fillet in the plated-through hole only (no zone I,II,IV or V present). Thus the strength of the solder joint is a function of the peel strength of the laminated copper on the top or bottom plus the strength contribution of the plated-through hole itself.

If the solder itself is the weakest link, then the critical area would be immediately adjacent to the lead. Therefore:

$$A_s = \pi \, D \, H = 3.14 \times 0.032 \times 0.0625 = 6.28 \times 10^{-3} \qquad (1\text{-}33)$$

If we introduce the multiplier into the tensile formula, we get:

$$X \times A_s = A_c \times \beta \qquad (1\text{-}34)$$

$$1/X = 6.28 \times 10^{-3}/(8.04 \times 10^{-4} \times 5.85) = 1.34 \text{ or } 134\% \qquad (1\text{-}35)$$

This means that the strength of the solder joint exceeds that of the lead by a factor of 1.34. If we take the strength of the lead as 31 lb, we get a solder strength of:

$$31 \times 1.34 = 41.39 \text{ lb} \qquad (1\text{-}36)$$

Therefore the critical area of the solder has a strength of 41.39 lb, and a well-made joint to this configuration would definitely not break in the solder.

The strength of the plated-through hole, then, depends on the peel strength of the plating from the inside of the barrel. This peel strength has two components. The first one is the peel strength of the laminated copper on both the top and bottom of the barrel. This component is predictable and is equal to the peel strength on zones IV and/or V according to the stress.

The second part of the hole strength comes from the adhesion of the plated barrel in the hole. This component makes a much smaller contribution to the overall strength. It is approximately 35% of the peel strength of the surface or laminated copper. Confidential test results on actual boards revealed this to be a conservative assumption. Little information is available in the literature, since these measurements are a function of the mechanical properties of the unplated hole, such as roughness, etc..

Table 1-3 reports zone—III as the sum of the contributions of the plated copper in the hole and the surface peel strength of the laminated copper. The table also reports the electrical conductivity of the solder in the joint, should it have to carry all the current.

1-10 THE VALUE OF THE TOP FILLET

Zone IV represents the solder fillet on the top of the printed circuit board (Fig. 1-6). In this analysis, we will assume a good fillet all around (360°) and a height of 0.005 in. (0.13 mm). In addition we will make all the assumptions used in discussing zone V to which this design is similar.

This should not be taken as an endorsement of top fillets. In many situations they are not essential to joint integrity, a conclusion which is borne out by the present joint analysis. In most cases, the solder is not needed to augment the plated barrel conductivity or strength. The calculations below provide further proof of this assertion.

1. **Electrical Conductivity of Zone IV;** This part of the fillet has the following conductivity if we take the critical area as the same values used for zone V. Then:

$$A_s = fHD\pi = 0.778 \times 0.005 \times 0.040 \times 3.14 = 4.89 \times 10^{-4} \quad (1\text{-}37)$$

If we add the multiplier X to the standard equation, we get:

$$XA_s = \mu A_c \quad (1\text{-}38)$$

$$1/X = 4.89 \times 10^{-4}/(8.25 \times 8.04 \times 10^{-4}) = 0.0737 \text{ or } 7.37\% \quad (1\text{-}39)$$

In other words the zone IV solder joint can carry 7.37% of the current of the wire. If we take the current of the wire to be 9.8 amp (for a temperature rise of 122°F or 50°C), then:

$$9.8 \times .0737 = 0.72 \text{ amp} \quad (1\text{-}40)$$

In comparison to the conductivity of the land (7.5 amps for the same

ZONE IV

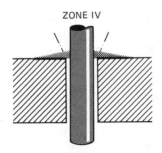

Fig. 1-6. A schematic of Zone IV.

temperature rise), this is an extremely small contribution. In addition remember that in order to have a top fillet, we must have zone -III. Thus it is easy to see why top inspection of solder joints is not required in many situations.

2. **Mechanical (Tensile) Strength of Zone IV.** Following the calculations for zone V, the formula (Eq. 1-24) would be:

$$A_s = 3.14 \times 0.040 \times 0.778 \times 0.005 = 4.89 \times 10^{-4} \qquad (1\text{-}41)$$

If we introduce the multiplier X, we can equate the area of the solder A_s to the cross sectional area of the lead and the strength ratio β (see Eq. 1-25):

$$1/X = 4.89 \times 10^{-4}/(5.85 \times 8.04 \times 10^{-4}) = 0.104 \text{ or } 10.4\% \qquad (1\text{-}42)$$

We determine that 10.4% of the strength of the lead is the strength of the critical area in the solder joint. Since this area has 31 lb UTS, we get 3.22 lb for zone IV. Therefore the strength of the critical area of the solder is 3.22 lb which is relatively small. Note that it is still larger than the copper peel strength of the appropriate printed circuit board circumference. The value for this peel strength can be taken directly from the calculations of zone V, where it is identical in shape.

Table 1-3. Zone Properties—Printed Circuit Board Solder Joint.
(Calculated from model)

| | ELECTRICAL PROPERTIES | | | MECHANICAL PROPERTIES | |
ZONE	Conductivity 120°F Amps	% – 2 OZ Copper	% – 20AWG Cu Wire	UTS Lbs	% – 20AWG Cu Wire
I	1.90	25.3	19.3	8.43	27.2
II	2.16	28.8	22.0	8.43	27.2
III	9.30	124.0	95.0	41.39	134.0
IV	0.72	9.6	7.4	3.22	10.4
V	2.88	38.1	29.5	12.91	41.7

1-11 HOLE-TO-WIRE RATIO

The hole size in the printed circuit after plating and the component lead-wire diameter have a large effect on a variety of solder joint properties. This effect is independent of the mode of production, whether by hand or by machine. Let us analyze these properties one by one.

Table 1-4. Effect of Hole to Wire Ratios Variations
on Zone—V[a]

CLEARANCE (D_{hole}-D_{wire})	f	CRITICAL SOLDER AREA (In^2)	CURRENT CARRYING CAPACITY (amp)	SOLDER TENSILE STRENGTH (lbs)
.000[b]	.890	2.24×10^{-3}	3.31	14.75
.005	.820	2.06×10^{-3}	3.04	13.56
.008	.778	1.96×10^{-3}	2.88	12.91
.010	.750	1.88×10^{-3}	2.78	12.38
.015	.685	1.72×10^{-3}	2.54	11.32
.020	.625	1.57×10^{-3}	2.32	10.34

[a] Using the dimensions on Figure 1-1, and a .020 inch fillet height.
[b] Not practical, for calculations only!.

1. *Mechanical Strength.* Sections 1-8 and 1-10 have already outlined the sensitivity of joint thickness to strength. Empirical data show that a solder joint is strongest when the solder cross section is between 0.003 and 0.004 in. (0.075—0.1 mm). (see Chapter 3, The Metallurgy of Solder, pp 92-93, Ref. 1-1).

Should the tolerance be too close, the flux cannot easily penetrate into the crevices and perform its function. In addition, it is difficult to displace it from those areas by the molten solder. There must always be an easy venting route for the flux and the gaseous materials formed. Experience has shown that close-fitting surfaces have poor mechanical integrity because they often have a large number of voids and discontinuities in the solder joint.

If the gap between the surfaces is large, the ability of the joint to withstand mechanical stresses is diminished because the bulk solder yields much more readily. This correlates with the ease of filling each hole during wave soldering.

Table 1-5. Hole to Wire Ratio—Effect on
Solder Joint Zones

ZONE NO.	ELECTRICAL CONDUCTIVITY	MECHANICAL STRENGTH	FILLETING IN WAVE
I	None	None	None
II	Yes	Yes	Yes
III	None	None	Yes
IV	Yes	Yes	Yes
V	Yes	Yes	Yes

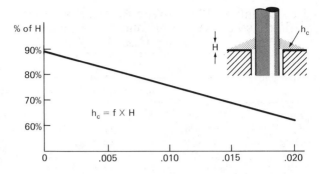

Fig. 1-7. Graph showing a critical solder joint section versus height.

The hole-to-wire ratio also affects the critical cross section of the different joint geometries. It is this effect which we had to keep constant in all of our calculations and which impacts every specific designs. Tables 1-5 and Fig. 1-7 show the effect of this clearance on our solder joint configurations. Note the rapid reduction in the critical area with the increasing hole-to-wire ratio (the smaller wire diameter in the same hole size).

Not all parts of the joint are affected by the hole-to-wire ratio. Zones II, IV and V are definitely affected, while zones I and III are not. The properties that depend on the ratio are strength and conductivity (see Table 1-5)

The reason for the effect on the electrical conductivity is the critical cross sectional area changes. Zones II, IV, and V decrease in conductivity with increasing clearance.

The effect on filling the joint in hand soldering is basically unimportant. Unless there is an interference fit, it is always possible to fill a large gap or hole with hand soldering techniques. This is not the case in automatic machine soldering, where surface energies, sometimes labeled *capillary rise*, dictate hole filling. These surface conditions also determine if the solder will stay in the joint after the board leaves the molten solder (wave, pot, etc.). These retaining forces are a function of the surface areas in the joint and their respective solderability. The opposing force which tends to drain the solder from the joint is mainly gravity. As such, it depends on the solder mass and the dynamics of the process. To summarize, these conflicting forces are:

1. Retaining forces that fill the joint
   a. Interfacial energies between the solder and the metals joined (lead wire and board). They are the function of:

Table 1-6. Optimum Hole-To-Wire Ratios. (Expressed as the Difference Between Hole and Wire Diameters)

$D_{hole} - D_{wire}$

APPLICATION TYPE	CLEARANCE		WAVE FILL	INSERTION	
	in.	mm		HAND	AUTOMATIC
For solder wetting	Min .002	.05	Yes	Impractical	
For mechanical strength	Max .006	.15	Yes	Difficult	
For ease of filleting	Max .018	.45	Yes	Yes	Possible[a]
For ease of insertion	Max .022	.55	Yes	Yes	Yes
For joints that may sag	Max .025	.63	Yes	Yes	Yes

[a] With good equipment only.

- • Surface area
- • Surfaces solderability (or type of flux)
  b. Cohesive force of the solder, a function of the environment:
  - • Flux
  - • Oil
  - • Air (dross)
2. Draining forces that empty the fillet
  a. Gravity, which acts on the molten solder, is a function of:
  - • Solder mass
  - • Vibrations
  b. Process dynamics, which vary with the equipment. In the case of a wave this is a function of:
  - • The wave dynamics at the point of exit
  - • The conveyor angle

In summary, then, we can see that for strength and machine filleting, we need very small gaps between the wire and the hole. For ease of component assembly, however, we would prefer larger tolerances. The practical compromise is stated in the following rule of thumb (Table 1-6):

1-12 ANALYSIS OF SURFACE-MOUNTED JOINTS AND THEIR METALLURGY

As fine-line boards become available, few if any large plated-through holes will be found on the boards. Thus our ability to anchor devices by their leads will diminish, and new design guidelines will be needed. This may take the shape of segregated functional design, mechanically secured components with screws and/or adhesives, and the redesign of the components themselves.

Just as switch technology changed with the advent of the keyboard, new devices will emerge. The old-fashioned toggle bears no resemblance to the membrane of today. Similar developments are in process for relays and transformers for low power. With time the use of all surface mounting on selected boards may be possible without giving up design flexibility.

The metallurgy of solder alloys and their reaction with the base metal are covered in Ref. 1-1. The metallurgy as it applies to surface mounting is discussed here.

The metallurgy of the component termination is a vital area that seems to be given little attention at present. Lead wire and board solderability are issues as old as the electronics industry. It is safe to say that up to now, the majority of unjustified soldering problems and costs are caused by unsatisfactory solderability. The danger in surface-mounted soldering, however, is much more acute since joint reliability is at stake.

In the through-the-hole interconnect, we have seen that there is a basic overdesign. In previous sections, we calculated the strength and conductivity of the through-the-hole connections. These calculations reveal that even single-sided boards have more solder joint strength than the average board. Double-sided board joint strength (with the plated-through hole) is 8 to 13 times greater than necessary. Under these conditions, it is easy to understand why termination problems have not been a major reliability issue. Even joints that were 93.2% defective could still function when the over-design factor was 13 times.

In surface mounting, joint design is dramatically weaker. Both the critical joint area and the amount of solder used are much smaller, and there is no overdesign safety factor. Preliminary calculations and tests show the need for 80% or better perfection per fillet. This is based on the assumption that the metallurgy of the system is stable.

The problems created by the soldering process are due to several elements. Here we will concentrate on the stability of the termination metallurgy. This can be further broken down into:

- The nature of the alloy zone created between the solder and the base metals for both the termination (with or without a lead) and the board conductor.
- The formation of intermetallic compounds and their properties.
- The solid state diffusion after bonding and its effect.

These properties are primarily a function of the solder type used. The standard solders for mass processing (wave, drag, and dip) of printed circuit boards have been the 63/37 (eutectic) or 60/40 tin-lead alloys. Both types are used with and without the beneficial addition of antimony and silver. These eutectic or near-eutectic compositions are vital because of their solidification parameters. In surface-mounted soldering, lower temperature bismuth and indium base alloys have joined this list. The metallic interaction of these alloys with copper, nickel, and iron is well understood. In reality seven elements (Copper, Nickel, Iron, Silver, Gold, Tin, and Lead) make up the bulk of the base metals used for board and device termination.

New techniques are used with surface mounting based on preplacing the solder and flux and then heating to wetting temperatures. The bonding materials are located in the joint in the form of solder creams and pastes or preforms. These techniques have a simpler solidification sequence, and near-eutectic alloys are no longer as important. This opens the door for the use of a host of new solder alloy combinations to meet different temperature and property needs. Here the metallurgy of the termination is more complex.

The problem is further complicated by the frequent use of silver (with and without noble metal additives) in the electrodes of leadless chip devices. These thick film materials are sensitive to the well-understood process of leaching or scavenging when the solder dissolves the finite small layer of surface silver, leaving a nonwettable high glass surface behind. There are several methods of retarding this self-defeating mechanism (Ref. 1-4). Here we should follow the practices of the established hybrid industry, which has learned to overcome these difficulties.

It is possible to overcome the scavenging problem by choosing a different solder alloy. However, it may be easier and more economical to have the chip producers stabilize the terminations. This is achieved with the use of a barrier plating such as nickel over the silver. The nickel does not dissolve or react with solder at process temperatures, and thus creates an effective separation. However, because bare nickel is hard to solder with milder fluxes, it must be further covered with a solderable layer like tin-lead. In this system the silver thick-film makes the termination, the nickel coating prevents scavenging, and the solder coat provides solderability and long storage life.

The term *intermetallics* has been misunderstood and widely abused. In pure metallurgical terms, it is an abbreviation of *Intermetallic compounds*, which are quite detrimental in soldering. Most novices, however, use this term to describe the process of wetting. This most unscientific use has created the impression that intermetallics are desirable. Nothing is further from the truth. For a list of intermetallic compounds formed with common solder alloys see Ref. 1-1, pp 136-138.

In the confines of small surface-mounted joint geometry, intermetallic compounds must either be dispersed or nonexistent. While they have improved electrical conductivity, their brittleness reduces their reliability under stress.

As this industry moves forward, many of these issues will be resolved. But special caution is urged in the selection of new untried metal systems. This applies to component and printed board producers, as well as to the assemblers. Be especially careful in applying unique solder alloy pastes over surfaces pretinned with another solder. Very low temperature melting interfaces may be formed, causing *hot-shortness*.

One of the most radical changes brought about by surface mounting technology is the configuration of the solder fillet. The shift from a large through-the-hole connection to a miniature lap or butt joint is not one of size alone. It involves a change in stress distribution, as well as the critical ratio between component weight and solder support. This is the area that needs most of the attention in developing the leadless chip carriers, and at present is the limiting factor for their growth.

For simplicity, let us look at each type of connection used with surface mounting and review some of the key features. These include the ability to accommodate TCE mismatches during thermal cycling, and the standard six design parameters for joint design (see Section 1-4).

1-13 THE RIGID LAP JOINT—I.E. LEADLESS CHIPS

The rigid lap is used with leadless chip components (see Figures 1-8 and 1-9). This is one of the larger joints in the surface mounted family because of the outside fillet. When this external solder is present, as a result of wave soldering, for instance, this is the most rigid configuration. It cannot compensate for mismatches of TCE, and this limits the length of the chip. If the amount of solder is controlled, as in top soldering with paste, for instance, the joint is similar to the compliant lap (see Section 1-14).

Note in the diagram (Fig. 1-8), the enlarged portion of the termination at

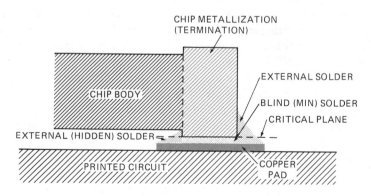

Fig. 1-8. Diagram of a rigid lap joint.

Fig. 1-9. Side view of a rigid lap joint made by double wave. (Copyright Manko Associates)

the chip's end. There are devices, in which this portion of the chip is at the same level, but they are rare today. The joint is shown with a small external solder fillet, as would be formed with solder creams. If the chip is on the bottom side and wave soldered, much more solder is present on the side, as shown in Fig 1-9. Note that the critical plane of the solder is hidden and defies inspection.

All the material oriented design parameters (electrical conductivity, mechanical security, and thermal dissipation) are satisfied when the chip is not too long. Since conductivity is small for these devices, there is no problem with the solder joints. The minimum and average joint strengths of this type connection are listed in Table 1-7. They are compared to a PTH to Wire connection (zones V, III and IV). In the calculations, the entire end termination overlaps the board pad. The degree of misalignment permissible in the industry will depend on the individual case. Heat dissipation is no problem because of the small currents involved and the proximity to the board.

The design-oriented parameters (manufacturability, inspectability, and repairability) are also easy to satisfy. They are a function of the assembly method, however, and become more complex when the component is glued to the surface, as in bottom mounting. The type of adhesive used, its volume, and its physical characteristics have a major effect on this solder joint.

Consider the ease of manufacturing for example. Here the use of the adhesive has consequences that go well beyond the accuracy of the as-

Table 1-7. Comparison Table: Chip Lap vs Pth.

SOLDER JOINT TYPICAL	CHIP LAP. MINIMUM	AVG.	PTH & WIRE .032 IN. DIA
Length, in.	.015	.03	.082
Width, in.	.04	.05	.1
Area, in.^{2}	.0006	.0015	.0082
% of PTH	7.32	18.29	100.00
Shear Sn63, lb	3.48	8.7	47.56
Tensile, lb	4.62	11.55	63.14
Pad peel, lb	.4	.5	3.14
Pad peel, <Tg lb	.0384	.048	.30144

sembly process itself. The geometry is such that thorough cleaning is often doubtful, and this limits the strength and type of flux used. In hard to clean assemblies, only milder rosin base fluxes seem suitable for most cemented chips, where flux residues are safe. As the strength of the flux is reduced, general solderability must improve, or reliability will suffer and touch-up rates increase. Companies with the advantages of water soluble intermediate fluxes (mislabeled as "Organic Acid", Ref. 1-5) will have to relearn to use weaker rosin formulations. It is also possible to improve cleaning by controlling the adhesive geometry (width vs height) in the cured stage. Today, several companies are successfully using organic fluxes with water washing on surface-mounted assemblies. The secret to their success lies in good design for washability, and process control.

Adhesives obviously also affect ease of repair. Special soldering iron tips make it possible to remelt the solder and simultaneously soften the glue by taking it above its T_g. In this weakened state, the component can be pried off the board (see Section 9-9). Fortunately, the chip components are relatively rugged, and replacement because of component failure is rare.

Inspection of lap solder joints is not simple (see Section 8-15) because the critical part of the joint is hidden. Remember also that it is visually impossible to inspect for entrapment of flux or solder balls under the chip device. Additional methods are required. In addition, silver-fired terminations are not always smooth, and the solder joint may reflect this rough surface.

1-14 THE COMPLIANT LAP JOINT—I.E. SMALL OUTLINE (SO) DEVICES

The compliant lap is used with leaded components in the *small outline—SO* category (see Fig. 1-10 and 1-11). This is a smaller joint than the rigid lap, and lead geometry controls many of its properties. Lead flexibility compensates for any mismatches in TCE, which explains its name.

The diagram in Figure 1-10 shows a typical cross section of this joint, with a minimum of solder in the fillet. Here, too, precise solder quantity control is possible only with top placement. Figure 1-11 depicts a wave soldered small outline dual inline package (SOIC), note that the amount of solder in the joint is excessive and uneven. This tends to counteract the flexibility of the compliant lead.

The strength of the minimum-solder joint depends on the quality of the heel. While the fillet should not rise more than two-thirds of the height to the knee, it must be above the heel. The length of the foot and the joint at the toe are less important, though they determine the critical plane. Any expansion due to mismatches in TCE cause stress concentrations at the heel, where the failure would start. This joint geometry is identical to that developed for the older flat pack package.

The material oriented design parameters depend mostly on the size of the foot-print, although the mechanical security is also a function of lead configuration and materials. For small, light devices, the material parameters are optimal when the entire foot overlaps the board pad. The degree of misalignment permissible depends on the type and weight of the component and the stresses in the individual case. It is obviously also a function of minimum spacing allowed on the board.

Table 1-8 summarizes the material-oriented parameters for a minimum

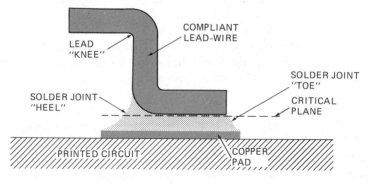

Fig. 1-10. Diagram of the SO lap cross section.

Fig. 1-11. The SO lap. View of wave soldered SOIC. (Copyright Manko Associates)

and an average solder joint. Note that the pad peel strength is half that of the rigid lap.

The design oriented parameters are more complex because of the variety of shapes being used. They depend on the location of the device on the board; the majority of top mounted SMD components are free floating, while the bottom mounted devices are cemented. This makes design parameters a function of the assembly method; complexity increases when the component is free floating (no adhesive).

Table 1-8. Comparison Table: So Lap vs Pth.

SOLDER JOINT TYPICAL	SO LAP MIN	SO LAP AVG.	PTH/WIRE .032 IN. DIA
Length, in.	.03	.04	.082
Width, in.	.02	.025	.1
Area, in.^{2}	.0006	.001	.0082
% of PTH	F 7.32	12.20	100.00
Shear Sn63, lb	3.48	5.8	47.56
Tensile, lb	4.62	7.7	63.14
Pad peel, lb	.2	.25	3.14
Pad peel, <Tg lb	.0192	.024	.30144

Consider the ease of manufacturing, for example. When an adhesive is applied, the consequences are similar to those of the rigid lap configuration. But the free floating devices have their own unique problems, which include:

- The method of preplacing the flux and solder in the joint, achieved mostly with the use of creams and pastes.
- The locating mode of the SMD component on the board, and the way they are secured prior to joint formation.
- The heating process used, which affects not only joint quality but also the balance of the assembly materials.

The repair of the compliant lap joint must also be divided according to the location of the device and the presence or absence of adhesive. The problems of multileaded component removal are complex because of the diminished size of each joint. The relatively fragile leads make component salvage rather difficult.

Visual inspection of these configurations is slow and rather cumbersome. Now more than ever, we can see the need for a universal automatic inspection device. Key parameters are wetting, the size and quality of the heel (visible), and the lack of voids inside the lap area (opaque).

1-15 THE COMPLIANT BUTT JOINT—I.E. THE "J" LEAD

The compliant butt joint is used with leaded components like chip carriers (see Fig. 1-12 and 1-13), where the lead is usually bent into a J shape. This is a narrow solder fillet and therefore very attractive for high density

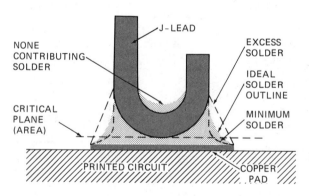

Fig. 1-12. Diagram of a compliant butt joint ("J" lead).

Fig. 1-13. View of compliant "J" lead on a chip carrier. (Copyright Manko Associates)

design. It is a joint configuration that promises to remain troublesome for some time. Here again, we rely on lead geometry and flexibility to compensate for any mismatches in TCE.

Note that the diagram shows an ideal lead bend, but that in reality the terminal shape may differ. This is by far the weakest fillet, and it cannot be strengthened by placing solder on top inside the curvature.

Because of the limited size of the solder fillet, the material oriented parameters are very small. At present, there is little danger to mechanical integrity because of the large number of leads per device. However, solder quantity control is vital; otherwise stress concentrations will fail the smaller fillets. Electrical and heat conductivity poses no problems because of the small currents used with surface mounted circuits.

The design-oriented characteristics of this joint configuration are even more critical. Device location and soldering must be extremely precise. Replacement is usually destructive to the component, and inspection is limited by geometry.

Table 1-9. Comparison Table: J Lead vs Pth.

SOLDER JOINT TYPICAL	J-LEAD MIN	AVG.	PTH/WIRE .032 IN. DIA
Length, in.	.025	.04	.082
Width, in.	.01	.015	.1
Area, in.^{2}	.00025	.0006	.0082
% of PTH	3.05	7.32	100.00
Shear Sn63, lb	1.45	3.48	47.56
Tensile, lb	1.925	4.62	63.14
Pad peel, lb	.1	.15	3.14
Pad peel, <Tg lb	.0096	.0144	.30144

1-16 THE RIGID LAP—I.E. LEADLESS CHIP CARRIER

The rigid lap is also a blind lap joint used with leadless chip carriers and similar components (see Fig. 1-14 and 1-15). This is by far the most troublesome joint in surface mounting technology. Here there are no leads to compensate for any mismatches in TCE, and we must depend on the entire material system for reliability.

Two distinctly different joint terminations are used in this joint design. One uses only the oblique bottom termination, which promises better reliability when properly designed. The other uses an additional area on the side of the chip carrier. This added terminal area is usually semicircu-

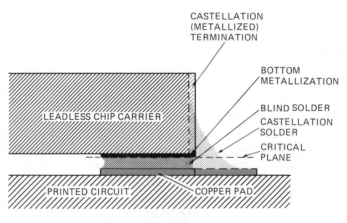

Fig. 1-14. Diagram of the rigid lap joint on an LCC.

Fig. 1-15. Bottom view of metalized pads (gold) on an LCC. (Copyright Manko Associates)

lar and is referred to as the *castellation*. The advantage of the added side joint is obvious, but its long term reliability is still not established. Non-uniform solder joints on the periphery of a *leadless chip carrier* (LCC) tend to cause stress concentrations and fail individual fillets.

When leadless chip carriers (both plastic and ceramic) are mounted to printed circuit boards, the system is not homogeneous. Many combinations in this category have failed in the solder joint after limited stressing. Heat cycling and/or vibration cause solder cracking because of the small fillet size. This has prompted many organizations to turn to leaded or pinned chip carriers for the near term, even though this results in loss of space. In very large devices [high input/output (I/O) count] or critical

Table 1-10. Comparison Table: LCC Lap vs Pth.

SOLDER JOINT TYPICAL	LCC LAP		PTH & WIRE .032 IN. DIA
	BLIND	CASTELLATION	
Length, in.	.04	.04	.082
Width, in.	.02	.04	.1
Area, in.^{2}	.0008	.0016	.0082
% of PTH	9.76	19.51	100.00
Shear Sn63, lb	4.64	9.28	47.56
Tensile, lb	6.16	12.32	63.14
Pad peel, lb	.2	.4	3.14
Pad peel, <Tg lb	.0192	.0384	.30144

applications, the added through-the-hole advantages may well compensate for the loss of density. This may be the preferred method until we learn to use the leadless carriers.

Note from Table 1-10 that we used the smaller LCC package geometry. In our rush to use larger numbers of I/O interconnects per device, we will further sacrifice the area available for this lap joint. Thus the design oriented parameters will further diminish. Also, note that the castellations add to the raw strength of the joint, doubling it in most cases. This is quite misleading because of the cantilever spring effect on the stress distribution.

In this joint, only the mechanical security is under scrutiny because the currents used are minute. Unfortunately, the use of stronger solder alloys alone cannot solve the problem; the right fatigue properties must also be present. Matching the compatibility of all the relevant materials offers a better solution, and such materials are being developed.

Manufacturing and repair problems are also complex. The blind nature of this design stretches our capabilities. Visual inspection is practically impossible, and incoming inspection plus inprocess control help to guarantee reliability.

1-17 PAD DESIGN FOR PLATED-THROUGH HOLES, SURFACE MOUNTING, AND TESTING

Pad design on printed circuits is not just a function of the space available and the insulation resistance. Easy soldering and the need for reliable joints, add to the requirements of pad configuration. If these requirements are disregarded, an unnecessarily large number of repetitive soldering faults, like bridges, will result. These faults, in turn cause expensive rework and may shorten solder fillet life.

The three major pad groupings each have their own guide lines as follows:

1. *Pads for Plated-Through Holes.* The pad configuration must be a compromise between several competing requirements. Board density and insulation resistance demand the use of the smallest outside diameter possible. Joint strength and reliability are best served if the pad is as large as possible to increase the copper peel strength. Easy soldering, however, requires a pad that is wide enough for repetitive wetting and not too wide for solder to drain away from the lead wire. Finally, we must consider heat transfer for wave soldering. This is aided by more area on the bottom (solder side) but is hampered by large ground planes inside or on top of the board. For details on repetitive wetting, see item 2 below.

The physical size of the pad can be made to vary from the exposed portion available for joint formation. A solder mask (resist) may be used to cover portions of the pad, providing a large pad with a small exposed area.

Pad size considerations must also include processing tolerances. This encompasses concentricity of the drilled hole, over- and underetching, mask and road map bleeding, and so on.

There are, unfortunately, no hard and fast rules on pad sizes for through-the-hole soldering. The following guidelines, however, are helpful.

a. Use round pads whenever possible. Oval pads are acceptable, but square ones are not recommended. This is due to the solder cohesive forces, which draw the solder into a sphere.

b. Make sure you have a minimum annular ring of 0.020 in. (0.5 mm) in all directions. This will provide the required strength.

c. Avoid larger pads with multiple holes. When this is unavoidable, separate pads with solder resist. Otherwise some parts of the common pads will not fillet properly, and icicles can form.

d. Isolate large ground planes from the pad by using *cart wheel spokes* for uniform heat transfer.

2. *Pads for Surface Mounting.* As in PTH design, layout density favors the smallest area, while solder joint stability indicates the largest pad possible. Manufacturing tolerances, including adhesive application and component placement, also dictate large pads. Pad dimensions for surface

mounting must, however, be divided according to the method of soldering into:

a. Bottom mounting for wave, drag, or dip soldering.
b. Top mounting with preplaced solder (cream, paste, preforms, etc.).

Let us see how processing effects these decisions.

a. When bottom mounting for wave soldering, the contact area with the molten solder should be as large as possible. This will eliminate the problem of lack of repetitive wetting. Because of the geometry of the components glued to the surface, there are areas of gaseous material entrapment (see Section 5-14). These cause skipping or misses on small pads. The larger the pad, the less likely the problem of wetting will be.

b. Top mounting by reflow (remelt) soldering using vapor phase, radiation, convection, or other methods has one additional requirement: Pad size is not as important for the soldering process as pad shape and uniformity. Since the pad dictates solder distribution and quantity, it also affects joint strength. Uneven joint ductility will transfer stresses to the stiffer joints, causing stress concentrations and joint failure.

Here we must worry not only about pad size uniformity, but also about the quantity of solder retained in the joint. Achieving geometric compatibility with the component terminations (accuracy of placement with the accumulation of tolerances) is no reason to make the pads uneven in size.

3. *Pads for Testing*. Automatic testing of bare and soldered boards is a desirable part of low cost manufacturing. It detects faults at the earliest possible stage and provides diagnostic data for rework. Common methods used to probe boards when using automatic test equipment (ATE), are as follows:

a. Use metallic probes (bed of nails) to contact the solder joints of through the hole or surface mounted configurations. Here large pad design is preferred. Repetitive contact, from board to board, with each joint is difficult and depends on joint height, lead wire location, and other geometry variations.
b. Use a separate test pad, away from the joints, which is always on the same plane. with this method, repetitive contact is much easier, provided the quantity of solder deposited is relatively uniform. Uniformity is enhanced by using round or at least oval pads. For recognition purposes, many companies use square inspection pads. These are more difficult for the solder to wet uniformly and are often skipped altogether.

c. Use edge contacts, like gold fingers. Here pad design is not relevant unless solder coated fingers are used. For wave soldering, to coat these fingers, they should be as long as possible. so we can get thin solder distribution through good drainage. It is also desirable to have them on the leading edge for the same reason.

1-18 FILLET CONTROL

Fillet control refers to the predetermination of the solder joint shape in all three dimensions through design. This is achieved by board layout and material selection, which makes it possible to continuously produce the same quality joint. Let us outline the sequence of fillet control evolution of a product.

The shape and size of the solder fillet must fulfill the design requirements (Sections 1-4 through 1-17). These requisites, in turn, become the workmanship or quality standards for the assembly. Each industry, product, and organization must set its own standards and should not follow the practices of others blindly. Overspecification (like unnecessary copying of government requirements) inevitably leads to unwarranted costs without increasing reliability. The hazards of underspecifying are obvious, but the hidden costs of field repair are often overlooked. Workmanship standards are the most logical compromise between cost and quality, and must be matched to the individual situation (see Section 8-2).

Once a fillet configuration is determined, design and production parameters must be merged to obtain fillet control. It is our aim in this section to analyze those factors which affect the size and shape of the solder joint. The purpose is to achieve a high degree of reproducibility with only an occasional touch-up. This way, we can utilize wave soldering, for instance, with optimum reliability. Repeated and/or excessive touch-up is not only very expensive but often degrade the quality of the end product (see Chapter 8).

Let us analyze the parameters involved. For the bottom side of the fillet only, these are based on several factors;

- Surface energies (surface tension, cohesive force, etc.)
- The process (wave dynamics, the use of oil, the incline etc.)
- The nature of the materials (base metal, flux, solder, etc.)
- The solderability of the surfaces

The wicking of solder up the plated-through hole and the formation of a top fillet are a function of these additional parameters:

- The hole-to-wire ratio (Section 1-11)
- The processing parameters affecting heat transfer (wave temperature, dwell time, preheating conditions on the top and bottom, etc.)
- The equipment (angle of impedance, conveyor speed, etc.)

It is not practical to consider the above factors in light of joint control only. These soldering parameters must be dealt with as they affect the process and the equipment.

There are several design parameters affecting fillet shape over which the designer has control. The arrangement of the bottom fillet is basically a function of;

- Length of the protrusion (component leads)
- Pad-to-wire ratio
- Drainage lines to and from the fillet

Let us look at these in more detail:

a. *Length of lead wire protrusion*. This refers to the amount of wire that extends below the underside of the board. It is often determined by the needs of the final assembly and is covered separately in the discussion of component lead cutting (see Sections 2-8 and 2-9). Here we will deal only with the parameters which affect the fillet itself.

It is easy to determine the minimum length of the protruding wire required, since this is a function of good inspectability. Experience has shown that the inspection requirements of the fillet are satisfied with 0.030 in. (0.76 mm). Preventing components from riding up in the wave is a problem only with nonsolderable component/flux combinations. In addition, solderable components mounted parallel to the conveyor direction may have a condition known as *leg up*. The first lead to contact the solder is drawn down, which can lift the second component leg up. If this lead wire is too short, it may not return into the hole, giving an incomplete solder fillet. To prevent this lifting, we require about 0.050 in. (1.3 mm) lead protrusion below the board to create mechanical interference with the upward motion. Sideways component movement, perpendicular to the wave, does not create a problem, since the wetting mechanism will pull the lead into the solder. Upon freezing, a desirable stress free fillet is created. Such stress free fillets cannot be formed when the component is tightly clinched to hold it in place.

Optimum lead length would give us just sufficient lead protruding out of the fillet for visual inspection. However, practical considerations dictate that the lead length below the printed circuit board should be between 0.030 and 0.070 in. (0.75–1.75 mm). This, of course, applies to good fillet control; leads are often much longer, reaching 0.100 to 0.125 in. (2.54–3.18 mm). When the lead is left too long, solder runs down the wire after leaving the wave. This robs solder metal needed to shape a good fillet, and increases icicling and bridging incidents.

b. *Pad to wire ratio.* In all cases hole-to-wire ratios must be maintained for easy filling. A minimum pad anchorage area is needed for strength as follows:

- Round leads and pads (straight through). Pad outside diameter (OD) should be three times the lead diameter, but not less than 0.025 in. (0.63 mm) wide, the width being half the average between the OD minus the inside diameter (ID). This will result in a strong and inspectable joint in zone V.
- Clinched lead and tear drop pad. The elongated part should be as long as the lead bend, which is a function of joint design. The pad should be 2.5 to 3 times as wide as the lead diameter. Thus the apron around the wire will permit the formation of a good and inspectable lap joint (zone I). The rounded portion of the pad should conform to the instructions above, if filleting is desired. (Consideration should be given to discounting the joint in zone II.)
- Flat terminal in the slot of an elliptical pad. The apron around the slot should have rounded corners and should be 0.025 in. minimum to 0.050 in. maximum in width (0.63—1.27 mm). There is no need to count on the fillet at the narrow edge of the slot, and the apron width there can be reduced to 0.010 in. (0.25 mm).

The fillet is also controlled by the lead length or height and its relation to pad size. The best results are obtained with a ratio of 1 : 1 to 1 : 2 between lead height and pad width. It is suggested that a ratio of 1 : 1–.5 be tried and an adjustment made according to the results. Remember that the materials and the process affect the fillet shape.

In general, if the pad diameter is too small, it limits the height of the solder fillet. If the pad is too large, solder drains away from the joint to give an uneven low profile.

The use of solder resist to change the size of the fillet is often a simple way to overcome design problems. It is possible to commit to the metallic layout of the boards and make minor changes on the solder resist screen. This is a relatively inexpensive method which provides full fillet control by adjusting the size of the exposed pad.

C. *Drainage Lines.* When exposed metal conductors go to and from a pad they can channel solder to the fillet or drain it away. This depends on the circuit geometry, the path direction, and the orientation over the wave. In this respect, large pads shared by several holes should be shaped rather than left to random geometry. For instance, a triangular pad with three holes near the three apexes is bad for fillet control. Here a cloverleaf arrangement of the same geometry would enable each fillet to assume its correct shape. Solder resist totally eliminates these hazards, but with exposed circuitry we must reckon with this fact.

While we are discussing pads, let us analyze the shapes available. The most prevalent design is a circular pad, which has the following advantages:

- The symmetry lends itself to easy location.
- The leads can be bent in any direction.
- It simplifies board layout and eventual engineering changes.
- It gives uniform looking fillets.
- It enables larger tolerances in case of misregistration of holes.

Other shapes are also available and are described in Table 6-4, of Ref. 1-1. None is as versatile as the plain circle. The round pad is also the ideal shape for fillet control. The square pad is very poor for fillet control, since the corners stretch against the cohesive forces of the solder.

To summarize, design and production conditions predetermine the shape and configuration of solder fillets in mass production. Beyond that, fillet uniformity depends largely on the quality of the materials and the solderability of the surfaces. Solder resist is very helpful in adjusting the pad for easy fillet control (see Sections 1-12 and 3-15). Solder resist is often located on parts of the metallic conductors without difficulties. The fillet itself must be easy to inspect and must fulfill design requirements for conductivity and strength.

1-19 BOARD LAYOUT FOR EASY WAVE SOLDERING

There are many excellent computer aided design (CAD) systems available for board layout. These help distribute the circuitry in an efficient manner over the allotted area. Baring any design changes, they generate an orderly layout that incorporates many soldering requirements. The program also generates the solder resist pattern, the size and shape of the pads, and the diameter of the hole needed.

The trouble begins when design or engineering changes have to be made. The best solution would involve the complete redesign on the CAD system of the entire board. This is seldom done; generally the change is squeezed in manually. It is at this point that soldering requirements are ignored.

The problem is even more severe when the board is laid out totally by hand. The complexity of this task causes many solder-oriented design violations. The guidelines published by the IPC (Ref. 1-3) are helpful in performing this task and should not be ignored.

Here are some tips to help the designer understand soldering needs. They are also useful in making the manufacturing personnel aware of design flaws. Remember that a single design error is repeated on every assembly made. It is often cheaper to redesign a board than to correct the fault in production.

- To avoid bridging with rectangular leads, lay out the board with all single and dual inline packages ((SIPs and DIPs) oriented perpendicular to the wave. In other words, make sure that the device travels parallel to the conveyor direction. In this fashion, the long side of the rectangular lead moves through the wave with a minimum of turbulence, minimizing bridging. Under no circumstance should components like integrated circuits be mounted in both directions.
- For controlled board transport through the wave, remember that the conveyor is a vital part of the solder system. It determines the depth of immersion and the uniformity of penetration into the solder. The designer must, therefore, make sure that the board is properly held in the fingers or pallet. The rules are simple:
  a. Select the direction of board travel over the wave. This should always be the narrow dimension of the printed circuit board for mechanical stability, to minimize board warpage during the solder heating cycle. In addition, make sure that the prints specify this direction in relation to the preferred orientation of the laminate weave, as indicated by the laminate trademark.

     Sometimes the direction of travel is dictated by other considerations, and the wide dimension of the board must be used. In that case, be sure to provide a stiffener (see Section 3-18) or support whenever the board is over 7-9 in. (17.5—23 cm) in the holding span.
  b. Leave 0.125 in. (3.2 mm) on the long edges of the board free of all components, joints or traces. This is the area where the board will have to be supported by the fixture or finger conveyor during soldering. Components interfere with the conveyor and require expensive fixtures. The solder joints give erratic results because of the turbulence caused by the conveyor moving through the solder. And the traces elevate the boards in the conveyor, lifting one side or both, making them ride high in the wave.
  c. An odd geometrical outline, other than square or rectangular,

must sometimes be used. To avoid expensive fixtures, consider leaving them in the web. They may be precut and returned to the web, or just held by small separators to be broken after soldering. This makes assembly, insertion, soldering, and cleaning easier. It is recommended for all small boards, to speed production and lower costs (see Fig. 1-16).

- To protect gold fingers, place them on the leading edge of the board. This will enable production to use a simple edge protector over the gold. This nonwetting rail prevents flux and solder from reaching the leading edge. It does not protect the sides or the receding edge properly. The rail can also be used as a stiffener to prevent board sagging during soldering (see Section 3-18 and Fig. 3-2).

  Gold fingers on all other sides require more expensive application of tape. Special conveyor fingers and channels can be designed to provide mechanical protection on the side of the boards. These are used only in machines dedicated to one type of boards.

- To avoid warpage due to weight distribution. Bending and twisting of the board during soldering is due to the softening of the material by heat above the laminate's glass transition point (T_g). With larger boards, the laminate structure cannot support its own weight even at room temperature. Once it exceeds the T_g temperatures, resin strength diminishes dramatically. At present, the glass transformation temperature of known laminates is below the melting point of the tin-lead solders.

  The proper weight distribution on the board thus becomes vital. Place all heavy components off the board if possible. Otherwise locate them as close to the supported edge as possible. In addition, try to distribute the weight uniformly on the board rather than concentrate it in one location. Use temporary or permanent mechanical stiffeners whenever these rules must be violated (see Section 3-18).

- To minimize heat damage due to uneven thermal requirements. Uniform heating of the board is important for high-quality results. Solder joint integrity depends heavily on the heat profile of the fillet. To obtain good results, then, isolate or avoid large heat sinks connected to pads. They make uniform heating of all solder joints difficult. This is especially true for those machines with a bottom preheater only. Machines that also have a top preheater can overcome more uneven top heat requirements.

  Uneven heating is caused by large components or large ground planes on the circuits. In either case, this problem cannot be overcome by higher preheating or longer dwell time. At the same time, we must be careful not to overheat or scorch low heat content areas.

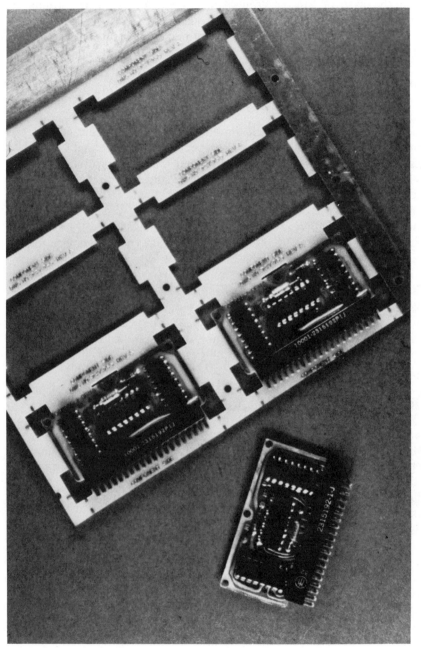

Fig. 1-16. View of PC boards in nest. (Copyright Manko Associates)

51

In the case of a large ground plane, etch out several sections around the hole. This will create a quasi pad, with several lands leading from it. The *cart wheel* that is formed conducts only a fraction of the heat through the spokes. This technique, of reducing the heat sinking problems, is used extensively in the inner layers of multilayer boards.

Large components can also create heat sink problems. If there is no practical way to increase the top preheat, it is necessary to slow the conveyor. A longer dwell time in the solder is usually helpful. Sometimes it is possible to modify inspection criteria around a problem component. If the joint is not needed for mechanical strength, relaxing the requirements for the amount of solder fill on the top may be possible.

- To lower the labor content of manufacturing (cost savings). One of the most neglected areas in our industry is value analysis. Designers and component engineers often select the least expensive device for their assembly in the mistaken belief that they save money for the company. These components are often *none wettable* which means that they cannot be wave soldered and/or cleaned inline. The cost of blocking the holes with a temporary resist, followed by hand insertion, soldering, and cleaning, often exceeds the purchasing savings by 200 to 300%. In addition, it encumbers the operation and defies full automation. It is mandatory that all nonwet devices be eliminated from future board designs.

1-20 BOARD ASSEMBLY FOR GOOD WASHING

The design of the assembly is important to the success of the cleaning operation. Board layout, component insertion mode, and the presence of other hardware determine the *cleanability*. In addition, the complexity of good cleaning conditions depends heavily on board and component compatibility with the process (see Chapters 3 and 6). Let us take the printed circuit assembly as an example.

The most difficult area to clean on the board is underneath components and subassemblies, especially for surface mounted devices. These areas are shielded from the cleaning fluid flow and thus cannot be readily flushed. The problem is compounded by the geometry of the gap created between the component and the board. This space may retain liquids or be entirely blocked. In either case, total residue removal is impractical. For details on the cleanability of this critical gap see Section 6-20.

To evaluate the cleanability of a printed circuit assembly one must consider many factors. Common-sense, however, makes it possible to evaluate any assembly in light of the cleaning equipment. An actual trial coupled with a cleanliness check will guarantee the results.

1-21 SOLDER RESIST (MASK)

A solder resist (or mask) is an organic coating applied selectively to the bottom of the printed circuit board. It restricts the contact area with the molten solder. As a result, only the exposed areas of the board can be wet. This is usually a permanent coating to be left on the end product. It offers many advantages, like fillet control, increased electrical insulation, and greater simplification of circuit design. Solder masks are essential for all surface mounting applications.

The material selected for a solder resist depends on the final substrate (rigid or flexible printed circuit boards), the cost of the assembly, and the environment in which it has to operate.

Solder resists are generally made of materials such as melamines, epoxies, acrylics, polystyrene, and polyimides, and sometimes contain polyurethanes and silicones. For further details, see Section 3-16.

It is essential that these materials form a relatively smooth, hard surface, or solder webbing will result (see Section 7-13). In addition, the materials must withstand soldering temperatures, as well as exposure to flux and cleaner solvents.

The least expensive materials are screened on and depend on a drying-curing reaction. The more expensive screenable formulations are thermally set. Both types should be used, with as little volatile plasticizing agents as possible. Low-temperature curing materials are also desirable, since the solderability and integrity of the printed circuit board are at stake. Ultraviolet (UV) -curable resist requires no temperature to catalyze the crosslinking and thus the curing.

These resist films must have good insulation properties, abrasion and thermal shock resistance, low moisture retention, good adhesion, and good chemical resistance.

Solder masks are mainly applied in one of two ways: screen-and-cure, or as a photosensitive film (wet or dry sheet) that is exposed and developed. The screen-and-cure method is usually cheaper but requires good housekeeping of the screens; otherwise the solder resist will bleed or migrate over the pads, causing solderability problems. To avoid resist spread, many designers try to pull the solder resist away from the metallic pattern to be soldered. They lose vital fillet control (see Section 1-18) but allow for bleeding, migration, and poor registration without suffering a reduction in solderability. With care bleeding can be kept to a minimum, although the heat of curing still is a problem. The UV-cured resists do not normally bleed, and are preferred for screened applications.

Wet or dry photographic film materials, on the other hand, are normally much more expensive. However, they allow precise pattern development due to the optical process used. They are presently the best type of system for fine line boards and thus for surface mounting.

The dry film resist is laminated to the board surface, and the pattern is exposed and developed much like a photographic film. The quality and precision achieved are better than those of any screening process. Dry film resist is recommended for all applications where solder mask is applied over tin-lead-coated boards. A similar film can be applied wet by curtain coating or similar techniques. Because of the late manufacturing stage of resist application, wet films are not very popular.

Of the standard platings used in printed circuit boards, tin and tin-lead are the only fusible alloys applied. They melt during the soldering process due to heat exposure in the wave. Any solder resist applied over such a liquid base is easily disrupted and often floats or flakes away. This creates a flux entrapment hazard and should not be used when nonrosin fluxes are applied. Because of its increased strength, the dry film resist is usually retained in place.

In all resists, there is hydrostatic pressure and surface tension of the molten solder trapped under the mask. This tends to destroy the coated surface uniformity, and as a result the appearance is crinkled. Crinkling of resist over fusible coatings is perfectly acceptable and not a cause for rejection in standard through-hole boards, provided potential flux entrapment is not a problem (this is true for certain rosin fluxes). For fine line boards, such as in surface mounting, this condition is dangerous. In closely spaced patterns, it is possible for the molten solder to be extruded from one land to the other, causing a short or a tightly adhering solder ball. To minimize this effect, it is customary to keep solder thickness in the plating to a minimum or to air-level the board (see Section 1-24). The ideal solution obviously is to use the solder mask over bare copper (SMOBC), or nickel plating.

The insulation characteristics of the solder resist are very important, especially in those cases where no conformal coating is used on the soldered assembly. In many ways, the solder resist helps to provide a partial conformal coating over much of the circuitry. The solder mask greatly reduces the amount of sensitive area that could be contaminated. By changing surface resistivity to volume resistivity, the effect of ionic contamination is reduced. This has prompted many designers to use solder resist even on the top (component side) of the board.

In addition, the solder mask reduces the danger of electrical shorting between metallic conductors and bare component leads. This increased insulation is also important because it allows closer spacing of electrical conductors on the surface.

The solder mask is also very cost effective. It reduces solder consumption and slows solder contamination. Solder resists usually pay for themselves by the savings they bring in solder consumption alone. In many cases, the added advantages make them outstanding improvements.

To summarize, the major advantages of the solder resist are as follows. They:

- Reduce solder consumption.
- Slow solder contamination.
- Simplify printed circuit design (bridging, etc.).
- Provide fillet control.
- Permit close spacing of conductors covered by resist.
- Provide a semi-conformal coating to the end product.
- Are usually very cost effective with increased quality.

Rather than talk about the disadvantages of solder resist, it is important to stress some production hints which eliminate difficulties in their application. These are as follows:

- Make sure the surfaces are absolutely clean for good solder resist adhesion and good resistivity under high humidity.
- Restore the solderability of metallic surfaces if it is lost during application and curing.
- Avoid air bubbling during screening to eliminate the dangers of webbing, flaking, and flux entrapment.
- Crosshatch large ground planes to minimize heat absorption and flaking.
- Find a balance between housekeeping (screening and registration techniques) and fillet control.
- Avoid getting too much solder under the resist (plated or fused) to minimize crinkling and solder shorts; use air leveling.

1-22 TEMPORARY SOLDER MASK

Temporary solder resists are often needed for production. They are usually used to block off metallic areas from the wave solder for future use. In some cases, components have to be hand soldered at a later stage because of cleaning or process limitations. Other surfaces, like gold plated edge connectors, must be kept solder free in order to function properly. In all cases, these resists must be removed either before or during the cleaning cycle, because they tend to trap some flux underneath.

The temporary solder resist basically fall into four categories as follows:

1. Soluble coatings. Various formulations are available which can be applied to the surface by standard techniques. While some are suitable for screen printing, most are applied by painting or dubbing on the surfaces to be protected. Later they are removed by the inline cleaning process. These materials can be water washable or solvent soluble. A fast evaporating solvent or an oven drying thinner like water is used to thin the material. Upon evaporation they leave a suitable resist behind. During washing, the stable heat-resistant filler is floated away, and only the binder dissolves.

2. Peelable materials. These are available in liquid and tape form. They are applied manually or by machine to the area to be protected. They are designed to be mechanically peeled off the surfaces. The nature of any adhesive residue left on the board is critical. It must be soluble in the cleaning process and can leave no detrimental residues.

3. Mechanical devices. In order to leave printed circuit board holes open for future use, one can load them with a nonwettable mechanical blocking material. Items such as wooden toothpicks, and especially molded dummy components with plastic leads, are used. These moldings are usually made to be reusable.

   The placement of these devices may become a high labor content operation unless the dummy components can be machine inserted or at least made a part of the standard assembly procedures.

   When gold plated fingers are located at the leading edge of the printed circuit board, it is simple to clamp on a nonwettable channel (made of tungsten, titanium, or high temperature plastics). This will prevent the contact with the flux and later the solder, thus leaving the gold plating intact. These channels have the additional advantage of helping to keep the board rigid. They are often used as mechanical stiffeners even if no gold plated surfaces exist.

4. Machine mounted devices. Mechanical guides built into the wave itself prevent solder from reaching connectors and gold plated fingers on the side of the board or in specific locations.

1-23 PROTECTIVE COATINGS

A protective coating is a temporary film which is intended to protect only the solderable surfaces during storage, handling, and assembly. It is normally applied to the entire board in recognition of the fact that a factory is not a clean environment. Protective coatings are mainly intended for copper surfaces; they make no real contribution to tin- or solder-coated boards. These coatings prevent heavy tarnish buildup but do not guarantee a tarnish free surface underneath. They are excellent for preventing dirt from settling on the areas to be soldered, thus aiding solderability.

Some of these protective coatings also contain an active fluxing ingredient, like organic acids, amines, or amine-halide activators. They are sometimes sold on the understanding that no additional flux will be needed in later soldering. This, unfortunately, is not entirely so, since the component leads have no similar coating. In addition, the liquid flux has many other beneficial properties for most processes.

The coating must be compatible with the flux, since it is not removed prior to soldering. For details on flux to protective coat interactions during wave soldering, see Section 4-7. The coatings fall into two categories, depending on the flux and cleaner to be used, as follows;

1. Rosin compatible coatings. These coatings are usually made of rosin or resins applied in a fast drying solvent. They soften in the flux thinner and/or melt during the preheating stage. As a result, they can mix well with the flux and permit its action on the surface. This flux action is necessary in order to remove minor films of tarnish that develop on the surface between solderability restoration and the application of the protective coating. In addition, it is obvious that air permeates through the coat during storage, and more tarnish is formed.

   These protective films have a limited useful storage life; thereafter, they either become neutral or hinder soldering and require removal. The quality and shelf life of the protective film depend on the plasticizer used. These beneficial additives evaporate with time or are lost during curing (overdrying). Once the plasticizer is lost, the coating becomes brittle and subject to crazing and flaking. It also becomes hard to melt and difficult to dissolve in the flux.

To obtain the best results from the protective film deposit, here are some hints;

a. Film thickness determines storage time, and vendor recommendations should be followed. One important parameter in thickness control is the solid content of the material when applied. This can be monitored by density control at the manufacturing location.

A convenient check on coating performance is the use of an accelerated aging test. Here the copper surfaces are exposed to steam and sulfides (see Test 1-1).

b. Rosin coatings are prone to overcuring, and the drying cycle recommended by the vendor is most important. In the case of plasticizer loss during overcuring, surface protection diminishes, and the coating becomes tenacious and will hinder the use of flux and solder.

It is, therefore, necessary to have available a usable stripping method which does not damage the assembly. Vapor degreasing in a solvent that boils near the melting point of the resin (approximately 170°F) will strip most of these materials. Strong solvents are also efficient but may damage the other board coatings.

c. Overaging of the film during prolonged storage also results in plasticizer loss and poor soldering characteristics. As with overcuring, the aged coating should be stripped prior to fluxing and/or assembly.

2. Acid-compatible coatings. These coatings may be simple cover films (like creosol-phenols) or *conversion coatings*. A conversion coating is a layer of metal-reaction products formed by the attack of a chemical on the surface. It is normally removed by the organic acid flux used.

The cover films are soluble in water and/or alcohols and easily penetrated by the flux thinner. Because of the strong nature of the organic fluxes used in conjunction with these coatings, the role of the coatings is more cosmetic. They also help to avoid gross contamination from poisonous materials such as sulfur and dirt. The quality of these films, and their ability to withstand harsh environments can be verified with Test 1-1. This is both a qualification test, and an inline process control check.

Test 1-1 ACCELERATED AGING OF COPPER WITH STEAM AND SULFIDES

This procedure is a modification of the standard aging test used for solderability checking (see Section 8-7). It involves a modification of the steam aging procedure called out in MIL-STD-202 Method 208, MIL-STD-833 Method 2003, and the IEC test. In this modification, the addition of ammonium sulfide is intended to check the quality of organic coatings over copper. It approximates a shelf life of 6 to 12 months.

In the standard test, a 2000 ml. Pyrex beaker with a watch glass (aluminum foil) cover is used. The test solution (the water) is kept boiling. Some means is provided to suspend the printed circuit board or components above the boiling liquid. The Pyrex beaker is heated on a laboratory type hot plate. For smooth boiling, some chips are included (Teflon coated or porcelain). In the modified test, a small amount of ammonium sulfide is added to the boiling water.

The samples are suspended about 1 in. (2.5 cm) above the boiling liquid for 15 minutes plus or minus 1 minute. The samples are then ready for the standard solderability test. This should be carried out within 1 hour of the completion of the accelerated aging procedure.

If the test is repeatedly performed, a stock solution of ammonium sulfide should be prepared and stored in a brown storage bottle. This solution is prepared by dissolving 5 ml ammonium sulfide (23%) in 1000 ml distilled water. The solution is good for 1 week. For the test itself, 200 ml of water is used, with the addition of 1 ml of the stock ammonium sulfide solution discussed above. The diluted test solution is good only for 1 hour of boiling.

1-24 FUSING AND LEVELING

Solder plating has historically been applied as a pattern-plated etch resist in the subtractive process. It was a fortunate selection because it also provided a solderable coating with an extended storage life. Solder plating as well as pure tin have also been applied to component leads, integrated circuits and similar devices. The solderability of these coatings and the process of fusing them, especially on printed circuit boards, is the subject of this section.

While solderability and storage are not the only motivations for tin-lead plating on printed circuit boards, we will deal with those aspects only. The quality of the electroplated deposit depends on the process used and emphasizes the need for good process control. The major points of interest in this respect are as follows:

• The solder plating must be applied over copper which has an anchorage area that will ensure wetting. During soldering, the plating on the surface becomes molten because it is a fusible alloy. The flux, however, does not reach the interface in question and cannot remove trapped tarnish. A simple solderability check prior to plating will ensure these results (Ref. 1-7).

- The electrolyte should be free of contaminating metal ions which might codeposit such as copper. The contaminants are introduced into the bath from low purity anodes, the plating chemicals, and the work when current is not flowing. They can be removed by plating (dummying) onto a corrugated cathode.

  The bath should also be free of nonmetallic particles (sludge) which cause rough plating. Constant filtration and bagging of the anodes help keep the solution clear.

- Control and maintenance of the organic brighteners used in the plating operation is necessary. The brighteners act to level the electroplating, making it more uniform and shiny. They are codeposited on the work, usually at the high spots due to electrophoresis. The metal fills the low spots by electro- plating, resulting in a shinier finish.

  With time, these organic inclusions deteriorate and the coating loses solderability. In addition, during the soldering process, the heat causes thermal decomposition, which results in outgassing and blow holes.

  Carbon treatment of the bath is normally used to remove old organics. This should be done periodically and before the addition of any fresh brighteners. Above all, keep the quantity of the brighteners to a minimum, following the instructions of the manufacturer.

In-process controls over plating can be established only if the operation is done in house. While the quality of the end product can be evaluated by solderability testing, this practice is time- consuming. For this reason, the industry has gravitated to the use of fusing, reflowing and leveling. By these processes, the plated surfaces are converted to hot, wetted coatings. They are designed to give the user 100% quality control monitoring of the incoming printed circuit boards. The advantages are as follows:

- These processes heat and melt the plated solder coating. This provides a 100% solderability check on all boards received.
- The melting process rectifies many plating deficiencies (expels organic brighteners, covers the edge of the conductor, etc.)
- Treated surfaces are shiny and uniform, and cosmetically more appealing than plated and etched surfaces.
- Some etching defects, like undercutting and solder slivers, are corrected by these techniques.

On the other hand, treating the plating by reflowing, fusing, or leveling is a costly operation and must be economically justified. In addition, it involves at least one more heating operation which may cause the plastic materials in the printed circuit board to deteriorate. The process is suitable only for the more expensive grades of laminate.

Depending on the position of the board during reflow, the solder redistributes itself and tends to "sag." This in itself is no problem, but must be considered as part of the operation. Let us define the various terms and the methods which are used:

The term *reflow* is often used in conjunction with the preparation of the printed circuit board in the unassembled stage. It is, however, more conventional to refer to reflowing whenever a solder joint is made by fusing two pretinned surfaces together, as in the case of surface mounting (see Section 5-23 in Ref. 1-1). The term "reflow" thus is not appropriate, *fusing* is the preferred terminology. In fusing we obtain a coating of molten solder, and the quantity is determined by the thickness of the plating before the heat is applied (see Table 5-6, Ref. 1-1).

The fusing processes are:

- Fluxing and immersion in a hot heat-transfer medium (oil, polyethylene glycol, etc.) to melt the plated surface.
- Coating the surface with a fluxing media and applying infrared radiation.
- Vapor phase condensation—an operation similar to vapor degreasing, using a high boiling solvent. The immersion in the vapor phase of the equipment causes uniform heating and reflow (see Section 5-11).

Solder "leveling" is a method that results in the removal of much of the plated alloy and leaves behind only a thin coating of solder. The basic advantage claimed for this process is the fact that the coating cannot mask any poor surface conditions underneath. It therefore yields a higher degree of reliability. The methods used for solder leveling are as follows:

- Mechanical force to remove excess solder.
- Hydrostatic pressure in the form of molten oil or salt impinging on the surface.
- The use of hot air impinging on the surface through special knives (air leveling).

Hot air leveling is a recent development that holds a great deal of promise. The fluxed board is totally submerged in the molten solder for wetting of all exposed surfaces. As the board is removed from the molten bath, two jets of hot air (above the melting point of solder) are applied

HOT–AIR LEVELING

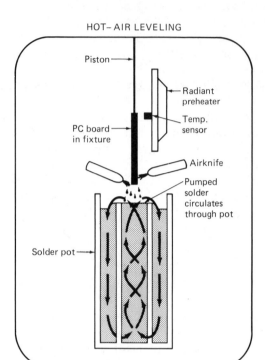

Fig. 1-17. Schematic of air leveling process. (Courtesy Hollis Automation)

from opposite sides, to remove excess solder, and clear the holes. Obviously, any poorly wetted solder will be removed by the air jets. This also makes it suitable for tinning of additive circuitry. In addition, any imbalance in the tin-lead ratio during plating is corrected by process. The principle of this equipment is illustrated in Fig. 1-17.

A less expensive process uses a hot reflow medium after the solder immersion. In all other respects, it is similar to the air leveling process (see Fig. 1-18).

Fig. 1-18. The "Lightning Machine" for hot reflow leveling. (Courtesy Alpha Metals)

REFERENCES

1-1. Howard H. Manko, *Solders and Soldering*, 2nd. Ed., McGraw-Hill Book Co. New York, 1979.

1-2. IPC-CM-78, "Guidelines for Surface Mounting and interconnecting Chip Carriers." The Institute for Interconnecting and Packaging Electronic Circuits (IPC), Evanston, IL.

1-3. IPC—The Institute for Interconnecting and Packaging Electronic Circuits. (Formerly the Institute For Printed Circuits) 3451 Church Street, Evanston IL 60203.

1-4. Howard H. Manko, "Selecting Solder Alloys for Hybrid Bonding," *Insulation/Circuits*, April 1977.

1-5. Howard H. Manko, "Organic Acid Fluxes—A Misnomer,"*Circuits Manufacturing*, March 1980, pp. 22-23.

1-6. MIL-P-28809, Printed Wiring Assembly, 1975.

1-7. Howard H. Manko, "Solderability A Prerequisite To Tin-Lead Plating", *Plating*, June 1967.

2

STORAGE, KITTING, ASSEMBLY, AND
PRESOLDER OPERATIONS

2-0 INTRODUCTION

A variety of operations take place between the receipt of components and printed circuit boards, and their use. In this chapter, we will deal with those steps that Affect soldering directly. This discussion should provide insight into a number of operations without outlining a precise sequence. It is feasible to perform only some or none of these presolder steps in house. The specific operations will depend on the process flow selected.

One major difficulty in this area stems from the division of control responsibilities. Purchasing should buy only what Engineering specifies, and at the lowest price. Receiving and/or Inspection should check all essential items, and those against the same engineering documents. Of course, Engineering should initiate the specifications with end product quality (reliability) and production ease (cost) in mind. Unfortunately, this chain of events is often neglected, and each department operates with little regard to the total picture. Production inputs are also kept to a minimum, beyond expediting to meet the schedule. This lack of coordination is costly in terms of unnecessary touchup and repair, and it reduced product reliability. If the U.S. electronics industry wants to remain competitive on a global basis, management must intervene.

When the incoming items reach storage, control changes hands again. The objectives must include such important items as stock rotation for *first in—first out* (FIFO), maintenance of a clean environment (bins), the right inner-packaging materials, and so on.

Getting an organization to work toward the same goal appears simple, but human factors often complicate this intentions. The reader is urged to study his own organization and, through discussion, to iron out most wrinkles. While no department is more important than another, each should be able to make its needs known. Value analysis is the best approach to the resolution of differences.

2-1 ENGINEERING SPECIFICATIONS FOR INCOMING MATERIALS

Each organization has different ways of generating these vital documents. This is usually the result of the the specific table of organization and historic developments within the company. The scope of this book does not allow us to delve into the intricacies of the process, but we will view the necessary ingredients for good specifications.

Here are the major elements needed:

- Electrical properties to fit the circuit needs (tolerances, mean time between failures (MTBF), etc.).
- Physical requirements to withstand the rigors of the anticipated use environment (thermal cycles, humidity, etc.).
- Physical size and shape to meet layout objectives (density, height, etc.).
- Ease of assembly to reduce costs (automatic insertion, in line soldering and cleaning, etc.).
- Solderability to eliminate costly touchup and repair.
- Washability to prevent costly manual operations where cleaning is anticipated.
- Inspectability to make visual inspection or automatic testing possible.
- Cost effectiveness based on a value analysis of the above factors and competition in the marketplace.

Note that the price per si does not enter into the consideration (see Section 10-2). A low cost component that must be hand inserted after soldering is usually much more expensive in the total picture (see Section 1-20). The added manufacturing cost outweigh the price differential with a suitable component. Consider the steps needed here:

1. Blocking the holes with a temporary mask.
2. Removing this mask after inline soldering.
3. Installing the component by hand soldering.

In a typical case, a single unsealed three pole switch (six leads) cost $0.84 to install this way. In addition the same switch caused maintenance problems in field service after approximately 1 year of use. The cost of a sealed, reliable switch was less than the price differential, and the manu-facturing operation became much more stream lined.

2-2 INCOMING MATERIALS AND THEIR INSPECTION

Incoming control is vital and can generate valuable information. Vendor quality performance should not be lost in columns of statistics. It must be tabulated and provide a feedback to Purchasing and management. Suppliers with poor track records should be warned. If no improvement is noted, they must be barred from making future deliveries. Remember the all-important value analysis concept: You must equate the unit purchase price with the incremental cost of rejection and/or production. There are four major groups of items that require attention in this category:

- Soldering chemicals and alloys.
- Printed circuit boards.
- Electrical components (with leads for through the hole (PTH) attachment).
- Surface mounted devices.

Let us review these in more detail:

1. *Soldering chemicals and alloys.* These are materials such as solder, flux, oil, cleaners, and dross reducers that are used in the soldering process. They should be purchased to specification and tested accordingly. Because of the proprietary nature of these formulations, the incoming test requirements should be written in conjunction with the vendor. For a detailed discussion of incoming tests, see Chapter 8 in Ref. 2-1.

When a number of different chemicals are used in the same process, it is necessary to check them for compatibility. Buying the complete line from one manufacturer may simplify this task. Obviously, relying on a single supplier is risky, and a second approved product package must be tried and held ready for substitution.

2. *Printed circuit boards.* They must be checked for solderability, as well as several other related properties. These include:

- Mechanical dimensions for a good hole-to-wire ratio.
- Width for stress free transport.
- Laminate thermal stability to resist measling and delamination during solder exposure.
- Coating quality and registration (i.e., resist, mask, protective coatings).

• Ability of packaging materials to preserve quality during storage.

The quality of the plated-through hole requires special attention. It can be examined and measured destructively by cross-sectioning. It is also possible to assess the hole quality by nondestructive means (see Section 8-9). Selecting the right board and/or hole for testing requires some knowledge, because the current distribution during plating results in the skimpiest deposits at the center of each panel and board. Normally, boards are manufactured in a cluster called a *panel*. This panel is cut into individual boards before delivery. Often a test coupon is designed into the center of each panel for a critical quality review. It is not unusual to have this coupon delivered still attached to a board for lot traceability.

3. *Electrical components.* Here we are concerned with devices that have leads for plated through hole or terminal attachment. These include axial and radial leaded components, single in line packages (SIPs) and dual inline packages (DIPs), and a variety of special components like switches, relays, transformers, and coils.

The solderability of the leads is naturally of primary importance; this is covered in detail in Section 8-7. Here too, packaging to preserve the finish integrity is a vital part of the program. But we cannot leave the selection of the lead wire or terminal finish entirely in the hands of the supplier. As with government specifications, we should avoid the use of hazardous coatings like gold and pure tin. Thus we must insist on buying only those components that are coated with an acceptable finish.

4. *Surface-mounted devices.* These are the components that are not anchored into a plated-through hole or around a terminal (see Chapter 5). They are held by small lap or butt joints to the surface of the board. They include leadless chips, small outline packages with gull winged leads, chip carriers with "J" leads, and leadless chip carriers.

Because of the reduced size of the solder fillet, termination metallurgy is very important (see Sections 5-4 and 5-5). It is also much more difficult to measure the solderability of these small terminations (see Section 8-7). We have not yet developed good accelerated aging techniques for these devices, and must study them more closely using laboratory procedures. In general, the metallurgical integrity depends on the presence of a diffusion barrier to stop intermetallic compound formation (for details, see Chapters 3 and 4 of Ref. 2-1).

In general, it is necessary to use common sense in setting up this program. It is not possible to anticipate all problem areas and establish a comprehensive, fail-safe procedure. With experience,however, specific areas of concern will become obvious. Then you can establish unique trouble shooting patterns to help keep your costs within reason.

2-3 STORAGE CONDITIONS VERSUS SOLDERABILITY

Storage of materials is an integral part of the logistics of manufacturing required to keep the lines going. It represents a compromise between good financial management and smooth production flow. On the one hand, a minimum of capital must be tied up in dormant stock. On the other hand, adequate supplies are needed to maintain production without depending solely on vendor deliveries. The problem is aggravated by prolonged delivery times for many items.

Companies usually monitor stock levels and provide physical control of the material to avoid theft. However, very little attention is given to the technical aspects of storage. We will cover only those parameters associated with soldering. They can be grouped into three categories:

1. Preserving solderability.
2. Avoiding moisture in boards
3. Storage of chemicals

1. *Preserving solderability.* This is a very important consideration when setting up storage controls. Let us briefly review the technical side of the issue. There are elements surrounding the work which affect solderability. They can be divided into those that are relatively benign and those that are detrimental:

a. Benign storage factors

- Oxygen
- Heat
- Humidity

These three elements actually cause the solderability of industrial finishes to deteriorate, but to a tolerable degree. For example, let us look at a tin-lead coated part. The solder coating is protected from the effects of oxygen once an initial tarnish layer is formed. While the oxides continue to build up, the rate of growth is very slow.

The formation of oxide is definitely accelerated by high temperatures, but these would be beyond human tolerance in the storage area. A chemical or metallurgical reaction such as oxidation would not vary much in the range we consider hot or comfortable.

Humidity also affects the tin-lead surface, but only if any contaminating chemical is present. For details, see page 14 of Ref. 2-1.

We can thus see that the aging parameters we call "benign" do affect the surface, but only to a limited extent. In our example, solder oxides are easily removable by even mild fluxes. The storage life under these conditions is strictly a function of coating thickness and the metallurgy of the system.

b. *Detrimental storage factors.* The list of such factors is long and involved. Many unusual circumstances can and do occur in industry which defy cataloging. The most common ones in the author's experience are as follows:

- Sulfur
- Chemical fumes
- Plastic fumes and silicone oils
- Dirt

These degrading contaminants react with the surfaces in two ways:

- The first two, sulfur and chemical fumes, react chemically with the surface finish. The resulting chemical products, like sulfides, are often not removed by the fluxes. Copper or silver and sulfur are good examples of this reaction. The same holds true for many other chemical fumes, which cannot be categorized easily. Chemical fumes can come from plating, etching, or the chemical laboratory, to name just a few examples of potential sources.
- The second group, plastic fumes, silicone oils, and dirt, coat the surface with an impenetrable layer. The flux, therefore, cannot reach the tarnish layer beneath and is rendered useless. Some types of silicone oils and greases are especially harmful because there is no effective way to remove them.

2. *Avoiding moisture in boards.* Every laminate can hold minute quantities of water, which varies with the resin in the laminate. Glass epoxy like FR 4 may contain moisture in quantities that are measured in parts per 10,000 of its own weight. Other resin systems, like polyimide, will absorb much more. In addition, crevices are formed during machining of the edges and drilling of the holes. Corrosive solutions may be absorbed into these cracks during wet processing like plating or etching. At a later stage, all this water generates steam at soldering temperatures and causes blow holes and other defects (see Section 7-17).

The amount of moisture absorbed is a direct function of the humidity in the air and the exposure time. It has been established empirically that a relative humidity (RH) of less than 50% at 65°-75°F is always safe. What is more, there are some indications that many boards can even be stored at 60-65% relative humidity in the same temperature range without trouble.

For higher humidity conditions, prebaking may be needed (see Section 2-12). Remember that the copper wall dimensions (thickness) have a large bearing on the generation of blow holes.

Attempts to exclude moisture with plastic bags and/or desiccants have failed. Such means only slow down moisture reabsorption by the boards as the humidity penetrates the bags. In general, most of the exposure to moisture occurs in the plant during assembly.

3. *Chemical storage.* Soldering and cleaning chemicals need to be stored using standard safety precautions. Suppliers furnish such instructions in both general and specific terms (flash point, acidity, etc.). It is also possible to obtain the standard U.S. Occupational; Safety and Health Administration (OSHA) safety data sheet with specific storage and medical data for your files. Finally, since regulations vary by state, county, and municipality, you must have copies of local ordinances to make appropriate provisions.

In general, both fluxes and thinners are flammable, and only small quantities should be stored in the production area. They must be stored in safety cans and placed in a special metal cabinet. The balance should be located in a special indoor area or an outdoor shed. If you purchase chemicals in drums or cans for economy, be sure that they are kept tightly closed. Full containers should be stored in a dry area or covered with canvas. Once a drum is tapped, it should be stored on its side to prevent water from accumulating on top and getting inside. Keep the drum tightly closed when not in use to stop moisture from condensing on the inside. If the drum is used up slowly, you may want to attach a simple air dryer to the intake valve.

The solder metal should be stored in a secure area because of its high intrinsic value. Keep it free of oil and water to prevent spattering when loading it into the molten reservoir.

Storage of spent chemicals and dross must also follow safety regulations. Some may be sold as scrap for recovery (dross, spent solvents, etc.). Others must be carted away for proper disposal at a relatively high fee. Consult your supplier and safety personnel for specific details.

2-4 GOOD STORAGE PRACTICES

A storage facility must also be managed so as to enhance the physical conditions that preserve solderability. The control aspects are simple and involve a good FIFO system. It will prevent the unnecessary aging of older parts while recent acquisitions are used immediately. Color coding by month and year offers one simple method of control. Stock cards, labels, or tags can be used to mark the date each lot was received. These should always be attached to the outside of transparent plastic bags, since most paper contains sulfur.

It is more difficult to achieve the right physical storage conditions. Remember that oxygen, heat, and humidity are always present and are considered benign to solderability. The effect of humidity on printed circuit boards may require segregated storage facilities. Thus a general storeroom requires no air conditioning beyond that required by the comfort of the employees. The printed circuit boards are best stored in an area with 55% relative humidity or less.

The storage area should be away from any laboratories or operations like plating and etching that generate chemical fumes. It is practically impossible to prevent such fumes from permeating the surrounding area. Only expensive positive air pressure techniques will keep the storeroom clean. The same holds true for plastic fumes and airborne dirt. It is a good idea to have solid walls and to avoid open wire mesh. This may have to be done only for electronic components and printed circuit boards.

Of the detrimental storage factors, sulfur requires special attention. It is present in all corrugated cartons and in most wrapping paper. Only special sulfur-free paper should be placed next to the components, and this is a relatively expensive grade. Thus it is necessary to place a barrier between the components and the sulfur source in the form of plastic bags. We must distinguish between the outside shipping container and the inside packaging material.

When parts are received in an unacceptable container, they should be repackaged. This provides low-cost insurance for troublefree soldering. It is also possible to implement a program that would establish rational package sizes. We must always endeavor to place no more than a few days worth of parts in one bag. This way the parts will not lay open to plant contamination for extended periods. Similar considerations apply to kitting.

Ideally, all parts in the storeroom should be placed in a reusable, washable plastic or metal tote box. The cheaper carton boxes, even those with a special coating of wax or plastic, are not recommended. The many cuts and edges in a folded carton give off sulfur, although at a slower rate.

In addition, the parts should be packaged in plastic bags to prevent their contamination by the environment. Oils, dirt, plastic fumes, and the like will settle on the outside of the bag, not on the components. A heavy plastic bag, 4 mils thick, is recommended. These bags may be reused until they become contaminated. Some have a "zip-lip" closure, while others are simply folded over and taped. Heat sealing is excellent but must be controlled to prevent plastic fumes from settling on the contents. Plastic bags offer the additional advantage of being transparent, allowing the contents to be seen. Some grades of static-free bags are opaque and require special handling. Remember to place any paperwork or shipping tags on the outside of the bag.

2-5 RESTORING THE SOLDERABILITY OF BARE BOARDS

The need for solderability of all surfaces during production is undisputed. A full definition of solderability, along with the test methods and solderability restoration procedures for incoming components, printed circuit boards, and so on, appears in Sections 8-6 to 8-8.

In this section, we deal with the inline processing of unassembled printed circuit boards with or without solder resist. Some of these boards will have a protective coating (see Section 1-23). If the boards have lost solderability, the protective coating must be removed.

The loss of solderability at this stage is normally due to reactions on exposed metallic surfaces. This may happen during production, in process storage, or during final storage. While fluxes are designed to overcome normal tarnishing, they often are not strong enough to remove the effects of environmental poisons. In addition, exposure is accompanied by sedimentation of various organic fumes and by accidental contamination. These airborne pollutants settle on the work in dry or wet form. Such foreign materials inhibit the later action of the flux, causing solderability problems. Finally, the solder resist curing temperature increases the danger. The contaminants are baked on the surfaces and become more tenacious.

The methods of solderability restoration at this stage can be roughly divided into three major categories:

- Mechanical removal
- Chemical treatment
- Tinning and leveling

Remember to clean the work by degreasing or washing before any of these steps are attempted. Surface contamination alone may be the cause of the problem, and this cleaning may be enough to restore the boards.

1. *Mechanical removal.* Use abrasion with pumice or a similar means. This is a very common form of treatment during board production. It usually involves passing the boards between two rotating brushes impregnated with abrasives. The process is simple and inexpensive, and yields a cosmetically attractive metallic surface. Unfortunately, brushed copper has been accepted by many uninformed people as having good quality. In reality, however, this treatment often cause the embedding of many parti-

cles in the surface. Serious solderability problems result because the abrasive is nonmetallic and thus nonwettable. The pumice drastically reduces the anchorage area between the solder and the base metal.

It is highly recommended that the use of mechanical abrasion on soft copper be eliminated wherever possible. The only corrective measure for the removal of embedded pumice is a chemical etch. However, this is a drastic treatment which also weakens the copper.

Brushing without an abrasive is also dangerous because of bristle smear (plastic or metal). This is sometimes used for the removal of undercutting overhang. The author has seen serious failures as a result of this practice. During the scrubbing, chemicals were trapped under the ledge and pinned down by the brushes. With time and humidity, these corrosive products leached out to the surface, causing failures (see Section 7-25).

The brushing of solder-coated surfaces, with and without pumice, is also undesirable. The reason, however, is different: Embedded pumice or bristle smear will float off the fusible tin-lead coating. This problem is found in the knee (inside edge) of the plated-through hole. This area, because of wetting geometry, has a very thin solder coat, and even slight abrasion can render it unsolderable. While this hole circumference is easily bridged on the bottom of the board, it may cause the solder to stop rising to the top. This condition is often misdiagnosed as a crack on the top fillet (see Section 7-29).

2. *Chemical treatment.* Reacting on the surfaces with chemicals is a much more reliable process. With this technique, etching liquids are used to remove the top layers of the metal together with the tarnishes and pollution particles. This exposes rough but very solderable surfaces. Unfortunately, they are not uniform and appear mottled and unattractive. However, this method is highly recommended to technically oriented organizations that are concerned with excellence and not with cosmetic appearance.

If the removal of conductor metal by etching is not desirable, there are many proprietory solutions available. These are designed to attack only surface tarnish without removing the metal underneath.

3. *Air leveling.* This process removes excess solder from the board by an external force. The solder on the board is rendered molten and removed, leaving only that solder which is well adherent or wet. The forces used may be centrifugal, acceleration, or hydraulic. Air leveling is a special technique developed for solder coating of additive PC boards. It is also used for the reflow of plated subtractive boards (see Section 1-24).

This method works because a strong flux is applied to the bare board. If the flux is not strong enough, chemical stripping may precede the retinning. Aged protective coatings can also be removed chemically. If the first tinning application does not give full coverage, it is possible to repeat the treatment.

Solder leveling of boards is an excellent repair method for overaged or defective printed circuits. The boards should be dry, and therefore pre-baking may be necessary. They are then fluxed with an aggressive flux and processed. If board solderability is not successfully restored, a second fluxing and leveling step will prove adequate in most cases.

2-6 RESTORING DISCRETE COMPONENT SOLDERABILITY

Components that do not meet the solderability requirement can be restored by methods similar to those used with bare boards. The problem is complicated, however, by the fact that most assembly houses do not have wet chemical facilities. In addition, electronic components are shipped in packages that do not always lend themselves to individual or batch treatment (on reels, molded holders, etc.). For this reason, individual unit retinning is becoming more and more popular.

Let us look at the three principal methods available, in the order in which they should be tried:

- Cleaning by washing or degreasing
- Descaling by wet chemical treatment
- Retinning

Here as with boards, cleaning is required before the last two steps— descaling or retinning.

1. *Cleaning*. Many device terminations do not solder because a layer of dirt coats the surface. The contamination prevents the flux from reaching the tarnish and promote wetting. Thus a good cleaning operation can restore solderability.

It is difficult to define an applicable process for each case. The diversity of equipment and processes is too great for universally valid instructions. One simply has to try what is at hand. The methods available are listed in order of preference:

a. Water washing with a saponifier.
b. Cold solvent dipping.
c. Vapor degreasing.

In all cases watch out for component damage, including:

- Attack on the outer casing or seal
- Removal of marking
- Swelling
- Entrapment of cleaning solution
- Electric deterioration

2. *Wet chemical treatment.* Many simple restoration liquid formulas are available for this purpose. One need only identify the metallic surface to be treated (e.g., copper, nickel, tin-lead). Thus it is easy to select the right chemical for surface preparation, which must be applied according to the manufacturer's instructions. Most of these chemicals are cold fumeless formulations, and the immersion time is seldom more than 1 minute. Thereafter, they require water rinsing and drying. Some common descaling solutions are listed in Table 5-1, (pp 174-175) of Ref. 2-1. In some cases, a protective coating may be applied to cover the freshly restored surfaces.

3. *Retinning of leads.* If desired, parts can be made solderable by using a strong flux and a solder dip. The aggressive flux residues are then washed away, and the parts are ready. This method is prescribed by many government contracts.

The tinning can be preformed on a dedicated machine with a rotary turntable and dip or wave modules. It can also be preformed directly on the production wave machine. In both cases, the fixtures holding the parts are of major concern. They require careful design and planning.

A single tinning process may not be adequate, depending on the metal-lurgy of the surface. A double dip is considered a must on soluble platings like gold or silver (excluding thick films). It is often advantageous on fusible alloy platings like pure tin and solder. In stubborn cases, a double dip is also helpful for bare base metals like copper or steel. In all cases, the first dip serves to strip the critical interface, while the second flux and solder application wets the surface.

2-7 RESTORING SURFACE-MOUNTED DEVICE SOLDERABILITY

This is a relatively new field in the printed circuit industry. The experi-ence gained in the last 30 years of *hybrid* manufacturing can only serve as a guide (hybrids is a term used for assemblies using a ceramic base sub-strate with thick film or thin film circuitry and surface mounting). This hybrid microelectronic industry is used to clean room processing stan-dards and scientific control of the operation. This is not the case in the printed circuit industry, which is poorly managed and controlled.

Surface mounted devices for printed circuit applications can sometimes be restored, but the handling problems are unique. The procedures are similar to those discussed in section 2-6. The cost is usually prohibitive because of the high labor content, and retinning makes the devices difficult to assemble.

The best approach is to avoid the need for any additional handling of these delicate and minute components. This can be achieved only through good engineering/purchasing specifications, followed by careful incoming inspection and an intelligent storage policy.

2-8 ASSEMBLY, LEAD CUTTING, FORMING, AND BENDING

In this section, we will deal only with the assembly of conventional leaded components. The assembly of surface mounted devices on circuit boards is relatively new and unique, and therefore it requires more detailed discussion (see Chapter 5).

Design and assembly requirements dictate the configuration of the solder joint (i.e., bent or straight through leads, etc.). The design of fillets has been discussed in light of the analysis of the five zones (see Sections 1-5 to 1-10). From this analysis, it becomes obvious that neither strength nor conductivity requires the clinching of leads. A review of industry practices clearly points to a shift in the direction of straight-through leads. Even components that are automatically inserted are only clinched to 30° maximum in order to approximate the straight lead. Let us, therefore, review the overall advantages of the straight-through configuration at the various levels:

- *Design.* Straight through leads conserve space (real estate) over clenched leads, simplify pad layout, and do not stress components during insertion. They are also conducive to the addition of surface mounted devices.
- *Assembly.* Assembly is easier when no clinch is required. This, however, requires inline processing, because the assembled boards need careful handling. The loose components may be jarred out, and the board cannot be turned upside down. Precutting and shaping the leads can greatly reduce this difficulty.
- *Soldering.* Perpendicular leads give much better uniformity, permitting good fillet control. This also causes fewer rejects.
- *Inspection.* Visual inspection of fillets on straight leads requires less time (up to 70% saving) over bend leads. The eye can detect faults easier and faster because of the uniform fillet profile.
- *Repair (replacement).* Repair of straight through solder joints in the

factory by trained operators, or in the field by unskilled personnel is simpler. A bent lead in the joint requires simultaneous solder melting and lead straightening. Remember the dangers of excess heat!

There is no doubt, therefore, that barring unusual circumstances, the straight-through lead is preferred.

Economic considerations should dictate the best insertion equipment to be used (manual or automatic, etc.) but should not influence the configuration of the fillet. The degree of automation involved is normally a function of the complexity and quantity of assemblies. A cost study comparing labor to investment provides a guide for equipment selection.

The subject of cutting component leads before and after soldering needs careful consideration. In general, component leads fall into two distinct categories according to the hardness of the base metal:

- Soft leads, mostly copper and copper base alloys
- Hard leads, usually iron base alloys

Historically, device terminations were made of copper, a soft and very conductive metal. They were also made of copper alloys like brass and bronze, which are harder.

In addition lead wires were made of iron alloys like Kovar, Alloy 42 (42% nickel and 58% iron), and the like, which are much harder. This last group was selected for its glass-to-metal thermal expansion matching properties (glass-to-metal seals). This TCE is also similar to silicon active devices used in semiconductor manufacturing. More recently, iron and steel leads have been used because of the cost savings.

The important areas to consider in relation to soldering and cutting are:

- Stresses on the component during lead bending
- Forces transmitted up the lead during cutting
- Corrosion potential of exposed lead ends after trimming
- Solderability and ease of restoration

1. Stresses exerted on the component body during lead wire bending and shaping may affect reliability. Cracked glass-to-metal seals, loosened encapsulants, and intermittent resistors are but a few of the recorded failures. The same consideration dictates great care in the force used during automatic insertion. The stress relief built into the lead wire bend is helpful only after soldering; it does not prevent damage during shaping and insertion.

2. Hard leads have more problems associated with cutting than soft

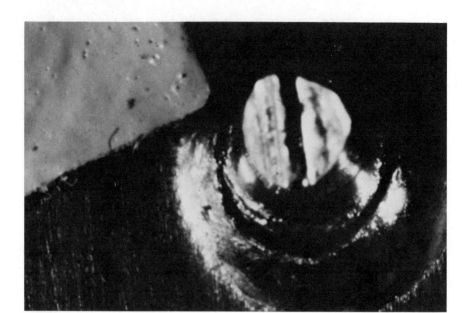

Fig. 2-1. Crack in a solder joint due to cutting after soldering. (Copyright Manko Associates)

materials. If a stress or shock is generated during cutting, it is easily transmitted along the wire. The force may cause damage to the solder joint and/or the component. Forces as high as 13 g have been recorded traveling up a Kovar lead when dull dykes were used. However, properly sharpened tools with uniform cutting motions do not cause such problems. Because of the transmission of forces through hard leads, solder joints have also been known to show signs of mechanical damage (see Fig. 2-1).

3. Corrosion considerations have caused worry about leaving the copper edge of the termination exposed. It has been argued that the galvanic potential poses a definite hazard. There is little danger, however, since galvanic corrosion depends on another ingredient—an ionic solution. The same electrolyte also causes current leakage on the board surface. The assembly would thus fail electrically before the corrosion mechanism could consume the metal.

This holds true for bare copper and copper alloys, which do not pose a threat to assembly integrity. However, the ferrous alloys have a tendency to rust in air without the electrolyte. The migrating rust has caused some bad field failures (see Fig. 2-2). It is recommended, therefore, that all steel

Rust

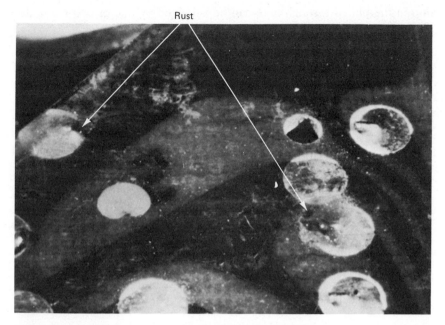

Fig. 2-2. The rust from these cut iron base leads migrated shorting out the circuit. (Copyright Manko Associates)

leads be resoldered after cutting. A magnet can help identify the magnitude of the problem on your own line. It is not unusual to find that 30-40% of the production leads fall into this ferrous category.

4. Solderability of the cut edge is a function of either lead material or finish. Let us review the effect of each on cutting. If the leads are cut close to the soldering time, even mild fluxes can promote wetting. When the raw edge of the termination is left exposed to the environment for longer periods, tarnish may form that mild fluxes can no longer remove. On the other hand, soft coatings like solder are smeared over the cut base metal, giving it some protection. Dirty leads create another problem; the dirt can be wiped over the cut surfaces, further retarding the effectiveness of the mild flux.

Component leads are manufactured to universal standards and cannot be influenced by the individual user. They are normally shipped loose, in packages, or taped on reels for automatic insertion. In this section, we will deal with the various assembly options in relation to lead cutting versus soldering, pointing out their advantages and disadvantages.

Getting from the standard supplied lead length to the final wire protrusion in the plated-through hole can be achieved by one of two methods. The options that are available for straight-through leads are:

- Precut, shape and insert
- Insert and cut

1. *Precut, shape leads, and insert.* This can be done for automatic assembly or manual insertion. Figure 2-3 shows a typical configuration for both modes of assembly. Note that the leads have all been shaped. Shaping is required for several reasons.

- Large heat producing components that must be kept off the board to dissipate the heat generated (resistors over 1-2 watt, etc.).
- Shaping provides a space under components for easy washing.
- Lead forming keeps insulation out of the plated-through hole (components that have an insulation meniscus on the lead like disk capacitors).
- Shaping with a spring lock action is also good for components that must be secured prior to soldering to prevent them from toppling off.

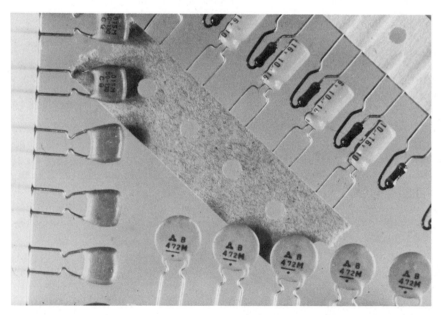

Fig. 2-3. Precut and shaped components. Note that some are taped for automatic insertion. (Copyright Manko Associates)

Shaping is the only suitable method for odd shaped components. It is done on small equipment in house for small and medium-sized production runs. Equipment is also available for mass production as an addendum to automatic insertion.

The precut length must be carefully planned and calculated. It is intended to end the need for trimming components again after soldering. This operation, unfortunately, is misused by many companies. It serves only as an interim cut, because it is too difficult to insert the full length lead wire. In that case, the lead is shortened just enough to make hand assembly easy, and a second costly cutting after soldering is required. In addition to the quality problems that ensue (see below), the economy of the process suffers. The author does not agree with the lame excuses that are often offered: "the precisely cut component leads are too short, and the component will fall out or rise in the wave!" Neither excuse is valid, because the shaping of the leads provides adequate mechanical security prior to soldering (see Section 1-18).

Cutting before soldering is usually done by hand tools or by precise jaws in equipment. After soldering, either hand tools or rotating blades are used, as described in the next section.

2. *Insert components and cut.* This popular production category involves various methods of holding the components while cutting. To understand the difference between them the issue of cutting before or after soldering requires serious consideration. Cutting after soldering should not be permitted for high reliability work because of:

- Danger of cracked solder joints
- Exposed ferrous lead ends
- Damage to components

For low cost assemblies, postsolder cutting is often used without resoldering. With the proper amount of care, the dangers listed above can be minimized or eliminated. No problems are encountered in such solder fillets when the following precautions are taken:

- The mechanical design permits cutting (correct hole-to-wire ratio, plated-through holes, etc.).
- No danger of rust development exists.
- The solderability of all surfaces is acceptable to prevent interfacial cracking.
- A good cutting device has been selected.
- The cutting edge is sharp and the motion correct.
- Careful high power inspection of suspect joints is feasible.

Metallic chips and debris left on an uncleaned board pose another hazard to the assembly. Careful cleaning after cutting is mandatory to prevent shorting.

In terms of the dangers outlined above, the inline cutting process with rotary blades is unique. It has the smallest mechanical damage impact, and the revolving surface tends to smear some of the solder coating over the cut surface, giving it limited protection.

There are process variations that can overcome many postsolder cutting problems. Their reliability levels vary widely. Let us review the list again in the order in which we will discuss them:

- Solder and cut only
- Solder, cut, and resolder
- Cold hold, cut, and solder

Each rationale of production must be evaluated against the potential hazards (reliability) and process costs:

1. *Solder and cut.* This operation may affect the quality of the work because of damage to the joints. It is used only at the low end of the cost and reliability spectrum for obvious reasons.

This process is used to trim when there has been no precutting or cutting was not precise. It is objectionable from the cost point of view because parts are cut twice. In addition, it often leads to excessive manual cutting, since operators will cut when in doubt (see Section 9-11).

This method is also used when the components are inserted in the length received. Rotary blades are positioned under the boards at a fixed distance (see Fig. 2-4). As the long leads pass the blades, they are trimmed to size.

2. *Solder, cut, and resolder.* This approach attempts to rectify any damage done by the first cutting operation.

- It repairs cracked solder.
- It wets the end of exposed leads.
- It tends to remove lose metallic debris.

The method has some disadvantages, however. Unless very heat-stable (expensive) grades of laminate are used, the first soldering operation tends to warp the board. This makes precise cutting impossible, and sometimes a board is cut and lost. In addition, the pad under the joint is above the glass transition temperature (T_g) as a result of the first soldering operation. The stress of cutting a hot board can cause lifted pads when the blades are dull or the laminate is inferior. Finally, solder consumption, as

Fig. 2-4. Closeup of leads being cut by a rotary blade (Courtesy of Hollis Automation)

well as the rate of solder contamination in the second pot, makes this an expensive operation.

The equipment used here varies; some processes use drag pots, while others use waves. A combination of drag and wave is also possible. A drag solder system is good in the first stage because it accommodates any size lead protrusion. Generating an equivalent deep wave is costly and creates a great deal of dross. On the other hand, the heat exposure in a drag system is much larger. Thus most systems use a wave for the re-soldering operation for the obvious advantages it offers.

3. *Cold hold, cut, and solder.* This process eliminates many of the disadvantages of solder-cut-solder. It provides mechanical anchorage during cutting below the T_g. There are no heat associated problems like additional board warpage before cutting. In addition, with this process pad peel strength is retained. The methods used vary according to the cold hold system as follows:

- A top weight is applied during cutting. This may be as simple as a bean bag or as sophisticated as a magnetic shaped and weighted cover that is automatically loaded and removed.
- A form-fitted plastic film (shrink-pack) is applied directly to the board. This is very effective when the components have a low profile

and are of uniform height. In this process, tall components become twisted and shorter components located nearby are left loose.

The adhering film may not allow flux fumes and expanding air to escape. This method has caused blow-holes and insufficient solder rise because of the back-pressure formed under the film.

- A top spray is applied that glues the components in place (spray webbing) and comes off during cleaning. This can be applied as a hot melt system or as an evaporating spray.
- A molten low temperature material that provides mechanical anchorage can be applied on the bottom.

The trade name for the last group of materials is "Stabilizers" (by Hollis Engineering), and the process is protected by a patent. Various wax-like materials are used that can be removed by the cleaning process. Some stabilizers are suitable for solvents and vapor degreasers; other organic compounds are water soluble. The higher melting materials reach 160°F and the lower products are in the 120°F range.

The process is simple. The wax is applied molten from a standing wave similar to a solder wave. There are two possibilities:

1. Use a standard flux, preheat the assembly to dry the flux, and then apply the holding compound.
2. Use a stabilizer and flux combination at the same time. In this process, the board is preheated first for good wax-flux penetration and then the composite formula is applied.

The holding material is allowed to solidify in order to develop its holding power. For the lower melting point materials, an air chiller is needed in line. Then the parts are rotary cut and wave soldered. The holding compound does not have to be removed before the soldering operation. It floats on top of the wave as a blanket or mixes in with the oil. The balance of the stabilizer material is removed in the regular cleaning modules.

Whichever method is used, one must watch for static problems in sensitive assemblies.

2-9 LEAD LENGTH VERSUS POSTSOLDER CUTTING

The length of protrusion of a straight-through lead below the board is important. It is controlled, on one side, by the Manufacturing department. Manufacturing desires long leads for ease of handling prior to soldering. They also fear that the component may ride on the wave and not settle back. On the other side is the designer, who wants to cram as much as possible into a narrow space. In terms of quality assurance, there is also a fear that long leads may bend after soldering to short adjacent circuits.

The dimensions recommended below are based on technical considerations only. For details on lead length versus fillet control see Section 1-18. These dimensions have been the standard at many well managed companies for a long time. They take into account all the points raised above, as well as the production cost.

- A maximum protrusion of 0.100-0.125 in. (2.5-3.2 mm) is used for good assembly practices and ease of inspection. It should be used whenever possible because of its simplicity and low cost.
- A minimum protrusion of 0.030 in. (0.76 mm) is used whenever design dictates close spacing. It constitutes the shortest inspectable lead under production conditions. In manufacturing terms this length is very expensive, difficult to inspect, and hard to achieve.

In general the following guidelines are recommended for lead length beneath the board:

Low cost—0.125 ± 0.030 in. (3.2-0.76 mm)
Medium cost—0.075 ± 0.030 in. (1.9-0.76 mm)
High cost—0.050 ± 0.020 in. (1.3-0.50 mm)

2-10 COMPONENT STRESS AND DAMAGE

The integrity of the solder joint cannot be considered independently of the electronic components. There are many areas where soldering and cleaning affect the reliability and life expectancy of devices. It is necessary to recognize these factors during component selection and to minimize them during assembly, soldering and cleaning. They include:

- Mechanical stress during assembly
- Heat effect
- Compatibility with soldering chemicals
- Stresses in place after soldering

Let us cover them individually:

1. *Mechanical stress during assembly.* This is normally due to the bending of component leads in preparation for insertion (see Section 2-8). In bent-over leads it may also occur during the clinching step. Both of these operations require forethought and planning. Remember, damaged components may fail only in the field.

Make sure that there is no stress on the area where the lead wire enters the component. In addition, it is important that the component be free to move after clinching. This will enable it to assume a stress free condition after soldering (see item 4 below).

For surface mounted components, where there are no long lead wires, similar precautions are indicated when the leads are formed ("J" and gull wing). For leadless chips, mass handling must avoid abrasion, since many chips have no outer shielding. The termination on all surface-mounted devices requires gentle handling.

2. *Heat effect.* Most components are heat stable enough for the rigors of wave soldering. This includes the heat exposure during prebaking of PC boards (see Section 2-12), the curing of surface-mount adhesives, the preheating before the wave, and the wave solder application. The heat effect is usually manifested in external damage to the plastic exterior and seldom in electronic functions. The heat exposure during cleaning and drying has the same effect. Note that in all of these operations the devices undergo uniform heating.

The most dangerous thermal damage occurs during touchup and repair with a soldering iron. The heat is more intense (higher) and is applied locally, causing thermal shock. Proper operator technique is vital and can minimize the damage (see Chapter 9). Touching the component lead wire with a hot tip above the board surface (on the component side) poses serious problems and should be discouraged.

3. *Compatibility with the soldering chemical.* The flux, oil, and cleaning agent must not attack the component in any way. Component marking, the external coating of the component, and its interior must be unharmed. The material or component supplier should be consulted for details on each unusual case.

4. *Stresses in place after soldering.* After the solder joint solidifies and cools, it is stress free. This is the result of the self-annealing properties of solder alloys. The solder, however, cannot relieve stresses set in the component or board during assembly (see item 1. above). If the component is free to float during soldering, it will assume a stress free position automatically. Bending components after soldering or cutting them may introduce new undesirable forces.

For the best results, solder only free floating components and do not move or realign them later. If a joint is mechanically disturbed, it should be reflowed to relieve any built-up stresses.

The unique conditions of surface mounting make all surface mounting devices sensitive to stresses due to TCE mismatches. For more details see Chapter 5.

2-11 CLEANING PRIOR TO SOLDERING

The need for contamination free surfaces prior to soldering is obvious. It must be stressed in all facets of the printed circuit manufacturing cycle. Foreign materials deposited on areas to be soldered will cause the following problems:

1. They hinder direct contact of the flux with surface tarnish. Dirt reduces flux efficiency significantly, causing solderability problems. The presence of silicone oil is a good example. This material is practically impossible to remove and is considered a poison to soldering. The dirt prevents wetting of the surfaces first by the flux and then by the solder.
2. Contamination which is not floated away during fluxing may cause problems in other areas, too. Most organic dirt compounds are not heat stable. Soldering heat causes them to decompose, creating gaseous materials. These contaminants, in turn, generate bubbling, blow holes, and trapped gas (voids) in fillets. Contaminants which retain liquids like water are especially troublesome in this respect (see Section 7-17). In addition, not all assemblies are cleaned after soldering. Contamination picked up before soldering will remain in place, with all of its associated problems. Some thermally decomposed dirt may not be washable, and, while functionally harmless, will cause cosmetic problems.

In an industrial environment, it is practically impossible to eliminate contamination while keeping production costs to acceptable levels. However, reasonable storage and handling conditions, coupled with good workmanship practices, can minimize dirt. The degree to which this can be achieved determines whether cleaning prior to fluxing and soldering will be required.

Economic and practical considerations dictate that precleaning should be used only when needed. Seldom is it necessary to include it permanently in the manufacturing flow. It is preferable to monitor and maintain cleanliness from receiving through storage to assembly and soldering. In those cases where precleaning of subassemblies (printed circuit boards, components, etc.) is necessary, follow the instructions given in Chapter 6. Availability of both equipment and material for precleaning play an important role in the decision.

In general, using a degreaser with a strong solvent is probably the best method. It will leave dry parts which normally absorb little if any cleaning liquid. This, of course, depends on the compatibility of the solvent with the assembly and the efficiency of the cleaner. If a suitable inline degreaser is available, the fully assembled but unsoldered boards can be passed through. This presumes that no components will be dislodged. Then the board can go directly into the fluxing station. A cool air knife is sometimes used before the fluxer. The air is designed to fulfill two functions. It removes any trapped solvents which might not have evaporated. The air knife also tends to cool the work so that it may be passed through foam fluxes without collapsing the foam head (heat causes bursting of the foam bubbles).

Water washing of assembled but unsoldered boards is not as easy because of the high pressure involved. The free-floating components cannot withstand the agitation. It is, however, an excellent method for cleaning the unassembled components or boards. They emerge very clean, and the process is milder than most of those using solvents. The boards, however, absorb a certain amount of moisture. Presolder baking may be required if they are to be used soon thereafter (see Section 2-12).

If no inline cleaner is available for this task, the use of a batch (kitchen type) unit should be considered. If components, for instance, are loaded into a closed wire mesh basket, they can be cleaned simultaneously to restore their solderability (see Section 2-6). Printed circuit boards can be stacked like dishes in the standard way (see Section 2-5).

For those who must prebake the boards, the precleaning stage is indispensable. Without cleaning, baking may cause the dirt on the board to harden, The boards will then lose solderability (see Section 2-12).

2-12 PREBAKING PRINTED CIRCUIT BOARD ASSEMBLIES

The baking operation just prior to wave soldering is intended to remove volatiles from the printed circuit laminate. These volatiles obviously come from various sources, but their action is generally the same. On soldering, they give off gaseous materials which cause blow holes and entrapped gas pockets (see Section 7-17). This gas evolution may create a force that ruptures the plated barrel and/or causes delamination.

The volatiles can be divided into several major categories:

- Surface liquids
- Trapped process solutions
- Organic volatiles
- Naturally absorbed moisture

1. *Surface liquids and moist films.* These are contaminants by nature and should be considered to be dirt. They are deposited on printed circuit boards and/or lead material during storage, handling, and assembly operations.

2. *Trapped processing solutions.* Such solutions residues as etching or plating solutions are the result of poor manufacturing practices. The quality and uniformity of the hole in the raw laminate are important, whether punched or drilled. If the laminate is damaged (cracked or fractured), the likelihood of processing solution entrapment is greatly increased. Unfortunately, such solutions are highly corrosive and cause problems from chemical bleed out. In the plated-through hole, they become trapped behind the plated barrel.

3. *Organic volatiles.* These materials result from the lamination, curing, and coating materials used in the fabrication of printed circuit boards. During these processes the temperatures, pressures, and chemical formulations may be slightly off the specified limits. This can result in a high volatile content. Multilayer and flexible circuits are especially sensitive to this problem. While this type of volatile is troublesome, it is the easiest to overcome by baking and leaves no permanent damage.

4. *Natural moisture.* The water content is inherent in all composite materials such as laminates, which absorb humidity. The quantity depends on the environment, because these materials attain equilibrium with the humidity in the air. It is universally true (except for polyimides) that the cheaper materials and laminates have a larger moisture content. The more expensive grades usually are made to absorb less water.

Prebaking the boards helps remove this moisture. As soon as the boards are exposed to room ambient, they will start to absorb moisture again. The rate and the quantity absorbed depend on the humidity in the area. Relative humidity below 55% at 70°F seems to cause no problems. Above 65% relative humidity and higher temperatures, water is reabsorbed relatively rapidly. Thirty minutes after the temperatures equalize with room ambient temperature, the advantages of baking start to be lost in a humid environment. Total equilibration of the humidity may take as long as 40 hours, but the rate is exponential.

If prebaking is required, a good precleaning operation is recommended (see Section 2-11). In most cases, foreign contamination has found its way to the assembly. Baking without cleaning may resolve the moisture problem but aggravate solderability problems. The contaminants may have hardened and become practically impossible to remove after baking. This hinders the flux action and, eventually, the quality of the solder joint.

Boards that were prebaked must not be allowed to reabsorb moisture from the room. They are therefore baked fully assembled, and should be soldered within 30 minutes of the time they reach room temperature. (Note: the rate of moisture reabsorption increases as the boards approach room temperature.)

The time and temperature of prebaking vary from case to case. They depend on the degree of dryness needed for the soldering process and the extent to which gas is evolving during soldering. The board materials and design are also important variables. The number of holes, the amount of metal not etched, the type and thickness of solder masks, and other factors all affect drying rates.

In general, boards are baked above the boiling point of water, 212°F (100°C) and below the glass transformation temperature (T_g). For various real and imaginary reasons, some organizations are afraid of these higher temperatures. Lower baking conditions for longer times are also possible. Table 2-1 summarizes the author's experience.

For fast thorough drying, use a temperature of 225–250°F (105–120°C) for 2 hours. Only rarely is this insufficient, and longer prebake times may have to be used. In many cases, a 1.5 hour bake may cure the volatile problem.

For those who do not wish to heat their assemblies above the boiling point of water, baking at lower temperatures for longer periods is possible. Remember that prebaking usually takes place with all components in place. Thus component temperature stability may become an issue.

For even lower temperatures which do not greatly exceed room ambient temperatures, vacuum prebaking is possible. A 2 hour bake at 120°F (50°C) is suitable. The vacuum in such an oven is normally generated only by a roughing pump and thus is not considered to be a hard vacuum (approx 1 torr or 1 mm Hg). One word of caution on this type of baking: Make sure that all gaskets in the oven are free of sulfur. The author had a bad experience with a replacement gasket which contained sulfur, and the fumes that were emitted completely destroyed the solderability of the assembly.

Table 2-1. Presolder Baking Temperatures.
(For volatile material removal)

TEMPERATURE RANGE (°F)	MINIMUM TIME (Hr)	AVERAGE TIME (Hr)	MAXIMUM TIME (Hr)	EQUIPMENT
225–250	1.5	2	4	Recirculating oven
160–180	3	6	16	Recirculating oven
120–130	1	2	3	Vacuum oven (1 torr)

Note that in the processing of multilayer boards and flexible printed circuits, various baking operations are performed. These are intended to cure materials, stabilize the structure, and so on. For obvious reasons, they do not obviate the need for presolder baking.

In general, boards should be free of volatile materials. Thus they require baking only when blow holes are discovered. Prebaking as a standard operation is costly and jeopardizes solderability. Only in humid parts of the country or during high humidity seasons can the routine baking of each lot be justified.

Plating thickness in the through hole also has a significant effect. It has been established that a copper plating of 0.001 ± 0.0002 in. (0.025 ± 0.005 mm) will prevent the natural moisture from escaping. Uniform, smooth holes are in the same category. Cracks, nodules, and similar defects in the plated-through hole also cause blow holes. For further details, see Section 7-16.

REFERENCE

2-1. Howard H. Manko, *Solders and Soldering*, 2nd ed., McGraw-Hill Book Co., New York, 1979.

3
SOLDERING AND CLEANING MATERIALS

3-0 INTRODUCTION

Soldering is a chemical-metallurgical process that is strongly material oriented. This imposes serious requirements on the selection of compatible material systems. The selection decision has to be made in the design stage; otherwise the process is encumbered with costly compromises. Appropriate material and equipment selection up front can lower manufacturing costs dramatically.

This book does not include a scientific treatment of the soldering process or of the materials involved. For a full technical discussion of the subject, the reader is referred to Chapter 2 (The Chemistry of Fluxes), and Chapter 3 (The Metallurgy of Solder) in Ref. 3-1. In the following sections, the materials are presented in the approximate order in which they are used in the process.

3-1 FLUX PROPERTIES

The flux is usually a liquid and is applied to the metallic surfaces to be soldered. The flux has specific properties as follows:

- It provides tarnish removal and mild cleaning action.
- The flux lowers the cohesive force of solder, aiding wetting.
- It protects the surfaces during heating from reoxidation.
- The flux moves readily out of the way of the spreading solder.

The soldering of printed circuit boards under production conditions without a flux is not feasible. Let us briefly review each property;

1. *Tarnish removal and mild cleaning action.* Metallic surfaces naturally tarnish in air. *Tarnish* is an all inclusive term for all the reaction products of metal in air. These include oxides, sulfides, and others, tarnish composition which depend on the impurities in the atmosphere. All engineering metals like copper, tin, and lead form tarnishes. Some of the noble metals like gold, and platinum are less prone to air reactions. Tarnishes are not wet by solder, and thus interfere with the bond formation.

Fluxes are specifically designed to remove tarnish from metals. The flux selection process thus considers the types of metallic surfaces to be soldered (see Chapter 4, "Designing the Solder Joint" in Ref. 3-1).

In addition, a surface collects pollutants from the air. These are mostly organic compounds of diverse sources like smoke, combustion emissions, and process fumes. This category also includes airborne particulats like dust and lint. These materials also do not wet and interfere with soldering. Fluxes are only partially able to remove small amounts of such materials. It is therefore advisable to protect our work during storage and processing from unnecessary exposure (see Section 2-3).

Finally, surfaces are contaminated with dirt. This comes from physical contact with other materials that hold the dirt. The contamination is transferred from one surface to another, ending up in the least desirable place—the work. Like tarnish and pollutants, dirt does not wet with solder. The nature of the dirt layer is often such that the flux cannot penetrate it. In industry, dirty surfaces are the biggest single source of soldering problems (see Section 2-11).

Selecting a flux can be a very scientific procedure, closely resembling a doctor's prescription of medication. The medical profession does not have the scientific knowledge to understand drug chemistry or the way a formulation affects a medical disorder. Yet the doctor links his prescription to generic classifications such as antibiotics, coagulants, and antihistamines. Furthermore, the qualified physician is certainly concerned with the side effects of each medication, and considers allergic and other adverse reactions.

The engineer must likewise select a flux to suit his specific assembly. Without delving into the proprietary nature of a formulation, he must choose a flux scientifically. The material must be matched to the application, without undesirable side effects. Chemical corrosion, electrical leakage, and attack on plastics and components are but a few of the problems to be avoided.

Considering the fluxes in this light, the ideal material would be completely inert at room ambient and yet active at soldering temperatures. Furthermore, flux residues, their pyrolithic breakdown products and fumes, should not harm the work at the operating ambient.

3-2 FLUX CLASSIFICATION

The best method available for the rationalization of flux formulations is chemical composition. The commercial approach of calling fluxes "corrosive" or "noncorrosive" is not acceptable. By definition, a flux must remove the tarnish from a surface which makes it corrosive to the base metal.

There are three basic categories of materials that encompass the soldering flux spectrum:

1. Rosin base fluxes.
2. Intermediate organic fluxes.
3. Inorganic fluxes.

The rosin fluxes have historically been associated with the electrical and electronics industries. They are made of a variety of natural and modified rosins. In addition, they contain a number of chemical additives labeled *Activators*. These activators are intended to give the flux more chemical strength for tarnish removal. Finally, they contain thinners to liquify them and some foaming agents where needed. This is the only family of fluxes that is covered by extensive government specifications. More details will be given later.

The intermediate organic fluxes have been around for many years. For the last 20 years they have become established in electronics. They are used by nearly half of the manufacturers who solder nonmilitary equipment. There are no government specifications to govern their electronic use.

The final category consists of the inorganic materials, which normally are the most aggressive. Straight acids like phosphoric and hydrochloric acids are used in conjunction with salts. Zinc and ammonium chloride make one of the strongest fluxes in this group. They are seldom used in electronic applications.

Our list would not be complete without mention of other states of aggregate besides liquids. Gases and solids, generally in the inorganic family, are also useful as fluxing agents. Reducing gases are common in the semiconductor field. Molten salts are used in applications like wire tinning. In general, however, they are more difficult to use and have not found their way into printed circuit board soldering.

Let us look at each major subdivision of fluxes.

3-3 WATER WHITE ROSIN (WW OR R)

Pure rosin makes an ideal flux. A solid at room temperature, it is both chemically inactive and electrically insulating. The same holds true for condensed rosin vapor and its metallic reaction compounds. Yet, when heated to slightly above its melting point, the rosin becomes active, being a mild organic acid. It reacts with a few metals used in electronic applications such as copper, silver, and gold.

Gum rosin is a steam distillate of pine tree sap and is graded by color. By designation of the American Society for Testing and Materials (ASTM), the purest grade of rosin is called *water white* or *WW* for short. In U.S. government specifications it is referred to as type "R". In industry it is also termed *nonactivated rosin*.

The major disadvantage of water white rosin is its inherent chemical weakness. A certain amount of presolder cleaning is mandatory in order to make it a reliable flux.

A number of solder joints are still being made with this flux, mainly in the telephone and telecommunications industries. It has also found many good applications in the more modern hybrid and microelectronics industry. Here the chemical activity of the flux must be restricted in order to prevent precious metal scavenging (leaching). In addition, the residues may not change the values of active and passive devices.

3-4 ROSIN FLUXES WITH ACTIVATING ADDITIVES

In an attempt to strengthen the fluxing properties of rosin, additives called "activators" are introduced. These materials fall into many categories of strength and behavior. In this regard, there is a substantial differences between the United States and other countries in theory and practice as expressed in national specifications. Rather than being classified from a historical view point, these fluxes are listed in order of increasing activity.

Such increasing strength, however, has little correlation with the chemical and insulation properties of the fluxes and their residues. The discussion to follow will also relate each material concept to the appropriate cleaning methods. For further details on cleaning, see Chapter 6.

3-5 MILDLY ACTIVATED ROSIN FLUXES (RMA)

This first category of activated fluxes can also be labeled "low activity rosin." The obvious intent is that full or partial flux residue removal is not required. However, the only truly "safe" flux for an application must be tested. Remember that contamination from other sources, like perspiration, can cause damage even if a safe flux is used. Let us explore the more important approaches to achieve this end:

1. *Mildly activated rosins in the U.S. philosophy as expressed in MIL-F-14256 and QQ-S-571 specifications.* These documents do not really consider the chemical composition of the flux. They only control the electrical and chemical requirements of the material before and after soldering. The manufacturer, therefore, is able to seek the most effective materials in this category.

According to this concept, the flux may contain any type of activator: halides, organic acids, amines, amides, and so on. The end result, therefore, is a flux formulation whose residues and recondensed fumes are noncorrosive and electrically insulating. In the opinion of the author, this type of flux can truly be considered harmless.

2. *Mildly activated fluxes in the British approach.* Here the amount of activator (all labeled as "chloride") are limited to 0.5%. This, in the opinion of the author, is a relatively weak specification, since it has little bearing on the quality of the work. It also provides no control over the type of activator employed. This enables the unethical manufacturer of fluxes to incorporate relatively hazardous materials into the formula. While it would meet the chemical requirements of the specification, electrically dangerous residues could be left behind. In essence, whole families of truly safe activators are eliminated.

3. *Halogen-free, mildly activated rosin.* This is a second flux philosophy which also relates to composition. The German DIN specification, in one of its two categories (DIN FS-W-32), classifies these fluxes as mildly activated. These fluxes are called "halogen-free", and the type of activator is limited to organic acids and similar materials. These, according to the specification, are relatively harmless.

Once again, the author does not agree with the philosophy behind the formulation of this flux. It limits the chemist to a small variety of chemicals, some of which have been proven to be rather dangerous to the work. This is especially so when the rosin is removed and the activators are left exposed due to inadequate cleaning.

In relation to cleaning, these three types of mildly activated materials are classified as both polar and nonpolar contaminants. They therefore require a double cleaning or a designed blend to remove both residues simultaneously. Only residues of the United States RMA flux can be considered harmless even in critical applications.

The mildly activated fluxes are much stronger than the nonactivated materials but are still relatively weak. They require solderability monitoring of all surfaces in order to achieve reliability and economy in soldering. These materials have found wide application in computer, aerospace, telecommunications, and military applications. They are uniquely suitable for surface mounting or touchup and repair where total flux removal is not feasible (see section 3-19 and Chapters 5 and 9).

3-6 ACTIVATED ROSINS (RA)

These fluxes are used for standard electronics manufacturing and mass production. They are also called "fully activated" rosin and were developed for applications where mildly activated fluxes were too weak. This group of materials constitutes about 85% of all rosin fluxes used in the United States.

Activated rosins in liquid form have only recently been included in the MIL-F-14256 specification. The category has been present in the core solder QQ-S-571 specification for many years. The German DIN classification F-SW-26 is very similar to this U.S. group of materials. The degree of latitude in activator selection is restricted by specification. These European materials are not quite as strong as their U.S. counterparts.

Activated rosins will leave residues which are not necessarily dangerous to many applications such as radio and TV assemblies. They have not caused any corrosion failures, but can conduct minute currents (leakage), especially when warm. Therefore, they are considered dangerous for critical applications.

In the removal of fully activated rosin fluxes, the use of both polar and nonpolar solvents is mandatory.

3-7 SUPERACTIVATED ROSIN (SRA)

The fluxes in this group consist of formulations that do not meet the government specifications. In their ultimate strength, they are very aggressive, and their residues are dangerous. They are so strong chemically that they will solder bare Kovar, nickel, some stainless steels, and similar alloys. The flux residues are definitely very active, and thorough removal is mandatory.

The need to use rosin in conjunction with high activity levels is dictated by the application or equipment on hand. A recent development in this area are the *Synthetic Activated* (type SA), which were designed to be highly cleanable in Fluorinated solvents. Some of the SA fluxes fall in strength into the RA category, although they are not covered by government specifications, others are in the SRA category. Many SRA fluxes are so strong that they can effectively compete with the next category of organic intermediate fluxes.

3-8 ORGANIC INTERMEDIATE FLUXES

Next we must consider the nonrosin organic base fluxes, or the *Organic Intermediate* category. These are also called *Organic Water Washable* or erroneously *Organic Acid fluxes*.

These materials, in concept, resemble the superactivated rosin fluxes, except that they are rosin free. Thus, cleaning no longer requires a nonpolar solvent, hence the name *Water Soluble*. There are no government specifications to cover this group, and therefore no restrictions on the amount or types of chemicals used.

We can best understand this family of fluxes by examining their history. In 1949, a U.S. patent (2,407,957), issued to J.E. Strader was assigned to the Battelle Memorial Institute. It describes a "self-neutralizing acid flux". The ingredients listed and the whole concept make this flux the grandfather of today's generation of organic acid fluxes. This formula also included one additional concept, namely, the use of a basic ingredient, urea, to neutralize the acids in the raw flux. A neutral flux, however, is still corrosive and conductive. Thus cleaning not only inhibits corrosion but enhances electrical performance as well.

The term "organic acid fluxes" or "O.A" for short, which is popular, is a misnomer. This flux category is made of ingredients such as organic acids, organic hydrohalides, amines, amides and similar materials. The term thus covers only some of the possible ingredients therein. The American Welding Society (AWS) in its *Soldering Manual* (Ref. 3-2), used a much more appropriate term for this flux category. Since the flux is stronger than a rosin base formulation and weaker than an inorganic salt type, the AWS labeled them "intermediates fluxes".

Since the average nonchemist views acids as dangerous substances, the choice of the term "intermediate" for these fluxes is more appropriate. While some materials, such as sulfuric or nitric acid, are extremely hazardous, many of the acids contained in these fluxes are present in our bodies and in many foodstuffs.

These organic acids include:

- Lactic acid (dairy products)
- Citric acid (citrus fruit)
- Oleic acid (natural fats)
- Steraic acid (animal and vegetable fats)

- Glutamic acid (protein)
- Benzoic acid (berries and other vegetables)
- Oxalic acid (spinach and rhubarb)
- Phtalic acid (medicine and perfumes)
- Abitic acid (major constituent of rosin)

In reviewing this list, we find that the dangers of organic acid fluxes may be unfounded. Still, management groups adopt organic fluxing methods reluctantly, because of the "acid" label. Note that rosin fluxes are truly organic acid fluxes; this adds to the problem of confusing terminology.

Organic intermediate fluxes justifiably are gaining popularity in industry because of quality and economic reasons. While management should use caution in introducing them into the line, they should not dismiss these fluxes out of hand. Being water washable, these fluxes are designed for efficient fluxing, easy removal, and economic washing. Therefore, when utilized properly, they can definitely solve many problems. It is unfortunate that it is too late to change the terminology.

3-9 INORGANIC FLUXES

The last category of fluxes is by far the most active and corrosive. Mostly based on acids and their salts, these fluxes are normally used only for structural soldering. This classification also includes fluxes made from basic ingredients or neutral salts. Most gases used as fluxes fall into this group, such as dry hydrochloric acid, hydrogen, and forming gas, which are used in the semiconductor industry.

The cheapest and oldest of these fluxes was called "killed spirits," and was made by dropping metallic zinc chunks into muriatic acid (HCl). The result is zinc chloride ($ZnCl_2$), which is a very temperature stable flux. To improve the formulation and lower its melting point, it can be mixed with ammonium chloride (salammoniac—NH_4Cl). This mixture is the base of most popular nonelectronic fluxes.

The white hazy film formed on the work is a zinc oxychloride which can be dissolved in a 1-2% hydrochloric acid solution. The mild acid is used in the first rinse and is followed by a number of plain water washes.

Other materials in this category are based on inorganic acids like phosphoric acid and its derivatives. Reducing gases, and dry hydrochloric acid gas are also used.

The tinning of lead wires during device manufacturing often requires this type of flux. Metallic leads like Kovar, nickel-iron, nickel-plated copper, or copper alloys under industrial conditions cannot be wet with a high yield using weaker fluxes.

Recently, during tinning, the temperature exposure of the device resulted in thermal stresses and some flux penetration. The chlorides in the flux were the cause of device failure in the field. Thus halide-free fluxes in this category are now being used by device producers for all tinning applications where there is no hermetic sealing.

3-10 SOLDER ALLOYS

Most solders are thought of as tin-lead alloys. Yet a large variety of other metal systems are suitable for low-temperature bonding. A *solder* is defined as a fusible alloy with a melting point below 800°F (425°C) (see Refs. 3-1 and 3-2). Soft solders are also available to cover the melting range of 100-600°F (40-315°C). They contain tin, lead, bismuth, indium, antimony, cadmium, and silver combinations. For those who are not familiar with the term, an *alloy* is a mixture of two or more metallic components (elements). For further information, see Chapter 3 in Ref. 3-1.

In the printed circuit industry, a certain amount of standardization has taken place. The most common solders used are the tin-lead alloys in the eutectic and near-eutectic compositions (63/37 and 60/40). The tin-lead system is popular because of its temperature range and it affords the best wetting for the price. Although there are many alloys with more desirable electrical or strength properties, none of them combines all these properties for the same cost of metals.

The question of which tin-lead ratio to use for a specific application is generally a compromise between price and metal fluidity. The cost of tin is over tenfold that of lead in today's metal market. Thus the price of the eutectic 63/37 tin-lead alloy (Sn63) is higher than that of 60/40 (Sn60). The eutectic also has slightly better fluidity and strength, as well as a sharp melting point. On the other hand, the 60/40 composition tends to bridge larger gaps, filling holes better if a large hole to wire mismatch exists. The bulk of the industry does not need the refinements of 63/37 solder and can use the more economical 60/40 alloy. The rule of thumb for wave soldering is *Use 60/40 solder for all applications except for multilayer boards, high aspect ratio holes, and surface mounting.* Government or costumer requirements may dictate the alloy for a specific contract.

A word about the addition of antimony is in order. The possibility of having an allotropic tin transformation at lower temperatures [Below 42°F (4.2°C)] exists. In this transformation the metallic tin changes to an amorphous (powdery) gray tin in a phenomenon also labeled *tin pest*. It was established that small additions of antimony have an arresting effect on this transformation. Thus the Federal specification QQ-S-571 calls for specific additions of antimony to the solder. In the United States no premium is charged for this deliberate addition, and it is highly recommended for general use.

There is nothing sacred about the amount of tin needed for wave solder alloys, other than the pasty range (see Ref. 3-1, pp 51-57). The author has developed a new group of alloys containing less tin but with larger antimony additions to improve its mobility and overall performance. These solders are labelled *functional alloys*, and are covered by a U.S. patent (Ref. 3-3). While somewhat less shiny than 60/40 tin-lead, they are easier to inspect, physically stronger, and economical.

The addition of silver in small amounts (2%) to the tin lead eutectic solder gives an alloy of 62/36/2 (SN62). This solder is costly because of the high price of silver. It is used to prevent the scavenging of metalized terminations in thick films and certain surface mounted components. It does not add to the quality of the solder in normal wave applications. There is, however, the danger of *silver migration*, a process whereby shorting occurs across a wet surface under electrical potential. This electromigration has been demonstrated in the laboratory for most metals but has not been reported for low silver-bearing solders in the field. There may be some pressure from chip component manufacturers to use such alloys for direct bonding of leadless surface mounted devices to printed circuit boards. See Section 5-5 for more details.

A word about bismuth alloys and indium solders for low temperature application and better slow cycle fatigue is in order. These low melting solders have found some special applications in surface mounting, and multi layer boards. They can be used at or below the T_g of many resins. For more information see Chapter 3, in reference 3-1.

3-11 GRADES OF SOLDER METAL

The quality of solder alloys is also a function of purity. In order to protect the consumer, many government and industrial organizations have issued purity specifications as safeguards. Unfortunately, these specifications are very broad and misleading. In reality, the level of solder contamination is an indication of the material source. The price, unfortunately, always increases with the purity. Let us look at the two basic grades of solders and the specifications covering them.

- Reclaimed Solder
- Virgin Grade Solder

1. *Reclaimed solder.* This grade metal is made from scrap tin and lead, which can be purchased below metal market prices. The proper refining of the scrap through regular metallurgical procedures is not cost effective. The only time such expensive processes as electro-refining can be used is when the scrap contains precious metals like gold. In other words, reclaiming pure tin and lead from scrap would cost more than extracting these metals from high-grade ores.

During World War II, the U.S. tin supply was greatly curtailed and its use had to be controlled. Reclaiming and refining of scrap became one of the major sources of solder. Federal specification QQ-S-571 and the ASTM standard therefore reflect the level of contamination permissible in a reclaimed and refined metal. This grade of solder is seldom used, however, in electronic mass production. The relatively high and inconsistent impurity content might cause serious problems by introducing additional variables. It is interesting to note that even manufacturers of radiators and refrigerators have switched from reclaimed materials. They use higher grades of purity in order to eliminate much of the costly rework and rejection in their automated setups.

2. *Virgin-grade solder.* The term *virgin grade* refers to alloys which are made of metal extracted from ore. The reader is reminded, however, that not all metal extracted from ore has high levels of purity. Additional controls over purity must be exercised in purchasing of these raw materials. The levels of contamination as specified in government and industry are not strict enough for the purity associated with this grade.

It is suggested that users of virgin grade solder write their own specification. This should be modeled after the QQ-S-571 outline for alloy tolerances, antimony content, and so on, except that the impurity range should be matched to the virgin grade, as outlined in Table 3-1.

The presences of tin-lead dross as well as other nonmetallic impurities in the solder is also important. There are no easy scientific tests for measuring the dross content other than vacuum or rosin melting. The observant operator, however, can gauge the dross content when melting down a fresh pot. Such nonmetallic additives as sulfur and arsenic, even in small quantities, have been shown to affect soldering adversely. No practical way to measure or control these impurities on incoming inspection is available, and expensive instrumental analysis is required.

3-12 SOLDER CONTAMINATION DURING USE

The incoming metal is not the only source of metallic impurities in a pot or wave. In the printed circuit soldering process, the work and molten solder are in contact. Several metals from the work can dissolve in the molten solder the way sugar dissolves in water. This action is associated with a detrimental effect on the process yield and joint quality. Before discussing these problems and their remedies, let us consider the source of the metallic contamination.

In a well-designed piece of equipment, the walls of the molten metal reservoir, the pump, and all surfaces in contact with the solder are made of a nonsoluble material. These include such metals as cast iron and stainless steel. They do not contaminate the solder unless a temperature over 800°F (425°C) is reached. Some units are also coated with porcelain, which is not reactive at all. The carriers, fixtures, and holders must also be made of compatible materials (titanium, tungsten, etc.). For further information on materials of construction see pp 224-227, or ref. 3-1.

Contamination of the bath during use can result only from contact with the work itself. This means that only a limited number of elements are picked up, depending on the nature of the work. In the dip solder pots used for tinning, this normally involves copper and zinc from brass, and so on. In wave soldering of electronic assemblies and printed circuit boards, it usually means copper, gold, and silver. In other words, a solder bath can only be contaminated with those metals with which it comes in contact and which are solder soluble.

To simplify the discussion, consider a single metal that contaminates the solder reservoir. Here a freshly charged bath contains only the elements present in the raw material (see Table 3-1). For best quality and economy, the purest solder available should be used. Once the reservoir has been charged with solder, the impurity dissolution starts. The amount of metallic pickup is a function of production volume for a set of standard soldering speed and temperature conditions.

Contamination pickup however reaches a steady state condition in the equipment. It does not climb indefinitely because of the dilution effect. From the outset, as the work passes through the pot, solder is dragged out on each joint. Therefore, the solder in the pot is constantly depleted and must be replenished with fresh solder. The solder which maintains the level in the pot comes in at a high purity value and thus constantly lowers the impurity level. Experience has shown that after a bath reaches equilibrium, the rate of metallic solder contamination is equal to the dragout and dilution with fresh solder. This creates a ''steady-state'' level at which that particular solder module operates. This contamination level becomes typical for the reservoir and the production line (Ref. 3-1, pp. 195-198).

When the contamination levels rise, the quality of the solder joints suffers and the process yield is reduced. There is no clear cut rule, unfortunately, to define the level of impurity that makes the solder unusable for a specific product. This problem defies mathematical or theoretical analysis because of the large number of variables involved. These include the soldering parameters (time, temperature, etc.) and the quality standards of the user, for whom the levels of shine or the roughness of the fillet may vary.

From a practical production-quality standpoint, this issue is not as serious as it may seem. Each organization can readily determine when the solder becomes a problem. At that point, two separate actions can be taken to prevent recurrence:

1. *Remedial action.* This is needed when the solder bath life is considered too short. The amount of contamination per unit work dissolved in the solder should be limited. This is achieved by either lowering the pot temperature or shortening the soldering time. It is also possible to expose less surface to the solder by incorporating a solder resist or masking part of the surface (see Sections 1-21 and 1-22).

b. *Preventive steps.* These are taken to eliminate future problems due to the rapid or gradual buildup of impurities. This implies a quality control warning system and a knowledge of the contamination levels that caused the problem. A general guideline can be found in Table 3-1. If you do have data about your own process, use it for control. Eighty percent of the impurity concentration found to be harmful to the operation is recommended as the solder change point. In either case, a quality check should be introduced which spells out a maximum allowable concentration of the element in the solder.

Production experience indicates that under control the same solder can be used repeatedly for many years. Even without being changed, the alloy never reaches detrimental contamination levels. Where contamination has proved to be the reason for production problems, changing the solder is mandatory. The ideal situation is a total changeover, which means shutting down the line for several hours. If this is impossible or undesirable, only half of the solder can be replaced, thus diluting the contamination level. This takes much less time and leaves the heating elements covered with liquid solder for efficient heat transfer. However, it also leaves a dirty solder module, and therefore should be done only when the routine maintenance schedule does not call for a cleanout. During solder changeover, there is a good opportunity to perform the appropriate equipment maintenance.

A word of caution is needed about cleaning the equipment in relation to impurities. After draining the solder and before recharging the equipment, remove all the nonmetallic dross and burnt flux adhering to the walls of the equipment. A relatively large amount of such material is formed on the inside of every solder wave or pot, even without agitation. No harsh metal but only a bristle brush should be used. This prevents exposure of the metallic walls of the equipment to the fresh molten solder. The heat scale formed on the equipment serves as an efficient contamination barrier, while shiny scoured surfaces might lead to iron contamination of the fresh solder.

In order to obtain the best quality and economy from a solder pot, Observe the following rule of thumb:

1. Charge equipment with the purest grade of solder available (virgin grade).
2. Operate at the lowest soldering temperature feasible for the operation.
3. Use the highest conveyor speed for the shortest time possible.
4. Mask off metallic surfaces which do not require soldering.
5. Replenish only with the best grade of solder available.
6. Eliminate clipping and drippings from molten solder.

To summarize, the quality of your work hinges on the purity of the solder. In the normal operation of the line, when trouble starts, you may want to follow the following corrective measures:

- Make certain that it is the contaminated solder, and no other parameter in the manufacturing operation, that is causing the problem. The unnecessary replenishment of solder is costly in terms of labor as well as materials.
- If the solder is at fault, change it completely, cleaning out the dross material, as described earlier.
- Cast the used solder into a shape that is easy to ship. It can be sold for a substantial fraction of its original cost.
- Record the contamination level in the discarded solder and recharge with the best available grade.
- Use one or more of the preventive measures listed in the rule of thumb above to minimize future contamination in the pot.
- Set up a quality control procedure to monitor the contamination level. Require a routine change of solder when it reaches 80% of the contamination level recorded.

3-13 LEVELS OF SOLDER CONTAMINATION

The results of an analysis of many industrial problems indicate that there is no one, clear-cut level of contamination that affects soldering. Problems depend entirely on the production line and its parameters—the flux, the time and temperature in the solder, the type of solder used, the type of joint created, the quality of the base metals, and similar factors. However, there are definite ranges in which trouble does occur, and the following discussion is a result of this accumulation of information.

Five basic contaminants are found in most solder pots (see Table 3-1 for details). Note the basic differences among the raw materials sold. Listed in columns 4 and 5 are the empirical observations over the last 20 years. The first figure in column 4 indicates the lowest level at which users reported problems. This level is considered very low and is not ordinarily a problem. The second figure in the same column shows the highest concentrations used before solder changeover. This is by far too high for most applications.

Column 5 shows the recommended level of impurities at which the solder should be changed. This should be observed even if no obvious trouble has been noted. It is a precautionary step to ward off increased touchup costs and reduced reliability.

It is interesting to note the large spread in the case of gold. A high value of 0.3% and a low value of 0.03% have been recorded. Because of the high value of gold, it is economically feasible to sell the used solder for its precious metal content. Once the solder reaches a breakeven point, the scrap value equals the cost of a new charge of pure solder.

Table 3-1. Contamination Guidelines for Wave Soldering.

ELEMENT	TYPICAL ANALYSIS		EMPIRICAL OBSERVATIONS		MIL-Std-454	TINNING[c]
NAME	VIRGIN GRADE	ASTM QQ-S-571	TROUBLE NOTED	RECOMMENDED POT CHANGE	MAX. LEVEL FOR POT CHANGE	
Copper	0.002–0.010	Max 0.080	0.200–0.500	0.300–0.350	0.300	0.750
Gold	<0.001	Max 0.08[a]	0.080–0.300	0.100–0.200[b]	0.200	0.500
Aluminium	<0.005	Max 0.005	0.005–0.010	0.005	0.006	0.008
Cadmium	<0.001	Max 0.001	0.008–0.020	0.005–0.010	0.005	0.010
Zinc	<0.005	Max 0.005	0.005–0.020	0.005	0.005	0.008

Source: Ref. 3-1, Table 5-5, with added information.
[a] Included in the "all other" category.
[b] At an economical breakeven point, where scrap value equals that of new solder.
[c] Figures reflect reconditioning (tinning) levels in the industry (not covered by any specification). They also apply to the solder in leveling of boards.

Column 6 shows the maximum contamination levels allowed in MIL-Std-454 (Standard General Requirements for Electronic Equipment, Requirement 5—Soldering). These levels reflect the maximum allowable concentration in a wave solder pot during use, not in the purchase of new metal.

The last column reflects to solder pots and waves used in tinning components as a preparation step for assembly and soldering (see Section 2-6). Note that these levels far exceed the contamination levels used in the final soldering. The reason is obvious: The thin layer of solder left on the tinned surface will become mixed with fresh solder in the final joint, and no ill effect is anticipated. This also holds true for board leveling (see Section 1-24).

While silver is also found as a contaminant in some processes, it is not considered harmful to the joint at levels as high as 2-3%. Other unusual situations may arise in which odd metals are used and dissolved. These must be studied one at a time.

3-14 CORE SOLDER FOR TOUCH UP

Solder for hand application is usually in the form of wire. When this wire contains the flux, it is called *core solder*. Using solid wire is not easy, because it requires the manual addition of liquid flux to each joint. Cored solder thus is in common use.

A spool of solder should be clearly marked with the following information (see Fig. 3-1):

- Solder alloy.
- Flux type and percentage.
- Applicable specification.
- Wire diameter.
- Manufacturer's name.

Let us discuss each in turn.

1. *Solder alloy*. It is customary to use the same alloy for hand soldering and touchup that is used in the wave. For economy, however, lower-tin and hence lower cost alloys may be employed. There are two quality assurance philosophies in this respect. One maintains that all solder joints must have a uniform metallic composition. The other prefers to be able to discern those joints that were touched up by hand. A duller low-tin alloy will serve this purpose while providing a cost savings (see Section 3-10).

For brevity, spools are often marked with codes taken from specifications. Thus 63/37 or Sn63 denotes the tin-lead eutectic, while 60/40 or

Fig. 3-1. Typical spools of core solder with label marking.

Sn60 refers to the 60% tin 40% lead alloy. The designation of 62/36/2 or Sn62 refers to a 2% silver addition to the tin-lead eutectic. The governing specification is often used to indicate the level of antimony (see item 3 below).

2. *Flux types.* The fluxes are normally indicated by abbreviations that follow the federal QQ-S-571 classification. Here type R is pure rosin, type RMA is mildly activated, and type RA is fully activated. Since this government specification does not cover all fluxes, vendors use designations of their own for organic intermediate fluxes (OA, etc.).

Flux percentage. Flux quantity is also critical and must be noted on the spool. Here again, specifications may offer a guideline or the manufacturer may device his own code. In general, a 2-3% flux is suitable for printed circuit board work.

3. *Applicable specifications.* To make sure that the alloy and flux are within specified limits, the corresponding specification should be quoted. This will ensure tin limits, the presence of antimony, flux strength, and so on.

4. *Wire diameter*. The size of the core solder is important for fillet control in hand applications. A large diameter (like 0.062 in.) is cheaper per pound, but shorter in length and usually higher in solder usage (large fillets). A fine wire (like 0.020 in.) is more costly per unit weight but much longer, therefore yielding more joints of controlled size. In most cases, the right wire is more economical overall. Obviously there can be too fine a wire (like 0.010 in.), which is very expensive and possibly wasteful.

The rule of thumb is: *Use a diameter that requires a joint feed length of 1/8 to 1/4 inch (3.2—6.4 mm) per joint.*

5. *Manufacturers name*. This is needed in order to identify the codes used. In a good operation, only vendors that have been closely evaluated will be on the approved list.

3-15 OILS FOR SOLDERING

Soldering oil is an archaic term that was coined when peanut oil and the like were used in the wave. Since then, a number of non-oil based materials have been adapted for the same purpose. Some of these are even water soluble, which contradicts the concept of oils. A better term might be "soldering fluids" or "tinning liquids," but "soldering oil" has become the accepted norm.

Let us review the functions of oil in the soldering operation, specifically in waves. These are listed (in no order of importance) as follows;

- To modify the surface energies of the molten solder. If the materials are formulated correctly, it is possible to decrease the cohesive force of the molten metal. This aids in wetting and improves the fluidity of the solder. Solder rise up the hole and solder draining (peel back) are positively affected.
- To exclude air from the solder process. This prevents dross formation and simplifies process control. Here the quality of the air and its tarnishing potential do not affect the results. The economic benefits of dross elimination are somewhat countered by the need to clean the oil and flux. Equipment maintenance, however, is facilitated, and is reduced in frequency.

Heat stability, the effect on surface tension, and the density or viscosity of the product determine the quality of the materials. While oils fulfill the requirements and are in the right price range, they are not water soluble. The best oils are in the petrochemical family, similar to crank case oil. There are a number of water- soluble materials in use; these are more costly. Some of these are actually high temperature, stable detergents.

There is a great deal of unwarranted concern that the addition of oil encumbers the cleaning operation. This conclusion has proved to be wrong, and several in-house evaluations have shown that cleaning with oil is actually easier than cleaning with no oil. The tests evaluated washing machine speed against ionic surface contamination and insulation resistance. These unexpected results can be explained by flux chemistry. The presence of a compatible oil reduces flux heat degradation during soldering. This improves residue solubility and retards the development of insoluble ionic residues. The presence of the oil also reduces the adsorption of flux ingredients which affect the insulation resistance of the exposed laminate.

3-16 SOLDER RESISTS (MASKS): PERMANENT OR TEMPORARY

In the normal flow of assembly, there are times when specific areas must be kept free of solder. For this purpose, a series of organic coatings have been designed. Termed *solder resist*, or *solder mask*, they are applied permanently (see Section 1-21) or temporarily (see Section 1-22). The permanent coating is solvent resistant, while the temporary mask can be washed off or peeled. Let us look at them separately:

1. *Permanent resists*. These are applied by the board manufacturer. They can be screened on or applied photographically and developed.

From the user's viewpoint, there are a few important factors to consider besides cost:

- The effect of the process on solderability.
- The behavior of the solder plating underneath.
- The inherent accuracy of the process.

These are covered individually below:

a. The solderability of the surface left exposed by the resist is critical. Curing with heat tends to tarnish the exposed pads, which is more detrimental to exposed copper than to solder plating. Heat also tends to bleed the resist onto the pads before full curing takes place. Obviously, any resist that migrates into areas to be soldered will cause solderability problems. The practice of pulling the resist pattern away from the land to avoid bleeding leads to poor fillet control in soldering and should be avoided (see Section 1-18). Surface mounting requires very precise mask placement because of the small footprint of surface mounted device terminations. Even in conventional boards, the lowest touchup rate can be achieved with resists that fully outline the pads. This can best be produced with photographic techniques, or with ultraviolet (UV) cured materials.

Screening, in general, can also cause serious problems due to lack of process control. Air bubbles in the material or resist starvation cause many breaks in the final coat. These increase the likelihood of flux entrapment, flaking, and webbing. The accuracy of screening is in itself quite limited.

b. The majority of double-sided boards used today have a solder plating on the surface. A resist that is applied over the solder does not prevent heat transfer and thus fuses (melts) underneath. The molten solder shifts under the resist in a phenomenon labeled *crinkling*, which in itself is acceptable. The shifting solder reaction is attributed to internal stresses in the resist coat, plus the cohesive force of the solder. With close spacing, the solder has been observed to run over the narrow insulation gaps, shorting adjacent conductors. In addition, when the resist is brittle and flaky, flux can be sucked into the crevices, which are not truly washable (see Section 7-30).

Solder resist that is applied over copper (solder mask over bare copper, or SMOBC), as in additive processing or for single-sided boards does not manifest the same problem. Some board manufacturers, who do not have additive facilities, strip the solder off the copper chemically. Others use air leveling to minimize the amount of solder and reduce the danger of crinkling and shorting out. Air leveling, however, leaves some solder behind, and thus does not solve the problem entirely.

c. Fine line definition and close spacing are best achieved by the use of photographic resist. While high in cost, it is well worth the benefits. Carefully applied UV cured formulations are second best; the short cold cure helps retain the screened configuration to some degree.

2. *Temporary resists.* These are applied by the board user during assembly and just before soldering. They are used to prevent solder wetting during wave soldering, then removed. They serve to:

- Keep a plated through-hole clear of solder for subsequent hand soldering.
- Keep the pad area free of solder for later mechanical assembly.
- Preserve gold-plated contacts.

There are a number of mechanical and chemical methods that can be used for this purpose. They are usually selected for their cost effectiveness.

a. Mechanical devices that are mounted in place and removed after the wave. Many of them are reusable. They include:

- Tooth picks.
- Dummy plastic or metallic plugs.
- Adhesive tape and precut shapes.
- Board stiffeners (see Section 3-18).

b. Chemical formulations that are applied by screening or hand dispensing (from a bottle, tube, syringe, etc.). They are then dried at room ambient, or heat cured. We can classify them by method of removal as follows:

- Solvent soluble.
- Water washable.
- Removed by peeling.

A word of caution is needed about cleaning with temporary resists. It is mandatory to remove the non-soluble resists prior to washing; otherwise flux residues may become trapped. This is especially true for the more aggressive fluxes that cannot be left on the work. Serious corrosion problems from this source have occurred in the field.

There are also methods of board design that help keep a single sided hole open. If the pad is split in the direction of travel over the wave, the solder will wet the two half moons but will not close over the apertures. This is particularly helpful for use with grounding screws and mechanical assembly.

3-17 CLEANING MATERIALS

Solvents, saponifiers, and water are described in detail in Chapter 6. They are an important part of the overall process and require careful selection in the design stage. If process modifications are needed later, always keep the cleaning materials in mind. Their compatibility with the components usually requires mild formulations. This will determine the type of flux family to use, as well as other factors.

3-18 STIFFENERS AND MECHANICAL SUPPORTS

The inherent heat of the wave and other soldering processes tends to soften the board material, especially when the temperature exceeds the T_g. In this structurally weakened state, the wider boards tend to bow and warp. An FR4 board of 0.062 in. (1.6 mm) thickness and a width of 7-9 in. (17.8-22.9 cm) is relatively immune to this problem, but wider boards are not. The safe width is obviously a function of board thickness and laminate quality. The addition of a metallic brace thus gives the board added rigidity. There are two types of metal supports:

1. *Temporary stiffeners.* These are metallic clips which are attached to the board when finger conveyors are used (see Fig. 3-2). They are normally placed across the leading and trailing edge in an effort to support the board during soldering. They must be removed prior to cleaning.

The stiffeners are often applied over the gold plated fingers to prevent their contamination. On the leading edge, the folded over design can prevent first the flux and then the solder from wetting the gold. This does not always work on the trailing edge, and therefore is not recommended there. Some high temperature plastic moldings are found on the market that prevent solder coverage, but they provide no mechanical support.

The metallic clips are made of tungsten or titanium, which are metals that do not wet. They come in a folded over design and a cheaper clip on variety. The latter type does not prevent gold fingers from being wet by solder.

2. *Permanent supports.* On very large boards, the mass-to-surface ratio dictates the use of a mechanical support even at room temperature. A

Fig. 3-2. Temporary stiffeners for PC boards.

complete frame around the board that also serves as part of the permanent chassis is quite common. A crossbar or two can be screwed into the board to serve the same purpose. Some laminated bus bars and connector blocks also strengthen the board.

With the use of such hardware, the materials of construction that are exposed to the solder are critical. The screws and edges that are submerged in the flux and solder can be wettable or unsolderable (depending on the need for electrical contact, etc.). Stainless steel is preferred, since it does not contaminate the solder. Silver plating is neutral; tin or tin-lead coated parts are also acceptable from that point of view. Copper and bronze (a copper tin alloy) contaminate the bath but can be tolerated. Zinc, brass (a copper zinc alloy), cadmium, and gold should be avoided since they interfere with the solder.

The use of a pallet to support the boards during soldering offers another solution to the warpage problem. These pallets can be metallic for long life and stability. The large heat content of such a frame, however, tends to rub heat during the soldering process. Nonmetallic pallets are less durable but have a lower heat content. Tempered masonite is an inexpensive and stable pallet material.

3-19 TOUCHUP MATERIALS

A discussion of soldering and cleaning materials would not be complete if we neglected the important area of touchup and repair. The materials used here are very similar to those used in wave soldering, but there are some important differences.

The term "touchup" is often used in a broader sense to include several other areas of hand soldering that come after the wave. One such an important operation is *add on*, where components are added by hand. These are either not suitable for wave soldering or nonwettable by the cleaner. Often component replacement during repair is included, which requires solder removal as well as solder addition. For further details, see Chapter 9.

Let us cover the materials in a logical sequence;

1. *Flux*. It is used in two forms: as part of the core solder (Section 3-14) or as an external liquid. The selection of the flux for this application is severely limited by the cleaning process to follow. For a variety of reasons, it is seldom feasible to channel the assemblies back to the cleaning process used after wave soldering. It is even more difficult to clean the board once nonwettable components are installed. A special in situ

cleaner is required which both supplies the liquid and sucks it off for repeated flushing (see Section 6-17).

When only a simple batch vapor degreaser is available after touchup, the use of a rosin flux is dictated. In other cases, complete flux removal is not possible, indicating the need for a mildly activated rosin (RMA) in critical applications. For further details see sections 3-3 through 3-7.

2. *Solder.* The solder is used in solid wire or core form. It is customary to use the same alloy as in the wave to ensure that the touchup has the same luster as the regular joints. There is a trend to use a much lower tin content solder in some high reliability companies, in order to mark a touched-up joint for easy identification at a later date. For further details on core solder, see Section 3-14.

3. *Cleaners.* These materials are used during touchup to remove the flux residues after hand soldering. They are fully described in Chapter 6.

The methods of bench application, however, bring these materials close to the operator, and safety becomes an important issue. There are two areas of concern: inhalation and skin contact. In that respect, isopropyl alcohol (IPA) is becoming very popular. It is the same material that is used in hospitals for skin disinfection prior to an injection. In addition to its safety, it is a low cost solvent, but being flammable, it cannot be used if smoking is permitted on the line. As a cleaner, it is both polar and nonpolar, and is the major ingredient in most rosin fluxes.

Water cleaning of both rosin (with a saponifier) and water-washable organic (intermidiate) fluxes is also becoming popular. There are no inhalation or skin contact problems, and the quality achievable is excellent. Small in-line washers (see Fig. 3-3), as well as in situ cleaners (see Section 6-17), are used for this purpose.

4. *Wicking braid.* These are also called "wick" for short, and are used for solder removal. Their action is based on capillary forces, which pull the molten solder up on the braided wire to form a wick. The braid is made of a solderable material and is prefluxed. The operator places the braid on the solder to be removed and touches the braid with a hot iron. The flux is intended to remove the tarnish and help establish a heat bridge. Once the solder melts, it runs into the capillary spaces and is wicked away from the joint.

The braid is usually made of copper, which is used bare or tinned. While the bare copper gives a simple color indication of the solder being wicked, it can deteriorate in solderability even though it is flux coated. The tinned wire does not show the extent of wicking, because there is no color contrast. The solder-coated wire is faster, however, and is less likely to lose its solderability. Remember that the copper braid is a large

Fig. 3-3. Small inline water washer suitable for post touch up cleaning. (Courtesy of Hollis Automation)

heat sink, and far more heat is required from the iron. For details see Section 9-7.

3-20 INSPECTION FLUX

This is a colored rosin (type R) flux used for marking defective joints during inspection. This flux promotes soldering during rework of the joint and stays in the flux ring until it is cleaned. In this way, the operator no longer has to search for rejects after the inspector has reviewed the work. Inspection flux reduces unnecessary touchup because the operator does not have to second guess the quality control personnel. In addition, it stops the inspectors from duplicating their effort and/or finding additional rework.

The fluxes are usually of the nonactivated rosin type (R or WW rosin). They are available in liquid form or in felt tipped pens. The color is nonionic and thus harmless. It does not disappear with the heat of the soldering operation, but washes off with the rosin.

This flux replaces the need for inspection labels, which can leave undesirable residues. Some companies insist on washing the residues off. Other organizations like to be able to trace a touchup later through the color in the residue ring left by the inspection flux.

REFERENCES

3-1. Howard H. Manko, *Solders and Soldering*, 2nd Ed., McGraw-Hill Book Co., New York, 1979.

3-2. *Soldering Manual*, revised ed., American Welding Society, Miami, 1979. (also the 1st Ed., 1959).

3-3. U.S. Patent 3,945,556: Functional Alloy for Use in Automated Soldering Process (Mar. 23, 1976).

4

THE SOLDERING PROCESS AND THE EQUIPMENT

4-0 INTRODUCTION

One of the more dramatic developments in electronic assembly was the concept of the printed circuit board. The growing complexity of electronic equipment made point-to-point wiring inside the chassis cumbersome, expensive, and impractical to repair. The printed circuit board solved these problems and made present day circuit density possible. This new technology and its later levels of sophistication (double-sided, multilayer, flexible circuitry, surface mounting, and hybrids) required a shift from hand soldering to mass production techniques.

In the beginning, the planar surface of the printed circuit board was simply fluxed and hand dipped in a solder pot. The process was not controllable and depended largely on the dexterity of the operator's wrist. While yields were low and the amount of scrap increased the cost, it was a significant improvement over manual iron soldering.

Dip soldering suffered from several problems. Manual dipping was difficult to reproduce. In addition, the flat surface of the board, laying on top of the solder, trapped gases underneath. These were mainly flux fumes, but they interfered with heat transfer and solder contact. As a result, the immersion time was long, 8-10 seconds, and heat damage was substantial. A skilled operator quickly learned to vent these gases by using a rocking motion that helped cut down on dip time.

The dross and burnt flux floating on the surface posed another problem. These had to be skimmed off continuously, since they interfered with the soldering operation. They slowed heat transfer and contaminated the boards, leaving insoluble and often conductive products on the work.

The historical sequence of mechanization of the dip process is not clear, as was established during wave patent litigation. Several innovations are discussed here in logical rather than sequential order. They depict the development of the wave while helping to clarify the process.

A homemade conveyorized dip system called *drag soldering* was used by an American radio manufacturer about 25 years ago. The system had a pallet mechanism which slowly passed the work through the flux and the preheater before lowering the board into a long solder pot. The fixture and board floated on the solder, being dragged along by the conveyor and lifted out again. Contact time was approximately 5-8 seconds, and the yield was good. Such equipment has since been perfected in Germany and Japan, and is now being imported into the United States. To alleviate the gas entrapment problem, some drag machines have a rolling motion during dipping to help the entrapped vapor escape.

Another machine, used by a U.S. television manufacturer, operated with a small solder trough which was raised out of the pot to meet the work. Called the *Slam Solder Machine*, it shortened soldering time to 2-5 seconds and had a higher yield, but required higher solder temperatures.

The most popular device became a wave solder machine, which is still used today. The solder was pumped through a horn and nozzle to the underside of the boards. When the wave impinged on the bottom of the board, no gaseous entrapment occurred. The contact time was dramatically reduced to 1-2 seconds, and the solder temperature was lowered.

A large number of variations on the same technology exist. Narrow waves, unidirectional waves, cascading waves, wide laminar waves, double "chip" waves for surface mounting (see Section 5-14), and other configurations are just part of the list. Differences also exist in the methods of dross prevention through mechanical design or the use of oil. The wave soldering line may also vary in the conveyor mechanisms and the impedance angle (horizontal versus inclined) and other factors.

Today the industry has a large selection of equipment and is well on its way to computerization of the manufacturing process. In this chapter we will treat the subject as units of operation, such as fluxing and preheating. There will be no detailed description of the equipment, as it relates to specific suppliers. The photos shown were provided by various manufacturers and in no way imply a recommendation by the author.

4-1 THE WAVE SOLDERING PROCESS

Wave soldering in its purest form consists of three simple steps, which follow the component assembly process. Some intermediate steps may also involve precleaning (see Section 2-11), prebaking (see Section 2-12), and similar processes.

Fig. 4-1A. View of a total computerized wave line. (Courtesy of Hollis Automation)

Fig. 4-1B. View of a total wave line. (Courtesy of Electrovert Ltd.)

The three basic steps are:

1. Fluxing
2. Preheating
3. Soldering

The process, however also includes the transfer mechanism (conveyor), ventilation, and other steps. Postsolder cleaning is described in Chapter 6.

Let us remember that wave soldering is a convenient way to attach components to the planar configuration of the printed circuit board. This is by no means the only application of the wave. It can be used to tin component leads, coat long lengths of wire and metal-strip, and so on. The use of the wave as a soldering tool is limited only by our imagination.

Since the soldering process consists of individual steps, we need to discuss them separately. As we cover each phase, we will dwell on the details involved.

4-2 FLUXING

The need for using flux is covered in Section 3-1, and in Chapter 2 of Ref. 4-1. We will concentrate on the methods used with wave soldering equipment.

The inline application of flux to the printed circuit board is achieved by one of several means. The selection of a specific process depends on the following parameters:

1. The length of the leads protruding on the underside of the printed circuit board.
2. Whether cleaning is required and by what method.
3. The quantity and uniformity of flux required; these determine the viscosity and density of the flux.
4. Material economy and housekeeping requirements, which can be translated into safety and maintenance of the equipment.
5. The cost of the equipment in relation to its function.

We will discuss the popular methods of flux application and point out the advantages and disadvantages relative to the variables listed above (see Table 4-1). In each case, the maintenance procedures recommended by the manufacturer must be carefully followed in order to utilize the equipment fully.

4-3 FOAM FLUXING

The foaming method is based on liquid fluxes which have a unique combination of surface tension and foam generating characteristics. The equipment consists of gas nozzles or porous stones immersed in a liquid tank and covered by an open chimney to channel the foam. The foaming elements are designed to break up the air stream into tiny bubbles. The foam that is generated pushes the flux up to meet the work (Figs. 4-2A and 4-2B). The boards pass over the foam orifice, and the flux adheres to the joint area in a uniform, controlled layer.

The vehicle to flux ratio is critical because it determines the reproducibility of the foam head, as well as the solid content. Flux foam formation also depends on viscosity. Thus control is needed over the following items;

1. The vehicle. Use only recommended thinners for adjustment.
2. Air pressure and air purity. Clean the oil and water trap daily.
3. The height of the liquid over the foaming elements. A height of 0.25 in. (6.35 mm) is suitable for most fluxes. Also, make sure that the stone is never exposed to air, or it will clog up.

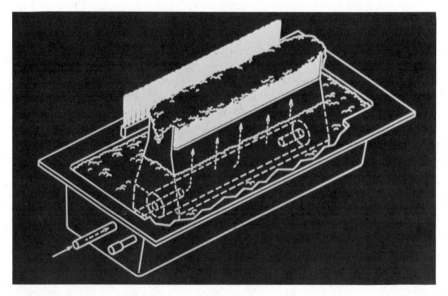

Fig. 4-2A. Schematic of foam fluxer with two foam support brushes, and a third wiping brush for quantity control. Note that such brushes tend to move the protruding component leads in the unsoldered board. This may disturb the component position, and require unnecessary touch up. (Courtesy of Hollis Automation)

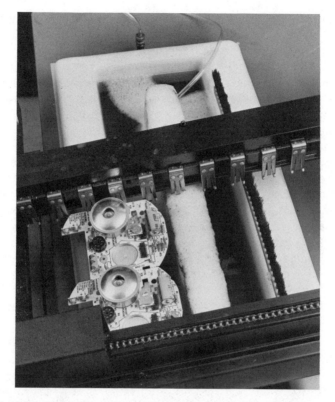

Fig. 4-2B. View of foam fluxer in action. (Courtesy of Hollis Automation)

4. The temperature. Both density and viscosity are temperature dependent. When measuring the density of the flux, make sure that the temperature of the sample being tested is equal to the temperature stated by the manufacturer. The density of fluxes varies with the temperature and may give misleading results unless adjusted. If the flux is too cool (normal for foam fluxes because of the latent heat of vaporization), hold the specimen under running hot water. If the flux is too warm, as in the case of wave fluxing, cool it with running cold water. The flux vendor may also provide a temperature correction chart or formula.

The foam flux is used mainly in automatic printed circuit soldering lines. It is especially suitable for rosin applications where no flux removal is anticipated. The solid content of the flux is kept low, and little or no residue is left.

The normal foam fluxer without any special support is often labeled a *free foam head* and reaches a height of roughly 5/8 in. (16 mm).

If greater heights are needed, a brush support can be devised on both sides of the foam chimney and the total depth can be extended to slightly above 1 in. (25 mm) (Fig. 4-3). This *supported foam head* requires assemblies that are not sensitive to the physical contact with the brush. Otherwise some short clipped components may be pushed out of the hole or "ride" in the hole.

Foam fluxing equipment is relatively inexpensive, since it has no moving parts. It is recommended where flux residue will be left in place and a thin flux coat is sufficient during soldering. When foam fluxing is used in an application that requires postcleaning, more viscous flux formulations are beneficial (up to 35% solids). Experience has shown that a skimpy flux layer bakes on during soldering and becomes hard to dissolve. A generous viscous flux coat also undergoes this change, but to a lesser degree, making cleaning easier.

The foam head is also sensitive to heat and may be depressed when a hot pellet is used. This happens when pellets are returned rapidly from the solder wave. It may be necessary to cool them in a cleaner soak. This

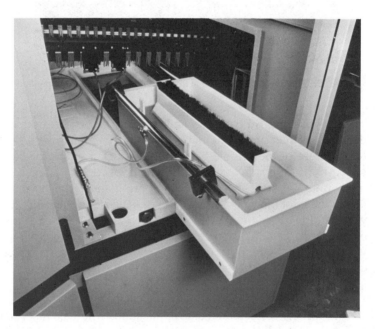

Fig. 4-3. View of foam fluxer pulled out for maintenance. Note the brushes supporting the foam and the air knife behind the fluxer. (Courtesy Hollis Automation)

would obviously also keep them clean. If this is not possible, it may be necessary to redesign the pellet to hold less heat (i.e., eliminating any heavy metallic cross sections that touch the hot solder, etc.; see also Section 4-17).

The maintenance of the foam fluxer is relatively simple;

1. The process requires oil-free, dry air. A simple filter and an oil trap are installed on the line to remove such impurities from shop air. Remember to service the filter and/or trap regularly.
2. The height of the foam requires uniform air pressure. It is wise to have more than one regulator. The first reduces the air from shop pressure down to approximately 5 psi (0.35 atm). The second, usually a needle valve, further reduces the pressure to the level dictated by the flux (0.5-1 psi or 0.35-0.7 atm).
3. Uniformity of foam also depends on a clean stone. The element should be soaked whenever it is not in use. An appropriate thinner or cleaner prevents the formation of solids in the pores by evaporation and clogging. During the operation, the foaming head should always be covered by the liquid flux to prevent crystallization.
4. Thinner evaporation causes poor material economy. The foam fluxer should be tightly covered overnight and when not in use. Maintain the specific gravity of the flux by adjusting it at least twice a day.
5. For good housekeeping, an additional wiping action is desirable once the work leaves the foam flux station. Here a final wiping brush is often installed; this brush must be kept soft and pliable to prevent component shifting. Air knives blowing in the direction of the foam head are much more gentle (see Fig. 4-3). The air stream is directed 1 in. (2.5 cm) behind the foam head to avoid unnecessary evaporation losses.

4-4 WAVE FLUXING

In this method, the liquid is continuously pumped through a trough, generating an exposed standing wave (Figs. 4-4A and 4-4B). The boards travel with no change in direction, while the flux is elevated to meet the work. The height of the wave may easily reach 2 in. (50 mm), which makes it suitable for long lead wires.

This method requires less stringent flux selection than foam fluxing, since there is no excessive evaporation. It requires no particular balance of surface energies or unique properties. The flux-to-vehicle ratio should still be carefully controlled, because it affects the uniformity of coating and cleaning.

This process is frequently used for high speed fluxing operations on automated lines, where foam regeneration may be too slow. It is always suggested when a generous coating of flux is desirable, as in the case of fast cleaning. It may be used with very low solid content fluxes when postcleaning is not desirable.

The quantity of flux deposited depends not only on its solid content but also on the wave construction. The most critical parameter in equipment quantity control is the angle at which the work meets the wave. Here the "drain back" is important for uniformity and materials economy. The wave should be kept as smooth as possible to make sure that the flux penetrates uniformly up the plated-through holes. It is generally difficult to keep smooth waves once they are pumped to a high crest. Optimum wave height requires discussion with the manufacturer prior to equipment purchase. The type of flux selected may also affect wave dimensions (width).

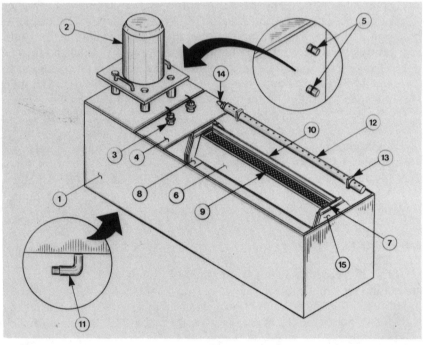

1- Pot	6- Nozzle Assembly	11- Drain Valve
2- Pump Assembly	7- Flexible Flap	12- Air Knife
3- Level detectors	8- Baffle	13- Bracket
4- Plate switches	9- Screen	14- Air inlet
5- Flux Recirculation Inlet and Outlet Port	10- Wave Deflector	15- Holding Plate

Fig. 4-4A. Schematic of wave fluxer. (Courtesy of Electrovert Ltd.)

Fig. 4-4B. Photo of wave fluxer. (Courtesy of Hollis Automation)

Maintenance of the equipment is relatively simple and follows good housekeeping rules. It is suggested that the fluxer be covered whenever it is not in use to prevent evaporation of thinner.

An additional wiping action by a free standing brush or air knife may be used to keep excess flux from adhering (see Fig. 4-3). This prevents dripping during the preheating operation.

The cost of this equipment is relatively higher than that of the foam fluxer. It is popular because of the trouble free nature of the process.

4-5 SPRAY FLUXING

This is a precise but messy method of application. It is suitable for odd shapes and very long leads. The *rotary screen spray fluxer* method is prevalent in the industry (Fig 4-5). While it is possible to build equipment on the principle of spray painting, the method is less often used. The flux cloud created tends to contaminate the whole area, especially the transfer mechanism. Spray fluxing, when used in conjunction with flammable or explosive materials, requires special safety controls.

A high degree of uniformity and quantity control can be achieved in rotary screen spraying. It is difficult to direct the spray into specific areas without the use of a mask. Rotary and flat masks are interspaced between the nozzle and the work in large volume automated systems. This reduces flux usage and can be coupled with a micro switch to activate the spray only when needed.

The rotary spray fluxer is shown in Fig. 4-5. It consists of a porous circular screen partially submerged in liquid flux. As the screen rotates, each cavity picks up a droplet of flux from the tank. An air jet is located near the exit, pointing toward the work. When the loaded screen passes over the air nozzle, the droplets are hurled against the work. Air pressure, screen rotation speed, and flux solid content determine the quantity of flux deposited.

One advantage of spray fluxing is the pure state of the flux, which is applied directly as received. There is no chance of contamination, and complicated flux controls are unnecessary. On the other hand, drippings and overspray must be discarded; this may be considered wasteful. The cost of the equipment is medium to high, and it is not always available. Many units are homemade, to meet unusual requirements.

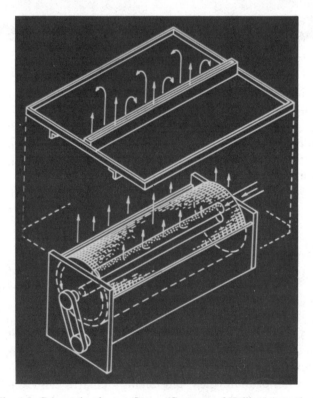

Fig. 4-5. Schematic of spray fluxer. (Courtesy of Hollis Automation)

Table 4-1. Comparison of Fluxing Equipment—Properties and Cost.

EQUIPMENT TYPE	FLUXING HEIGHT (in.)	(mm)	EQUIPMENT COST	MAINTENANCE COST	MATERIAL USAGE	COST RATING
Free foam head	5/8	16	Low	High	High	High
Supported foam	1.0+	25+	Low	High	High	High
Wave fluxer	2.0	50	High	Low	Medium	Low
Spray fluxer	Any		Medium	Medium	Medium	Medium

Note: This comparison assumes similarity of construction materials (i.e., stainless steel versus plastic).

Table 4-1 summarizes the equipment properties and compares them to the cost. The expense of flux monitoring has been included for controlled processes. Flux volume usage for the same square footage of boards is also given. The reader is urged to compare these data for a value analysis (Section 10-2), before investing in equipment.

4-6 FLUX CONTROL AND FREQUENCY OF CHANGE

Careful control of the flux during use in all types of equipment is vital. Flux density and temperature dependency were covered earlier. The frequency of change and other monitoring parameters are covered here.

As computer control over wave soldering becomes more widespread, the vital flux control variables will become material and quantity oriented. The quantity changes with the hardware and is simple to adjust. The material parameters are more sophisticated; for example, flux activity may be monitored by ionic content measurement. The pH may be very misleading as a reference point, depending on the formulation. Other chemical checks and titrations may be feasible, but only with close coordination with the vendor.

The frequency of flux changing is covered by the following rule of thumb:

Continue to use the flux as long as it maintains its clarity and color!.
The two key items here are:

1. *Turbidity.* This indicates the pickup of foreign materials, such as dirt from the work or dross from the finger conveyor. It may also mean the presence of water and/or oil from the compressed air.
2. *Color.* All fluxes change color when they react with metals like copper. Rosin fluxes are also sensitive to oxidation and activator breakdown, and become dark brown.

Change foam fluxes at the end of a production week, regardless of the number of shifts. This only if the turbidity and color rules are monitored. If the equipment is used only infrequently, filter the flux before returning it to a special in-use storage container. Discard the flux after 40-50 hours of use.

For wave fluxers, the frequency of change is much lower. The flux may remain usable for many weeks or months. An activity check may be advisable on a bimonthly basis.

Spray flux is usually in a closed system and does not readily pick up dirt. Use common sense in changing the flux.

None of the above guidelines are intended to supersede density control or the regular maintenance schedule recommended by the equipment vendor. In addition, flux material data sheets should be read carefully.

4-7 THE PREHEAT STEP AND ITS REASONS

This part of the printed circuit soldering process is the least understood yet one of the most critical parts of the operation. This step is a combination of chemical and physical reactions. The chemistry is mostly flux oriented, although thermal degradation is also a chemical process. The physics consists basically of a delicate thermal balance, plus the mechanical effects of temperature rise.

In discussing the reasons and advantages of preheating, we will separate the horizontal from the incline conveyor. The two types affect the various properties of the printed circuit board differently

The reasons for preheating are:

- Volatile evaporation
- Flux activation
- Reduction of thermal shock
- Effect on soldering speed

1. *Driving the volatiles off the surface of the printed circuit board.* These volatiles are basically the flux thinners and, by design, are easy to evaporate. In this regard, the water-based fluxes are the most difficult to handle. They usually retain enough moisture to spatter during the soldering process. This volatile removal requires some natural convection of the saturated air underneath the printed circuit board. With an incline conveyor, this is the natural draft due to the heat generated. Remember that only limited venting by updraft is available through the holes in the printed circuit board.

With horizontal or near horizontal conveyors, it is recommended that additional air movement be generated. Preheating equipment using convection (the movement of warm air) is employed.

Preheating unfortunately, is not enough to remove volatiles like moisture that have been absorbed into the printed circuit board laminate. Thus the addition of convection equipment does not replace the need for prebaking (see Section 2-12). Only flux volatiles from the underside of the board are removed.

One can anticipate a certain amount of "sizzling and spattering" on the wave even after preheating. This is directly related to the amount of volatiles still left in the flux. In some formulations, overdrying is also not desirable because it makes the flux relatively immobile. In this state it interferes with solder wetting because it is not easily replaced on the surface during the wave application.

2. *Flux Activation.* Some fluxes, like rosin, depend on heat to become active. Other formulations become active as soon as they come in contact with the base metal; however, their action is intensified with preheating. In the rosin family, the flux is designed to provide a relatively inert material at room ambient. In the flux categories of water white (R), mildly activated (RMA), and activated (RA) rosin, the residues are often left on the board. In these cases, it is necessary to gain the full strength of the flux by preheating the boards to the *activation temperature*. Most rosin-based fluxes become active at around 180°F (82°C), and this should be the minimum preheat temperature at the bottom of the board. As we will see later, higher temperatures are required for other benefits (see Table 4-2).

The organic acid fluxes, to some degree, and inorganic acid fluxes do not require heat for activation. In these cases, the flux is active even at room temperature and starts removing tarnish on contact.

3. *Reduction of thermal shock.* The thermal gradient between room ambient and soldering temperatures is enough to cause serious damage to most nonmetallic assemblies. The materials used in printed circuits are not subject to physical cracking due to thermal shock. The major concern is the distortion which occurs and is often referred to as *warpage*. Warpage is mainly due to the difference in thermal expansion between the underside of the hot board (exposed to solder) and the cooler top. As a result of the differential expansion, there is bowing of the printed circuit board laminate. This depresses the center and pushes it even further into the wave, while the sides curl up and may not contact the molten solder. Warpage is further complicated by the random location of holes drilled in the structure and by the uneven distribution of component weight.

By preheating the printed circuit board, the thermal gradient between the top and bottom is greatly reduced. When the proper equipment is

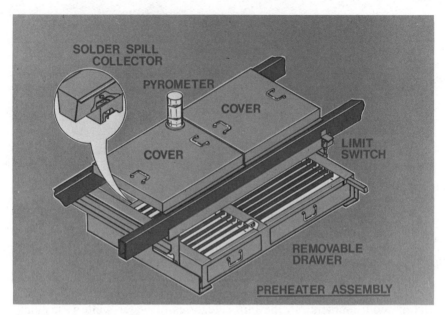

Fig. 4-6. Schematic of preheaters. Note the covers which help uniform preheating, and save energy. An additional top pre heater may be needed to increase heating rates for multi layer, and thick boards. (Courtesy of Electrovert Ltd.)

Table 4-2. A Rule of Thumb for Best Preheat Temperatures.

PRINTED CIRCUIT BOARD				
TYPE	THICK		TEMPERATURE RANGE	
	(in.)	(mm)	(°F)	(°C)
Single sided and flexible	all	all	175–200	80– 90
Double sided	max. 0.063	1.6	210–230	100–110
Multilayer (up to 4 layers)	max. 0.063	1.6	220–250	105–120
Multilayer (over 4 layers)	min. 0.093	2.4	230–270	110–130

Note: Measured when leaving preheat station, on top of board. Temperature taken on insulation, in between metallic conductors. Holds true for average speed, and component density. Remember that large ground planes, heavy component population, and other heat sinks require more heat.

132

used, the heat soaks slowly through the board. Table 4-2 gives a rule of thumb for best preheat temperatures.

4. *Effect on soldering speed.* In the process of wave soldering, the heat required to raise the surfaces to the wetting temperature (approximately 475°F, 245°C) comes from two sources. First the preheating step and second the wave solder application. These work in tandem to supply the heat necessary for the joining process. The higher the preheat temperature, the less heat is required from the wave. This can be translated into less time in the solder wave or higher production speeds. The temperature of the wave is normally fixed. For further details, see Section 4-10.

4-8 ADJUSTING THE PREHEAT EQUIPMENT

The sequence in which we adjust the preheat temperature as a function of the four preceding considerations is as follows:

1. Select an appropriate preheating temperature from Table 4-2. You must also consider the shape and size of the assembly at hand.
2. Test to see if this temperature dries the volatiles by listening for signs of excessive spattering at the wave. Then check the soldering results to make sure that the flux is active. In general, the preheat should span at least one third of the temperature difference between room ambient and wetting temperatures for the solder alloy used.
3. If the preheat temperature fulfills all requirements, adjust the other machine parameters (soldering temperature, impedance angle, conveyor speed, etc).
4. Testing by trial and error. Increase or decreases the preheat temperature and study the effect on production rates and the incidence of touchup.

In general, the top side temperature should be as high as possible, without exceeding the glass transition temperature (T_g) of the laminate. Since the bottom of the board is hotter than the top, caution must be exercised.

The top side should reach between 220 and 250°F (105-120°C) for minimum warpage. This holds true for the average 0.062 in. (1.6 mm) printed circuit board with a normal heat sink population on the top. If thicker printed circuit boards with heavy heat sinking are soldered, it may be necessary to provide top preheating for good results (see Fig. 4-6).

The preheat conditions presented in table 4-2 have been found to apply to the average production process . They have been established empiri-

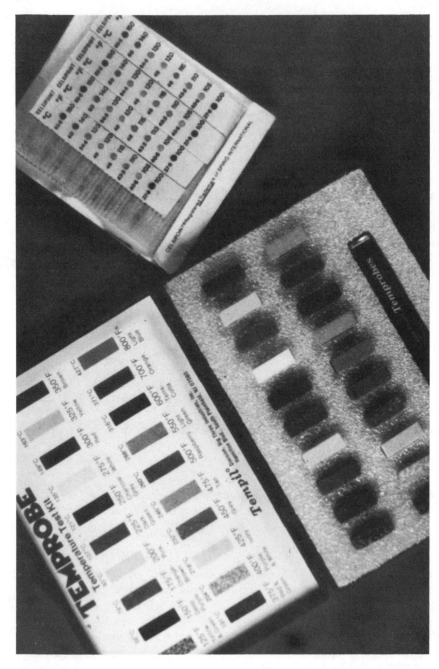

Fig. 4-7. These probes and similar materials melt at the indicated temperature. Use small shavings on the board, which should melt as they leave the preheater. The use of the thermal strips on paper is not as meaningful, and more expensive.

cally and are measured on the top side of the printed circuit board, on the insulation (not the conductors), just as the assembly leaves the preheater.

The simplest method of measurement is to use a temperature- sensing stick (which melts at a predetermined temperature) or a heat- sensitive pigment (which changes color). When using the stick, one simply shaves off a small section with a sharp blade and places it on top of the board. If it melts after leaving the preheater, the designated temperature has been reached (Fig. 4-7). The molten material is not harmful and washes off with the rosin flux. Remember to place the shavings on the insulator between the metallic tracks, not on the conductor.

The use of thermocouples for this application has some inherent problems. Thermocouples give misleading results because of board heat content variations due to layout. It is also difficult to embed them precisely on the surface of the printed circuit board. In addition, they are useful only when the board is new. Repeated use of the same printed circuit board gives varying results because the board material cures and ages. Thermocouples can be successfully fastened to the surface with aluminum conductive adhesive tape. If the same size tape segment is used to hold the thermocouple to the same area of fresh printed circuit board, reproducible results are possible, although the heat characteristics of the tape are unknown. The use of thermal paper strips has the same unknown factor.

Thermocouples, however, have been used successfully inside of multilayer boards. They can be placed accurately in critical areas of the inner layers and laminated in place during assembly. This is an easy way to get a temperature profile of the inner layers in a multilayer board as it travels through the process.

4-9 PREHEATING EQUIPMENT

The equipment used for preheating in wave soldering falls into three general categories, depending on the mode of heat transfer:

- Bottom radiation
- Bottom convection
- Top radiation

Since any method of heating is adequate for this process, the type of equipment and its method of construction are not critical. The parameters which guide equipment selection are uniformity of heat, energy efficiency, ease of control, ruggedness, maintainability, and cost.

For bottom preheaters, it is important to remember that flux dripping is unavoidable. The flux which drips onto a hot plate may bake on and result

Fig. 4-8. Self-cleaning exposed element preheater, note the changeable aluminum reflector. (Courtesy of Hollis Automation)

in nonuniform performance. The units which have exposed heating elements that ash themselves clean are therefore preferred. The reflector in these modules can be easily lined with replaceable aluminum foil (Fig. 4-8). Such units are in essence self cleaning. Any flux dripping onto the element will burn itself clean, while flux dripping onto the aluminum can be replaced periodicaly. When a hot plate is used for this application, it idles at a lower temperature, and therein lies the problem. Flux dripping cakes on the surface and does not ash. Uniformity of preheat is thus affected, and the maintenance is much more difficult.

One energy efficient, easily controllable unit should be mentioned. Quartz lamps with reflectors can be mounted on the top and bottom, and switched on only as needed (Fig. 4-9). Their fast cycle and self cleaning property make them economical. The reflectors behind the quartz lamps are also readily replaced when dirty.

4-10 THE WAVE SOLDER STEP

This is the last unit operation in the chain—the application of the solder itself. The process involves direct contact between the work and the

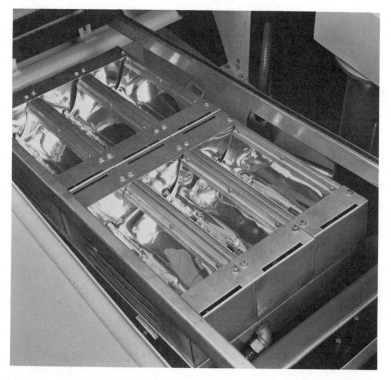

Fig. 4-9. The radiant quartz heaters can be mounted top and bottom. They are normally used with a plain reflector-cover. Since they reach operating temperatures very rapidly, they are used only when required, and save energy. (Courtesy of Hollis Automation)

molten metal. It can be broken down into two distinct physical events as follows:

1. *Final heat transfer.* A step needed to raise the surfaces to wetting temperatures. This is a function of:
   a. Solder bath temperature (490 ± 10°F for Sn63).
   b. Wave contact length (1-3 in. or 2.5-7.5 cm).
   c. Conveyor speed (dwell time 0.7-2.0 seconds).
   d. Wave dynamics.
2. *The supply of molten solder.* A step needed to provide solder for wetting and filling the gap. This is a function of:
   a. The solderability of both surfaces.
   b. Design (hole to wire ratio and fillet control).
   c. Wave dynamics.

As indicated, both parts of the process depend on the wave dynamics, in other words, on the shape of the solder being pumped. A large variety of wave designs can be found; all are variations on the same technology.

The solder wave is useful not only for printed circuit boards but for numerous other solder applications. Component lead tinning, component manufacturing, hybrid circuit assembly, and continuous wire tinning are just a few examples. We will concentrate, however, on printed circuit boards only.

4-11 SOLDERING AND WAVE DYNAMICS

This process analysis explains the impact of the various wave zones on the printed circuit board. By understanding the principles, the reader can avoid common problems and utilize the equipment more effectively. The aim is to achieve high quality joints at the lowest possible cost. To obtain this goal, it is necessary to reduce and control all variables, provide the right board design, and have appropriate materials. This discussion is also intended to facilitate in troubleshooting and the evaluation of line problems.

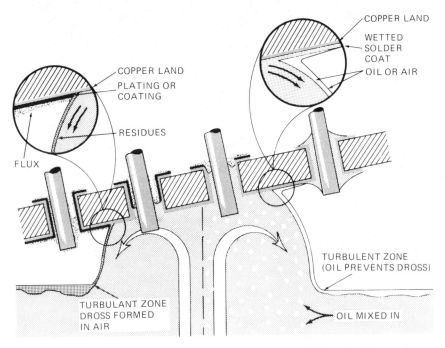

Fig. 4-10. Schematic of wave depicting the three major impact zones of the wave on a board. Note that zone I (entrance) and zone III (exit) have been enlarged in the circled areas.

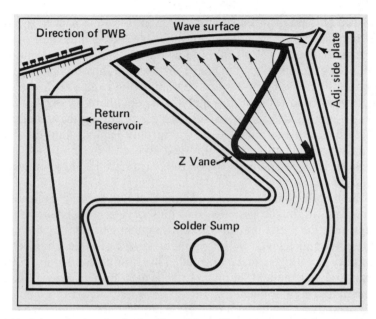

Fig. 4-11. Schematic showing the location of the baffle used to control flow behaviour in a wave. (Courtesy of Hollis Automation)

Figure 4-10 gives a schematic view of a solder wave with a printed circuit board passing through it. The wave has been arbitrarily divided in two. The left side shows no oil being pumped, while the right side shows a typical case of an oil mixed in. In reality, both sides are obviously the same; they either contain oil or are dry.

The solder is always being pumped from the bottom, where there is no dross. It usually passes through a number of baffles to ensure uniformity of flow to the top (Fig. 4-11). These baffles may be of iron and must be tinned; others are made of stainless steel and require no wetting. The area where the solder falls back into the reservoir is termed the *waterfall* (Fig. 4-11). It is in this turbulent area that dross is formed whenever there is no protective cover on the solder. The dross normally floats on top of the reservoir and will be discussed in detail later.

The interaction of the wave and the printed circuit board can be divided into three zones. These are discussed in Sections 4-12 to 4-14.

4-12 ZONE I—POINT OF ENTRY

Let us consider the bottom of the printed circuit board or substrate as it enters the wave. This is depicted in the enlargement on the top left side of

Fig. 4-10. This point of entry is at the most dynamic part of the wave. Here the solder flows rapidly down the wave, while the board moves in the opposite direction. This differential motion creates a washing action that removes the flux from the board. In addition, any organic layer such as a protective coating or surface contamination is also flushed away.

Flux removal is total on the metallic lands where the solder wetting occurs. Some of the more viscous fluxing materials may cling to the laminate between the conductors, where only laminar fluid flow conditions exist, complicated by the physical layout of the lands and component leads. Thus some flux is left on the bottom of the board between the metallic lands.

On the metallic surfaces, we indicated that flux removal is immediate. This occurs within the first few fractions of a second needed for heat transfer until the metal reaches the wetting temperature. Physics teaches us that given the correct wetting conditions of cleanliness, temperature, and affinity, wetting is instantaneous. Thus the denser solder displaces the flux, and an excellent thermal bridge is established between the wave and the metallic components.

Soldering to a base metal under these conditions is simple and follows the rules of wetting. The mechanism is similar yet different for a metallic coating over the board or the components. These coatings are usually electro deposited, and are applied to provide solderability and/or shelf life for the surface. Three types are possible:

1. *Fusible coatings*. Most of these cover layers will melt at soldering temperatures (tin, tin-lead solder, etc.).
2. *Soluble coatings*. Others will dissolve in the molten solder (gold, silver, etc.).
3. *Stable coatings*. Few usable metals are neither soluble nor fusible at soldering temperatures (nickel, iron, etc.).

In the wave, the washing effect of zone I has different effects on these coatings. The mechanisms is:

1. *Fusible coating*. If the coating will melt in the hot solder, the wave movement will wash the fused layer away replacing it with fresh solder. The original plated layer is thus seldom distinguishable in a metallurgical cross section of the bottom circuitry. Since most such coatings contain only tin and lead, they do not contaminate the solder pot.

2. *Soluble coating*. If the coating is soluble, the sequence of events will depend on the solution potential of the metal and its thickness. Given enough time in the wave, it will be dissolved and washed away. These observations apply to soldering operations of average speed and tempera-

ture. It is conceivable that with fast production parameters some of the original coating will be left. In the case of hefty components with a relatively thick silver plate, silver can still be observed on the surface after wave soldering. This is attributed to the large heat sinking of the component, coupled with the heavy plating, which will retard total solution. After solution, these metals become an impurity in the solder.

We should include copper in this category because it is also soluble in solder. However, the copper on the boards or in the lead wire is very thick. Total removal, similar to that of gold or silver plating, is not likely. Because of its solubility, copper is the major contaminant in the solder alloy (see Section 3-12).

3. *Stable coatings.* Nonfusible or nonsoluble coatings like nickel behave like the bare base metal discussed above. Only here no solder contamination is picked up from the work.

4-13 ZONE II—HEAT TRANSFER AND SOLDER RISE

If we had to wet only the bottom of the printed circuit board, the wave solder operation would be complete shortly after the point of entry. However, there are also component leads with a substantial heat content that need to be soldered. In addition there are plated-through holes which must be wet and filled with solder. All this occurs in zone II.

The rise of solder through the hole in the laminate must be analyzed more closely. Let us start with a plated-through hole, which is more complex. This will help us understand the process in a single-sided board (Fig. 4-12).

The rise of solder in a plated-through hole depends on a specific sequence of events:

1. The flux must cover the entire surface. This may occur during the fluxing stage, and the flux may be transported even up to the top pad. Sometimes this happens only just before the solder rise, when the flux rides on the crest of the solder.
2. Sufficient heat for adequate fluxing action is needed. At this stage, volatile evaporation should be complete.
3. Heat transfer must raise the lead-wire and the plated-through hole barrel to the wetting temperatures.
4. The solder must rise and fill the hole, displacing the flux.

In the correct sequence of events, the climbing solder is preceded by a solder film with a small dihedral (wetting) angle close to $0°$. If this sequence is disrupted, serious fillet problems may ensue.

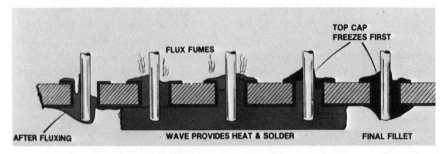

Fig. 4-12. Schematic of wave impact on a double sided board. Note that top cap "should never be allowed to freeze while still in the molten wave. Such premature freezing can cause chemical entrapment.

The importance of good heat transfer increases, the higher the solder must rise in the hole. Hydrostatic pressure pushes the solder only partway up the through hole. The board edges can be only partially submerged, or solder will spill over the top. The solder rise in the last part of the fillet is due entirely to wetting forces, which, in turn, depend on temperature. Here the heat balance is greatly affected by board design, component type, and component density. Multilayer boards require special attention in this regard, and their design must take heat transfer into account. Any heavy metallic interlayer requires special preheating and the use of a top preheater. There are ways of reducing these thermal demands by etching a quasi pad around the hole, connected to the ground or power plane by two to four separate lands. This resembles the spokes of a wheel and is referred to as *cart wheeling*. When these thermal considerations are disregarded during design, there is increased danger of delamination.

In addition to the heat supplied by the wave, there is the heat which the assembly absorbed during the preheating cycle. This additional heat is critical in good fillet formation. A board that is already warm can pass more rapidly through the hot wave. Soldering speed is thus reduced, which also decreases thermal damage. Remember that thermal degradation is accelerated at higher temperatures. The greatest damage is done during exposure to the wave. The assembly can tolerate the lower preheat temperatures for an extended period with few if any ill effects.

Under normal conditions, the rising solder will spread over the pad and continue partway up the lead to form a perfect fillet. However, soluble and/or fusible coatings in the barrel, on the pad, and around the lead will not be washed away, as in zone I. The impurities will usually be concentrated in the top layers and adjacent to the surfaces from which they emanate. In the case of gold leads, these surfaces may appear quite gritty as a result of the intermetallic compounds which are formed.

Because of the importance of heat transfer, we refer to this part of the wave as the *heat transfer zone*. It lies between the point of solder to metal contact (several fractions of a second after wave penetration) and the point of wave exit. Remember that as we approach the point of exit, the upward push of the solder due to fluid dynamics decreases in importance. Wetting is by far the dominant force in the solder rise through the plated through holes.

It is obvious that the top side of the board is the coolest part of the assembly during processing. The top of the solder fillet is the first area where solder freezes to form a cap. In the right sequence of cooling, cap solidification is very beneficial. It provides the mechanical structure which prevents the solder from sagging after the board exits the wave. This is especially important when there is a large mismatch between hole diameter and lead size. Without a cap, the solder will drop out of the hole because of its weight, forming an icicle (see Section 1-11). Premature freezing of the cap also causes many quality problems in terms of blow hole entrapment and icicle formation. Some of these problems will be described in Chapter 7.

In discussing heat transfer, let us review the sequence of events in the solidification of the fillet (Fig. 4-13). Here the thermal conditions may differ in some cases, and our comments apply to the majority of solder joints:

1. The top of the board is usually the coolest zone on the board, and the top of the fillet freezes first.
2. The bottom of the board is the next zone to cool below the solder-melting temperatures. Thus the second joint extremity solidifies next.
3. The center of the joint is normally the last to freeze, and stays molten for the longest period.

This thermal profile can be deduced from metallurgical studies of solder joint cross sections. It is also the cause of the small round *freezing vacuoles* found in the body of the fillet, which are often mistaken for gas entrapment. These solidification cavities are common in most cast structures, where the center solidifies after the skin in the mold. The cooling metal shrinks in volume, creating this "piping" effect.

While this freezing pattern occurs naturally in most applications, it also triggers the best joint quality. When this sequence is disrupted because of

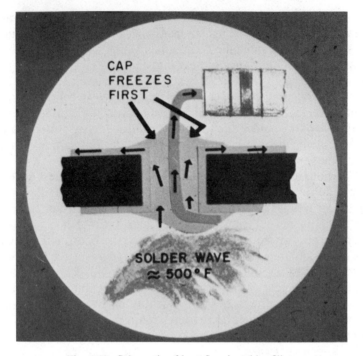

Fig. 4-13. Schematic of heat flow in solder fillet.

unusual circumstances, solder joint imperfections occur. A prolonged top molten condition, for example, has been observed to cause joint sagging, icicles, incomplete fillets, and sometimes complete solder dropout. Premature freezing on the bottom has caused blow holes and pin holes on the top of the fillet. This cooling sequence also explains the phenomenon of blow holes on the bottom of the board. See the discussion in Sections 7-16 and 7-17.

As a practical guide, when setting up a wave, the author uses the following rule of thumb:

For proper fillet freezing, adjust the heat balance to allow the top of the solder joint to solidify within 3/4–1 in. (20–25 mm) from the point of the wave exit.

The wave exit in this case is the point of solder separation, or peel back.

4-14 ZONE III—POINT OF EXIT (PEEL BACK)

In order to obtain the best wave solder results, we seek to obtain a high degree of uniformity in fillet configuration. This makes inspection easy and dramatically reduces unnecessary operator touchup. To obtain such uniformity, we must control the forces shaping the underside of the fillet at the point of wave exit. These forces basically fall into two categories:

1. *Surface energies.* These forces are predictable and can be subdivided into:;
   a. The interfacial energy between the solder and the base metal. This is affected by solderability but is independent of the environment.
   b. The cohesive force of the liquid solder, which is greatly affected by the second phase with which it is in contact (flux, oil, or air).
2. *Hydraulic forces.* These are often random; they depend on the following factors:
   a. Wave design, which is related to turbulence, direction of flow, and the quantity of solder being pumped.
   b. The impedance angel of the conveyor, which varies from horizontal to a positive incline.
   c. Board land configuration (exposed metal) and the distribution of the thermal load.

The best point of wave exit, therefore, would correspond to a location where these hydraulic forces can be neutralized. This can be achieved by withdrawing the fillet from the wave at a static location. This location is found where board travel speed and direction are similar to solder flow and velocity. This ideal exit point is normally adjusted by the equipment manufacturer for standard speeds only. Any drastic deviation from these parameters merits adjustment by the user.

By exiting at this point, a wave application subject to liquid flow forces has in essence been converted to a static dip process—an operation that is well understood and controllable (Ref. 4-3).

To further control the solder fillet size, we can increase the favorable withdrawal geometry at the point of exit. By inclining the conveyor, we can promote the solder *peel back*. The angle used varies with the wave configuration and is normally fixed by the equipment manufacturer. It ranges from 3-5° in dry waves to 7-9° in oil-intermixed applications.

We have already discussed the upward force due to the depth of immersion in the solder. While it is a driving force in fillet formation, its importance is limited. It is more important to have the correct sequence of solder feeding and heat transfer. It is necessary, therefore, to correlate the depth of immersion with board thickness and type. A good rule of thumb is outlined in Table 4-3:

In other words, submerge single-sided and flexible boards approximately one third of their height. Plated-through hole boards up to 0.062 in. (1.6 mm) thick should be immersed approximately two thirds of the way.

Table 4-3. Suggested Depth of Wave
Immersion versus Board Thickness.[a]

BOARD TYPE	BOARD HEIGHT (in.)	(mm)	IMMERSION RANGE LOW	HIGH
Flexible[b]			Kiss	N/A
Single Sided	0.062	1.6	Kiss	1/3
Double Sided	0.062	1.6	1/3	2/3
	0.093	2.4	1/2	3/4
Multilayer	0.062	1.6	1/2	3/4
	0.093	2.4	5/8	3/4
	0.125	3.2	3/4	7/8

[a] For an average thermal load only. As fraction of board height.
[b] When not attached to a rigid board; otherwise use the total board thickness as guide.

While multilayer boards with more than four layers (or more than 0.062 in., 1.6 mm thick) should be immersed three fourth of the way. Very thick boards may have to be immersed even deeper (Fig. 4-14).

These guidelines refer to the average thermal mass on a board. Heavy components or large ground planes may require somewhat deeper immersion, coupled with more preheat and slower speeds. The reverse is also true; light thermal loads need less depth of wave penetration.

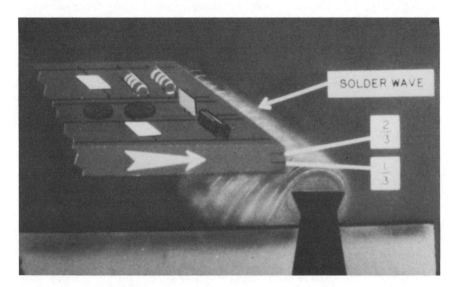

Fig. 4-14. Schematic of board submersion in wave—see text for details.

Remember that the preceding is only a guide; each case is unique. Much depends on wave shape and design, and smooth waves allow deeper penetration. Board warpage, conveyor finger or pallet design, and mechanical stiffners on the leading edge also affect the immersion depth.

The depth of submersion can be calibrated in one of two ways. The operator can bring a trial board just before the wave and adjust it by eye. There is also a more scientific method, using a glass plate with a calibration grid. This is placed in the conveyor in place of a board, and the wave print against the transparent glass is measured. For more details, see Section 4-16 and the Lev-Check (Hexacon Electric Co.).

Fillet formation in single-sided boards is much simpler (Fig. 4-15). The solder must still rise up the sides of the leads, but to a lesser extent. Fillet solidification is much more rapid and uniform. This is due to the smaller mass and geometry of the single-sided configuration. The effect of the solder's cohesive forces on fillet shape is more pronounced.

The reverse is true for multilayer boards, where solidification is very complex. There is a more critical hole diameter to board height ratio, which is further complicated by the higher heat content of the multilayer boards. Here the heat balance becomes a major factor, and top preheating is usually beneficial. Note that ground planes inside the multilayer board act as large heat sinks. It is recommended that cart wheel etching around all holes be incorporated in the design.

Flat cable and flexible circuit boards defy generalizations, since heat and joint requirements vary from one application to another. Their similarity to single-sided boards during freezing is generally the best way to describe the process.

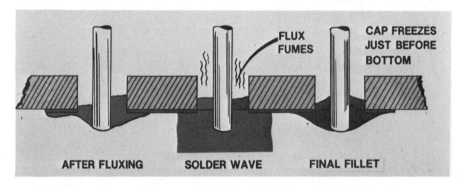

Fig. 4-15. Schematic of solder impact on single-sided board. Note that the shallow fillet should be drawn up on the lead. Here the top freezes just before the bottom.

4-15 DROSS FORMATION AND THE USE OF OIL

Dross is a metallurgical term used to describe nonmetallic waste products like oxides and sulfides. These products form on top of a molten metal as a result of interaction with the air. In addition, any nonmetallic particles included in the melt tend to float up to the surface (depending on their mass to surface ratio). In soldering, the dross from pure metal is usually a combination of tin oxide and lead oxide in a ratio closely resembling that of the parent metal. In practice, this floating mass of tarnish also contains other metal reaction products like sulfides and organic residues such as burned flux and the like.

Dross has been reported to contain the metallic impurities picked up during soldering. This, unfortunately, occurs only when the impurity concentrations exceed their solubility limits. At that point, most systems develop intermetallic compounds, some of which float to the surface. With the normal impurities picked up during electronic soldering (copper, gold, silver, etc.), the dross is not a purging mechanism. The level of impurities detrimental to the joint formation is far below the solubility limit. See Sections 3-12 and 3-13 on impurity levels in wave soldering.

To clarify the wave soldering operation, we will define and use two practical terms—*dry dross* and *wet dross*. Dry dross consists entirely of nonmetallic components, as described earlier. It is normally found on still pots, drag machines, and similar equipment. Dry dross may be mixed in with flux, oil, or any other liquid material. The term "dry" refers basically to the absence of any metallic solder.

Wet dross in contrast, contains metallic droplets of molten solder suspended in the dross. It is normally formed as a result of mechanical agitation. In wave soldering, droplets of molten metal are splashed and are retained in the floating dross. Because of their location, they are unable to coalesce with adjacent droplets. In addition, their small size (weight) prevents them from rejoining the parent metal underneath the dross by gravity. Wet dross often appears as a spongy metallic mass floating on the surface. It can be mechanically squeezed to release much of the parent metal, a practice no longer recommended because of operator exposure to lead poison. With the use of a solder blanket and/or oil, wet dross can be substantially converted to dry dross. This releases most of the valuable trapped metal and allows it to join the parent solder.

A thin film of dry dross without organic material assumes a color which depends on the solder temperature and exposure time. This color varies with increasing temperature from a dull gray through a yellow gold to gun metal blue, then violet, and finally a dull brown. When a sill pot is carefully skimmed, dry dross normally comes off powdery, with little parent metal entrapped. This dry powder must be handled with care because of its health hazard.

Lead is a poison to the human body and must not be introduced during soldering. It has been established, through years of experience and measurement, that molten solder does not generate dangerous fumes. At electronic soldering temperatures, it stays well below permissible inhalation level. This holds true for molten reservoirs like pots, waves, or hand soldering with an iron. The dry dross, however, when carried by the air, can be inhaled in quantities above safety limits. To avoid such inhalation, safety precautions are advisable, such as the use of a simple aspirator.

Solder can also penetrate the body by ingestion of the metal or its compounds. Employees should be instructed to wash their hands thoroughly prior to eating, drinking, or smoking. Lead traces are retained on the skin of the hand whenever solder is handled. This is true for bulk metal, spools of hand wire, tinned parts, and so on. Fortunately, lead is not absorbed by the skin, but we must prevent its ingestion. Eating, drinking, or smoking at or around work stations should be strictly prohibited (Ref. 4-5).

Dross quantity depends largely on equipment design, solder velocity, and the height of the *waterfall* (the height from which the solder falls from the crest of the wave). Soldering temperature and alloy composition are also important. So is the quantity of flux used and the presence of any other organic materials (wax, oil, etc.).

There are several ways of reducing dross formation in wave soldering. The first of them is strictly mechanical:

1. Wasteful dross formation can be limited by wave design. This can be achieved by minimizing the turbulent area. Figure 4-11 depicts one mechanical design where "side plates" with a controlled solder return shortens the waterfall height.
2. Another mechanical way to reduce dross is through the use of glass beads. These float on top of the solder by buoyancy and are rotate slowly by the falling solder in the waterfall zone. Thus they mechanicaly break up the wet dross, until the system gets overloaded with powder. At that point, they are simply removed with the dross and replaced with a new charge.

The second method is the reduction or total elimination of dross through the use of a chemicals:

1. Solder blankets exclude the tarnishing atmosphere from the molten solder in the reservoir. They do not, however, protect the exposed

solder on top of the wave or any exposed areas. Dross formed on the wave may be chemically reduced back to metal but this process requires aggressive chemicals. These are not normally found in blankets because of the proximity to the work. Remember that they do not just float there, but give off fumes that may be redeposited on the work. In addition, they may splash up and deposit on the boards.

A blanket is basically a heat-stable material that is liquid at soldering temperatures. The materials used, in order of increasing chemical activity, are wax, rosin or resins, and oil, all containing some proprietary inhibitors.

Another function of a solder blanket is to prevent wet dross formation. If the material has the right surface tension, it prevents the suspension of solder in the dross. The mechanism involves a partial suspension of the dross itself in the blanket, followed by surface reactions. The solder blankets float on top of the metal reservoir and are not pumped with the solder. If the waterfall turbulence at the bottom of the wave is large, some of the blanket is pulled into the solder and may end up on the work.

2. More aggressive chemicals are used to break up the wet dross when no blanket is used. These are not applied to the entire surface, mainly because of their cost. The operator periodically skims the dross to one side of the pot and applies a cup full of this material. It must then be worked thoroughly into the dross with a spatula. There are no dangerous lead fumes given off during this treatment, which is very cost effective.

The third method of dross inhibition relies mainly on atmosphere exclusion in the reservoir and the flowing solder.

1. This is accomplished by pumping oil together with the solder through the wave. There is a monomolecular layer of oil on the surfaces at all times, which also enhances solder wetting. The pumping of oil together with the solder is achieved by using the same impeller and providing an oil feed from the surface or a fresh reservoir. This method which is depicted on the right-hand side of Fig. 4-10.

2. A second, older method relies on the introduction of droplets of oil by the falling solder in the turbulent area. The oil is immediately sucked into the impeller of the pump and pushed back up to the surface In this method, more air is introduced into the oil and a heavy sludge is formed.

3. Oil is also introduced into the wave by other means. It can be

pumped directly to the base of the wave by a separate unit and allowed to come up to the surface together with the solder. It can also be pumped by a separate impeller to the back of a unidirectional wave and is carried forward by the solder. In the last two methods, care must be taken to heat the oil to the operating temperature prior to its use.

4. A separate oil wave, immediately after the solder wave, is also used to help reduce icicling and bridging. Here the oil is naturally carried over the solder wave in front.

The reduction of dross is of great economic concern. The dross constitutes a loss, since it is a conversion of usable metal to a nonfunctioning form. While this scrap has some intrinsic value, it must be shipped to the smelter, which further reduces its value. In addition, dross formation increases the need for machine maintenance. Thus dross reduction can be equated with actual savings.

Oil has three other major advantages in addition to minimizing dross formation:

1. It lowers the surface tension of the solder in the direction of improved wetting. As a result, it is possible to solder at temperatures of 20-30°F (11-16°C) lower than those without the oil. Lowering the wave temperature is not always an advantage from thermal supply considerations. Improved wetting, however, minimizes the formation of bridging and icicling, due to the shallow fillets that are formed.

2. The second big advantage of oil lies in the area of fillet formation at the point of wave exit (zone III). Since the oil excludes the air at this particular interface, solder cohesive forces are greatly reduced, while no dross skin forms. A balance of surface energies is established between the wetted copper of the printed circuit board, the solder wave, and the oil layer (see Fig. 4-10, top right). This is by far a more uniform and controllable configuration than it would have been if exposed to the outside environment. If air is present, a dross skin is formed, retarding the cohesive energies. As a result, the degree of fillet uniformity with the use of oil is much higher.

3. There is one final advantage whose value has been proven in marginal solderability situations. The presence of the oil in zone II (heat transfer stage) seems to provide an additional fluxing action after the original flux is gone. This is especially important if wetting was marginal, as in the use of RMA rosin fluxes. It also helps where assemblies are not really solderable. It occurs when the oil is

pumped with the solder, and not when it is introduced only as a blanket.

The use of oil, however, implies a mandatory cleaning operation after soldering. The nondrying oil residue mixed in with flux and dross should not be left on the work. This, to the author's thinking, is a blessing in disguise, because it provides an economical contribution toward the achievement of clean work. Today the desirability of trouble-free clean assemblies has been well established. Presolder processing, handling, and storage contaminants are removed together with the soldering chemicals such as flux, oil, and wax. New test procedures and government oriented cleanliness guidelines are now available to evaluate cleaning efficiency (Ref. 4-4).

We must not end this discussion with the impression that the use of oil is the only possible method. Vendor limitations and product design may often preclude the use of oil. Dry waves are also very versatile and can yield excellent results. In the author's opinion, however, oil is well worth considering.

4-16 THE CONVEYOR

The method of transporting the printed circuit assembly through the process requires planning and much consideration. It starts in the design stage, with specific rules that relate to the conveyor (see Section 1-19). It also requires the most thorough investigation during equipment selection and purchasing. Once the decision has been made, it is too late, and manufacturing ease (cost) as well as quality suffer.

The conveyor fixes the precise location of the board relative to the molten solder (depth of immersion—see Table 4-3) and thus influences fillet shape. It also determines the movement of the work through the molten metal, affecting the process dynamics. Finally, it controls the thermal balance in preheat and soldering.

As such, the transport must be smooth, have constant speed, and stay level and stress free. Let us look at these and other requirements individually:

1. A vibration in the conveyor will upset its relative position to the solder and the fluid dynamics in the wave. In the extreme, it can cause solder skips and disturbed joints (see Section 7-5). Vibrations can be generated externally; check;

- Conveyor drive.
- Individual fingers catching.
- Solder pump.

- Ventilation motor.
- External building vibrations.

An uneven motion of the conveyor can cause the same results. Check for friction, lubrication, and other interference.

2. Conveyor speed is essential for uniform contact with the materials (flux, solder, oil, air knife, etc.) and the thermal balance. The more elaborate machines provide a digital speed control or readout. The simple machines require manual calibration. For speed variations check:

- Conveyor drive.
- Voltage variations.

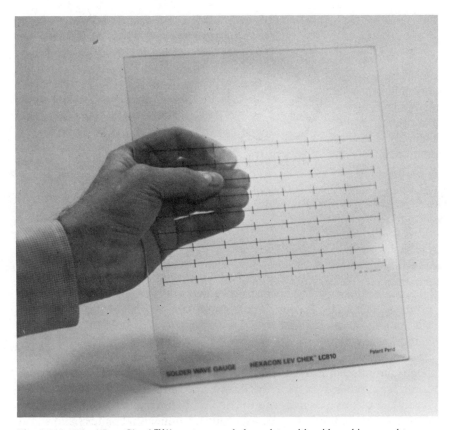

Fig. 4-16A. The "Lev-Check™"—a tempered glass plate with grid markings used to measure and adjust the depth of immersion in the flux and solder. (Courtesy Hexacon Electric Co.)

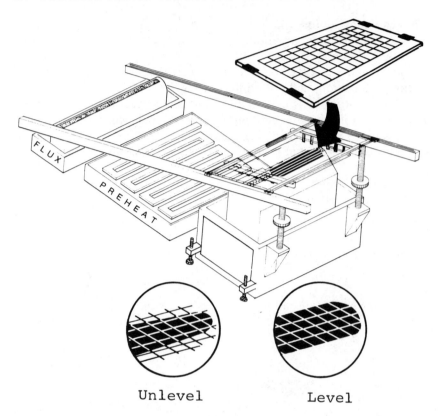

Unlevel   Level

Fig. 4-16B. A schematic of the "Lev-Check™" showing how it is used in checking the conveyor position over the liquid from front to back. (Courtesy Hexacon Electric Co.)

Damage due to variations in speed is mainly heat oriented and may not be dramatic.

3. The conveyor must be kept level from left to right with the liquids in the process. It is easy to adjust the conveyor position relative to the fluxer and the solder wave. It is practically impossible to change the liquid layer in a standing wave of flux or solder, since liquids seek the horizontal levels.

The adjustment can be made with the use of a tempered glass plate. The plate is mounted in place of the work in the conveyor and passed through the system. While it is over the fluxer, and later the wave, it is visually possible to observe or measure the depth of immersion. The width of the depression in the liquid is used as a gauge. Commercial plates have a grid marking (Figs. 4-16A and 4-16B), which makes it easy to measure the width. Plain glass enables observation only.

4. The stresses built up in the work due to thermal expansion must be absorbed by the transport system. Otherwise, the board which is above its glass transition temperature will warp. It usually bows in the center and often *submarines* under the wave, with solder running over the top.

Make sure that cold boards are loose when placed in the conveyor. They will expand as they heat up, and must not get into compression. This holds true for finger conveyors as well as pallets. If warpage persists, check to see if your board width is within manufacturing tolerances.

4-17 FIXTURES, PALLETS, AND FINGER CONVEYORS

There are two major methods of holding the printed circuit board in the conveyor. First, there is the fixture or pallet, which may be universal and adjustable, or made specifically for a particular shape (see Fig 4-17). Second, there are the *finger conveyors*, which hold rectangular boards by the edges and are adjustable only in width (Fig. 4-18). Each has its own

Fig. 4-17. Typical adjustable pallet for wave soldering. (Courtesy of Hollis Automation)

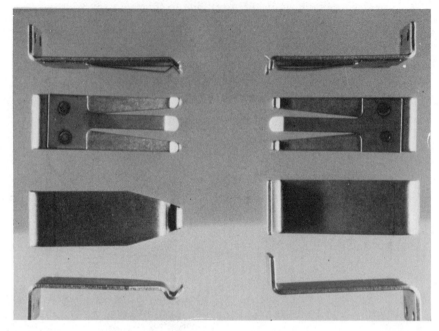

Fig. 4-18. Variety of fingers for conveyors. The top left finger is the most popular, since it lets the solder reach nearly to the end of the board, although with some turbulence. The angle of the tip however, causes variations in height between narrow and wide boards, this is corrected by the finger on the top left (note it has a 90° bend). The bottom left finger is used at 3/4 in. centers to hold boards that tend to warp, or are already warped before soldering. The bottom right finger is suitable for pallets, special connectors on one side, etc.

characteristics as follows;

1. *Fixtures and pallets.* These usually ride on chains, pulled along by their own prongs. They have the following advantages:

- Hold odd shapes.
- Suitable for multiple boards.
- Useful also for insertion.
- Simple and rugged transport.
- Good for low to medium volume.
- Intermix board sizes.

They are not cost effective for medium to large volumes if finger conveyors can be used. The labor needed to load and unload pallets is great.

They must be handled and stored carefully to avoid physical damage. They interfere with the total cleaning process and must be unloaded before degreasing or washing. They must be cleaned, however, to avoid tackiness and contamination of the fluxer. When hot, they tend to depress the foam in the fluxer, resulting in skips or uneven coverage.

Fixtures and pallets can be made from nonmetallic materials like tempered masonite, phenolic laminate, etc.. They are also made from metallic materials like anodized aluminum, Teflon-coated steel, and titanium. When selecting a material of construction, remember that it must have the following properties;

- Nonwettable with solder.
- Resistant to flux.
- Does not warp in heat.
- Be a small heat sink.
- Easy to fabricate.

One major advantage of pallets is their ability to provide complex support and stress relief. They are often used even with finger conveyors, held by the edge like a board. This combination of pellets and finger conveyors is useful when heavy components like transformers are mounted on the work.

2. *Finger Conveyors.* These are permanently attached to the conveyor. They have the following advantages:

- Low labor content.
- Automatic pick up and delivery.
- No heat mass.
- No volume limitations.

Finger conveyors are very cost effective from small to large volumes and fit into conveyorized manufacturing systems. They can only handle boards with parallel edges and cannot mix width sizes. The fingers must be kept clean and adjusted.

Many shapes are available (see Fig. 4-18). Pay attention to these properties:

- Thermal expansion relief.
- Board height control.
- Accommodation of width tolerances.
- Turbulence created in wave.

Special configurations have been developed to allow the use of connector blocks at the holding edge, on one or both sides. Others protect gold fingers at the sides with the use of special channels. It is also possible to support long boards in order to avoid warpage.

Cleaning the fingers on the conveyor helps keep the system in balance. It stops contamination of the flux station while keeping the fingers free of tacky residues. It is especially helpful when oil is intermixed in the solder.

4-18 THE ENCLOSURE AND EXHAUST

Wave soldering and cleaning equipment is available in two basic forms: modular units and preengineered systems in their own enclosure. The modular approach lets the individual select the specific units for the machine. The conveyor, fluxer, preheater, and wave are the basic parts needed. The user must then provide the exhaust and possibly isolate the equipment from the rest of the plant (Fig. 4-19).

Obviously, all of this is provided by the enclosed pre-engineered system. But a certain freedom is also forfeited, because it is very difficult to add to or modify such a unit. The enclosure creates only part of the restriction; other elements are also fixed in place and cannot be adjusted later.

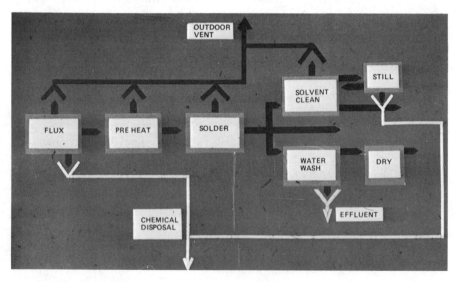

Fig. 4-19. Schematic of vapor and chemical handling in a wave soldering operation.

Let us look at the reasons why a machine requires exhaust and list the advantages of an enclosure:

1. The exhaust is required for to remove the following types of fumes:

- Flammable flux volatiles.
- Pyrolytic breakdown fumes.
- Lead-containing dross particles.
- Oil vapor, if any.
- Restricted cleaning solvents.
- Water vapor (moisture).

2. The exhaust is also needed for thermal management:

- Uniform machine output.
- Operator comfort.
- Reduced air conditioning load.

3. The advantages of the Enclosures include:

- Lower energy use.
- Less exhaust volume.
- Reduced noise levels.
- No separate room required.
- Aesthetic appearance.

When selecting an enclosure or building your own, certain points must be considered:

- Does not encumber production.
- Process visible, well lit.
- Easy to maintain and service.
- Simple to inspect and control.
- Readily tied into a larger system.

The choice of either approach is a function of individual needs and cost. Modular units should be located in a separate room. Wall openings like windows can serve as input and output ports for an inline system. Modular units may also benefit by a separate room, where chemicals and other supplies can be stored.

4-19 COMPUTER AIDED MANUFACTURING (CAM)

The proliferation of low-cost minicomputers has had an impact on wave soldering and cleaning, too. As the engineering community learns to use data processing equipment to control manufacturing operations, few repetitive processes will escape automation. Soldering and cleaning machines with micro chip controls are readily available today (see Fig. 4-20). Stand alone computer systems can also be added to established lines.

Still, the availability of hardware and software does not automatically ensure success in soldering, as measured by less touchup or repair and coupled with higher reliability. The computer must be viewed as an additional instrument in the system which has to be fitted in. In doing so, we must develop a systematic approach to soldering and cleaning. This by itself will soon result in cost savings by forcing all relevant disciplines into line. These extends from design and layout, through material selection and control, to storage and handling, and of course assembly.

In an ideal flow (Fig. 4-21), the steps necessary to utilize the ability of the computer fully start at the beginning. The solid lines on the bottom represent the process itself, while the dashed lines show the information fed to the computer. (Note that one solid line indicates the computer's output in the form of hard copy reports.) Looking at the first element, we realize that we must start with controlled materials. This function can also become computer directed and extends beyond inventory control. It would include the following soldering and cleaning-oriented items (see Section 2-2):

1. Soldering chemicals and metals must be analyzed for conformance to incoming specifications.
2. Printed circuit boards and other substrates must be inspected for the quality of plated through holes, the solderability of exposed surfaces, resistance to soldering heat, adhesion and compatibility of the solder mask with the flux and cleaner, and so on.
3. Leads on components must be tested for solderability and compatibility of surface finishes. The component body must be stable in the cleaning environment.
4. Surface mounted components need special scrutiny for termination stability and solderability. Compliant lead geometry must also be checked to ensure stress absorption in service.
5. Washability of all components mounted on the board must also be guaranteed throughout the assembly process. A sufficient gap underneath flat configurations must exist to avoid entrapment.

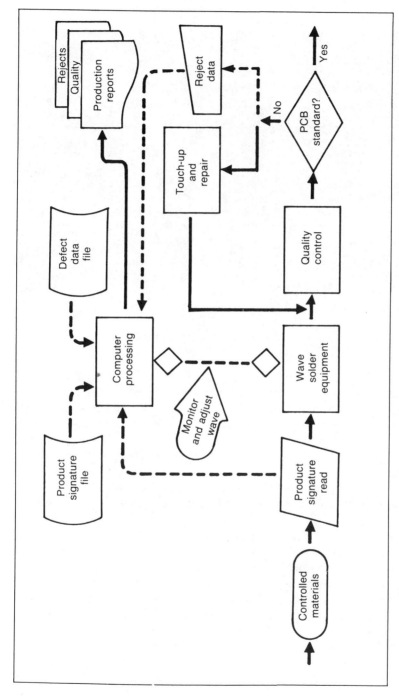

Fig. 4-20. Flowchart of proper computer utilization. Solid lines follow the process itself and the hard-copy output of the system. Dashed lines represent information fed to the computer.

Fig. 4-21. Printed circuit boards with bar graph to identify the "Signature."

6. In addition, the normal checks for electrical and mechanical accuracy are necessary. This includes the hole-to-wire ratio, mask registration, and similar board characteristics. It also includes the stress relief on components, insulation meniscus on the lead, and stand-off wherever needed.
7. Finally, the assembled board must be checked for proper component insertion by type and orientation; for the size and shape of the adhesive spot on surface mounted devices; for the application of temporary resists or stiffners where needed; and so on.

These presolder controls are not really part of the soldering process, but they do affect the yield. In any survey of soldering defects, such faults must be culled out in order to get a true picture.

The second element of the system focuses on the equipment control itself. A machine is seldom dedicated to a single printed circuit board; instead, it must process a variety of sizes and thermal loads. Sometimes groups of boards are treated in different ways (some may not pass through the cleaning process, while others do). Thus the computer must know which boards are being processed. Let us review the functions needed for this process:

1. The first computer input is the *product signature*. This signals the computer such important factors as board width to adjust the conveyor width; thermal requirements to set preheat temperature settings and conveyor speed; board height to control the depth of immersion; cleaning parameters, if any, and so on.

The board signature can be printed directly on the printed circuit as a bar code (Figs. 4-22A and 4-22B), mechanical notches, or in any other machine recognizable form. It can also be fed manually by the operator through a keyboard.

Once the information is received by the computer, optimized soldering conditions are retrieved from a *signature file* stored in the computer memory. Obviously, this file had to be fed in by the user earlier. The computer can now make the appropriate wave machine adjustments needed to process the boards.

2. The second function of the computer is the continuous monitoring and adjusting of the solder machine elements. These elements include:

- Flux level, density, and activity.
- Preheat temperature.
- Wave height, flow, and temperature.
- Solder metal level, and oil flow.
- Conveyor speed, chatter, and angle.
- Operation of special attachments like the turbulent chip wave, hot air knife, and so on.

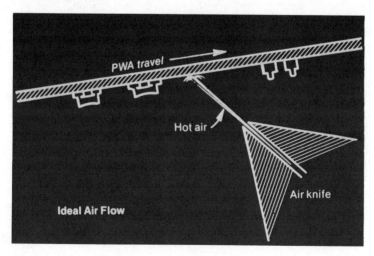

Fig. 4-22. Schematic of air knife mounted behind the wave. (Courtesy of Hollis Automation).

- Washing temperature, spray rates, liquid levels, dry cycles, chemical composition, or additives.

These and similar parameters must be continuously monitored while feedback circuits control the equipment. Analog computers have been used for years in this process control. Most of today's minicomputers have analog to digital interfaces suitable for this function. The problem lies in finding appropriate equipment to install on the wave for the critical parameter control.

The third level of the computer system is very crucial. It is intended to generate the feedback data required to keep the process trouble free. This step still requires human inspection; however a sizable effort is underway to automate the solder joint evaluation.

Following this process, the boards are compared to existing quality standards. Good assemblies are counted and passed through, while defective boards are shunted to a touchup station. The type of defect and its location are fed into the computer before they are corrected. The computer processes this information continuously, and identifies possible corrective action by utilizing the *defect data file*. When any repetitive failure pattern is noted, the microprocessor is programmed to take specific action.

In order to achieve this remarkable feedback capability, the defect file must first be established. There are some general categories of rejects that are universal, but most systems must set up specific data banks for their own boards. The defects are usually classified into the following groups:

1. Process faults like flux skips, solder misses, and board flooding.
2. Material problems such as solderability (wetting), blowholes, and webbing.
3. Spatial (board location) defects like unfilled holes (hole-to- wire ratios), insufficient rise (weak knees), and repetitive bridging (design oriented).

The defect data file also holds specific corrective instructions for each problem. Some machine controlled parameters can be instantly corrected, but the control should be left to the operator's discretion. Let us follow such a case:

The computer receives the input that two boards of a specific lot have been flooded with solder, which is a catastrophic failure. The probable cause of this defect is one of the following, in order of descending probability:

1. Too deep in the wave (immersion depth).
2. Too much side pressure from the conveyor.
3. Board warpage before soldering.
4. Rough wave.

The computer should issue printed instructions (hard copy or on a CRT screen) to the operator in one of two ways as follows:

1. Stop sending any additional boards over the wave and back out any boards that are already in the preheater.
2. Signal that the system is on hold.
   Direct the following checks:
   a. Depth of immersion should be $1/x$ board height
   b. Width of the board should be X".
   c. Is the board warped?
   d. Is the wave rough?

It is up to the operator to take corrective action from this list. Further instructions can be stored in memory for each item. For example, if the board was too wide, thermal expansion during heating will exert pressure between the unyielding conveyor and the board. As a result, the printed circuit will collapse downward, submarining the board in the solder. A conveyor change without a dimensional check of the board is not possible. If the operator establishes that the board is wider than the permitted tolerance, he may request further instructions on item b. The computer may than display the following additional massage:

b. Check board width on sample of 10 printed circuits; If they are within 0.00X in. of one another, feed new value for temporary use. A larger variation will require board sorting. Notify Incoming Inspection of problem and get instructions for future lots.

This solves the immediately problem and should trigger long-range solutions.

The fourth and final series of events in the computer use involves the acquisition of data and the generation of valuable reports. These must be used to prevent any future repetition of problems. The corrective actions may include design changes for recurring layout problems, specific vendor rejection because of a poor quality history (vendor profile), new assembly instructions or other inline processes.

The outline of the software described above must be generated by the user. No such programs are available in the marketplace today. This is understandable because there is no real standardization within this industry at present. This is also the reason why the mere purchase of a computer-controlled machine does not solve all the problems. The machine is only a part of a system, which must function properly to achieve the desired results.

4-20 THE HOT AIR KNIFE FOR DEBRIDGING AND INLINE STRESS TESTING

Among the recent wave soldering innovations is the hot air knife. Like most such developments, it was an answer to a stubborn problem in the industry. It later found additional applications as industry learned more about its capabilities.

The problem originated with a large volume, high density printed circuit where good design rules had to be ignored. In addition, competitive pressures dictated a solder-and-cut operation. This meant a tangle of many long, closely spaced leads on the bottom of a wave soldered board, which obviously contributed to numerous bridges. With the development of the patented hot air knife, it was possible to eliminate the bridging problem.

The hot air knife works by concentrating a jet of hot air, with a temperature above the melting point of solder, to blow off excess solder and bridges. It is mounted immediately behind the wave and is directed at the immerging solder joints (see Figs. 4-23A and 4-23B). The hot air removes

Fig. 4-23. A view of the computer-controlled Hollis GBS™ (Guaranteed Bridgeless Soldering) system, utilizing the hot-air knife after the wave. (Courtesy of Hollis Automation)

Fig. 4-24. Wave soldered SOIC with typical bridge problem.

the unwanted bridges before the solder solidifies. Only the solder that is bonded by surface tension (wetting forces) remains on the board. Properly filled solder joints are not affected by the process, because the surface tension exceeds the force of the air pressure. In addition, the direction of air flow will not push solder up the hole even if it is molten.

The process, however, has some exciting and unexpected advantages. By blowing superheated air at each solder joint after it is made, it is

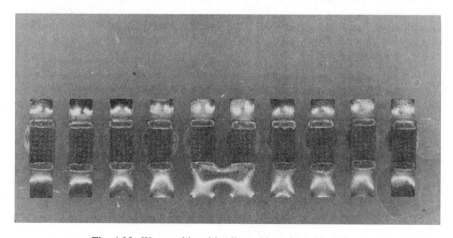

Fig. 4-25. Wave soldered leadless chips with bridge short.

possible to stress test each fillet during production. If either the lead wire or the board pad has solderability problems, the knife will expose the culprit. In this fashion, it is possible to stress test 100% of all the solder joints without leaving internal stresses or damaging joint integrity. A fillet that withstands the air knife is thus a most reliable joint. This is an exciting yet simple method of ensuring high reliability in wave soldering.

The hot air knife process is simple to control but requires careful exhausting of the solder (lead containing) particles. By adjusting such parameters as air temperature, air velocity and volume, impedance angle, and proximity to the wave, the amount of solder removed can be closely controlled.

The process holds special promise in surface mounting, where bridging of closely spaced chips, or high count compliant leaded devices are wave soldered. Figures 4-24 and 4-25 show typical bridges on an SOIC and a leadless chip. It is also possible to avoid bridging on leaded chip carriers even though the leads extend in all four directions. The airknife opens the door for the attachment of most active devices by the wave soldering process.

REFERENCES

4-1. Howard H. Manko, *Solders and Soldering*, 2nd ed., McGraw-Hill Book Co., New York, 1979.

4-2. Howard H. Manko, "Understanding the Solder Wave and Its Effect on Solder Joints," *Insulation Circuits*, January 1978.

4-3. T.Y. Chu, "A General Review of Mass Soldering Methods," *Insulation Circuits*, November 1976, pp. 73-75.

4-4. MIL-P 28809—Military Specification, "Printed Wiring Assemblies."

4-5. Federal Register: *Occupational Exposure: Proposed Standard—Lead*, October 3,1975.

5

SURFACE-MOUNTED SOLDERING TECHNOLOGY

5-0 INTRODUCTION

In the second half of the 1980s the fastest growing development in the printed circuit industry will be the use of surface-mounted devices. The concept of surface-mounting is not new; it was developed in the late 1950s for the hybrid industry. *Hybrids* are thick and thin film circuits on ceramic substrates with surface-mounted components. In addition, flat pack devices were introduced in the 1960s and have been surface-mounted on printed circuits ever since.

There are many undisputed advantages to surface-mounting that propel the industry in this direction: higher electronic speeds due to shorter delay-lines; more density per cubic inch, which increases the possibility of circuitry sophistication; fewer external interconnections, which increases reliability; and last, but not least, lower costs for materials, processing, and labor.

One unscientific force pushing U.S. industry is competition from Japan. The Japanese have successfully applied surface-mounted devices to printed circuits for many years. In fact, however, surface-mounting was innovated and created in the United States. Only the application to printed circuits, rather than ceramic substrates, is a more recent innovation that was neglected in this country.

There are several lessons to be learned from the first stages of the hybrid industry in surface-mounting. During the early development period of thick and thin film technology, the importance of material compatibility became obvious. The metallurgy of the terminations had to be compatible with the solders used. A close match in the *thermal coefficient of expansion* (TCE) between substrates and devices was also vital (see Section 5-3). Board laminates have a dramatically different TCE than ceramics and/or the silicon of the active devices. This large mismatch in TCEs was the reason we neglected to think of surface-mounting for printed circuit boards.

The hybrid industry has had advantages that the printed circuit board industry does not enjoy. It had the time to develop their technology without the pressure of competition. In addition, the staff and equipment used are of a high caliber, with scientists and engineers running the lines. Fabrication is done under clean room conditions, and often under temperature and humidity control. These conditions are very different from those in the printed circuit assembly and soldering shop. Such controls had not been necessary for conventional boards, but this does not mean that the U.S. industry can afford to rush into surface-mounting without preparation and forethought. Proper design (see Sections 1-12 to 1-17), material selection (see Section 5-4), and equipment are needed. Let us look at the additional areas that need planning for successful surface-mounting on printed circuit boards.

5-1 WHAT IS SURFACE-MOUNTED TECHNOLOGY?

Surface-mounting is the technique of attaching components and devices only to the surface of the board. No holes or terminals are used in this process; only the board pads are soldered. If any plated-through holes are used, they serve as via or interconnect holes.

The pad size corresponds to the footprint of the surface-mounted device, or SMD for short. These footprints are very small and allow a high density of component population. To utilize fully the advantages of this miniaturization, special fine line boards are needed (see Section 1-3). These boards are pushing today's state of the art in circuit manufacturing; they are not yet readily available.

A variety of components in the surface-mount configuration are already available. These will be described further in this section. But we are far from being able to design sophisticated circuits using only surface-mounted devices.

For this and other practical reasons, in the near term, surface-mounted devices will be soldered on conventional boards with lower densities. In the beginning, the total component density will be increased by the addition of surface-mounted components to the unused bottom of the board. This trend will be furthered by cost pressures and production capabilities. Later, more and more boards, or portions of a board, will be dedicated to surface-mounting only. There is no rush to move into a dedicated technology in this area.

Several types of surface-mounted devices are already being applied to printed circuit boards. They can be divided into two major groups:

1. Leadless devices in which the body of the component has metalized areas to serve as terminations.
2. Leaded devices, in which the lead shape is conducive to a surface connection.

The devices themselves fall into four additional subdivisions, according to shape and function, as follows:

- Leadless chips (see Sections 5-12 and 5-13).
- Small-outline compliant leaded components (see Section 5-15).
- Leadless chip carriers (see Section 5-16).
- "J" leaded chip carriers (see Section 5-16).

5-2 THE WHY AND WHEN OF SURFACE-MOUNTED TECHNOLOGY

The drive to the use of surface-mounting has several forces behind it. Each is important, yet all are interrelated. Let us discuss them individually:

- Cost reduction in materials and processes.
- Increased density and speed.
- Reduction of interconnects.

1. *Cost reduction in materials and processes.* This cost reduction stems from several factors, not just a reduction in size. These factors are listed in no order of importance:

- Material savings in devices (smaller size, shorter or no lead wires, lighter for shipping, less bulk to store, etc.).
- Simpler assembly and processing.
- The ability to use less complex and cheaper boards for the same density. (i.e., single-sided versus double-sided versus multi-layer).

2. *increased density and speed.* The possible increase in density is obvious. These components can be mounted on the top and bottom of a board, with little change. Their physical size also contributes to compactness and shorter delay lines (increased speed)
3. *Reduction of interconnects.* In terms of reliability, the number of interconnections in a circuit always poses a problem. The fewer external

joints produced at the user level, the better. This is why there is such a strong trend toward ever-larger scales of integration.

The status and future of surface-mounted technology is a complex situation. First, the implementation of this technology is not simple. Few reliability data have been accumulated to define its use at the higher end of the printed circuit industry. It must be used with new and more costly laminates, specialized assembly techniques, and so on. By nature of any new developing trend, the cost justification for small and medium-sized runs will not exist for several years. Therefore, we can expect to see this technology grow in the following areas:

• Mass production at the lower end of the industry.
• High-technology applications like computers.
• Miniaturized and/or high density circuits.

1. *Mass production at the lower end of the industry.* Surface-mounting of devices for mass production is already underway. Japanese industry has had a jump on the rest of the world for several years. However, the trend in the United States, and Europe, is now in full swing.

2. *High-technology applications like computers.* These can justify the increased investment because of the circuit speeds gained. The U.S. computer and telecommunication industries are leading the penetration into surface-mounting.

3. *Miniaturized and/or high density circuits.* The high density, high reliability industry is also using parts of this new technology. Here density, rather than economy, is the driving force. Some defense applications have used flat pack and similar surface-mounted devices for years.

5-3 THE FUNDAMENTAL PROBLEM

The long delay between the use of surface-mounted devices in the hybrid technology and the printed circuit industry is due to a basic problem. Surface-mounting requires relatively small solder joints. Since solder, by nature, is not a strong material, the forces exerted on a joint in use become important. These include:

• Thermal cycling.
• Vibration.
• Mechanical shock.

1. *Thermal cycling.* The thermal expansion and contraction dictate a close match in coefficients of expansion. An electronic circuit may experi-

ence a temperature excursion as large as several hundreds degrees. If there is a thermal coefficient mismatch, sizable forces are generated in the solder joints. These forces can cause failure after a limited number of cycles because solder is very strain rate sensitive.

The hybrid industry is using a substrate that is very similar in TCE to the component materials. Printed circuit laminates, however, are drastically different, which creates a problem. Only small components can be directly soldered to the board where their short length prevents undue stresses. In addition, devices with ductile leads can also be mounted safely on the surface, because the lead absorbs the stress.

A search is now underway for new laminate materials to match device expansion properties. Until they are developed, there will be a physical size limitation on direct bonding. Compliant leaded devices are obviously excluded.

2. *Vibration.* Vibration is a less severe hazard to surface-mounted joints, even though the moduli of elasticity are quite different. This is because of the solder ability to withstand high frequency stresses well. In addition, only a small device-to-weight ratio per joint has to be supported. In extreme cases, the component can be held down further by an adhesive.

3. *Mechanical shock.* The situation with mechanical shock is similar to that of vibration. Here too solder is capable of withstanding reasonable levels of a one-time force.

5-4 MATERIAL CONSIDERATIONS

Surface mounting obviously requires careful selection of the soldering chemicals used. It also affects the laminate and the coatings applied. We will discuss the changes from standard practices which are required for soldering:

1. The selection of the laminate must include the following considerations:

- Flatness and the tendency to warp. A flat mounting interface is critical for accurate device placement. It becomes more critical as the size of the component increases. In addition, any changes in the contour of the board after soldering will result in stresses on the connections and possible failure. Most permanent dimensional changes are associated with heating cycles during manufacturing. Every time the board is heated above the glass transition temperature (T_g), its structure seeks to relax and eliminate stresses. These internal stresses originate from such mechanical operations as drilling, routing, and punching. They can also be caused by wight distribution of the mounted components, hole location, etch patterns, and so on. Finally, we tend to handle and/or store the board by supporting it at the edges only. This causes the structure to sag under its own weight, as well as that of the mounted components.

- Dimensional stability relative to registration during and after manufacture. This problem involves the same rational as flatness and warp. It is not enough to generate precise artwork if the the laminate structure shifts. Because of the small size of the surface-mounted components, the tolerances are much more strict.

- With the obvious reduction in pattern size, surface insulation properties must also be considered. Any material savings on the laminate may be negated by costly conformal coatings applied at the end of the process.

Thus, we must learn to select the best material for the application. Small leadless chips require little change, and low-cost materials suffice. The larger the surface-mounted device, however, the more critical this becomes, and only the more expensive laminates may be applicable.

2. Board finish is another area of concern. While it must be carefully selected for good solderability and high surface-mount yields, this is not enough. Solder resists are also beneficial for fillet control, but only if they do not shift. Thus, a solder mask over reflow solder becomes a problem, while resist over bare copper (SMOBC) is ideal. For details, see Section 1-21.

3. The flux and its cleaning philosophy are much more critical than in standard boards. The crevices under the components, with and without adhesive, may become traps for hazardous residues. The flux must therefore be either safe if left on or totally removable. Cleaner selection is obviously also critical; the solvents may not interact with the materials used (see Chapters 3 and 6 for details).

4. Selection of conformal and other top coatings requires a careful review of their coefficients of humidity and thermal expansion. If they get underneath a surface-mounted device, any mismatch will strain the connections. A semidry spray rather than a dip may minimize this problem.

5-5 THE SOLDER ALLOYS

It is too early in the development of this branch of technology to know which solder alloy will become the norm. The traditional tin-lead alloys have many advantages and may continue to dominate the market. However, there is a need for several other solders with the following properties:

- Solder compositions with better strength and fatigue properties. The diminishing size of the fillets, coupled with the growing dimensions and weight of the devices, dictate their use.
- Alloys with higher melting points for differential (piggy-back) soldering. The melting point is selected so that consequent soldering with lower-temperature solders will not remelt the fillet. This is important for mounting components on the top of a board first, and then soldering the balance of the assembly in standard fashion. Vapor phase reflow, infrared radiation, and hot air convection are some of the suitable technologies for such applications (see Section 5-11).
- Solders with lower melting points are also becoming of interest. They minimize heat distortion of the boards, component damage, and other problems. The close proximity of the solder to the devices during surface-mounting makes these alloys attractive.

The use of nonstandard alloys is not as simple as a substitution in the present system. They have different wetting potentials on base metals, making flux selection more critical. They also require different surface preparation when fusible alloys other than tin-lead are used. The problem arises from the fact that the molten solder mixes in with any fusible finish coated on the printed circuit board or the components. This changes the solder composition and behavior in the joint. Thus, a high temperature solder like tin-antimony, when used on a tin-lead reflowed solder-plated board, will no longer be just tin-antimony. The new mixture will contain lead and will have a lower melting point. Since there is no way of guaranteeing that the mixture will be uniform, any strata within the structure may be very low melting. Unfortunately, most of the desirable alloys for soldering are not suitable for electro-plating, making their use more difficult.

Many of these special alloys have not been traditionally wave soldered. Some high temperature waves are available, but not every alloy can be used this way. As a rule of thumb, only eutectic or near-eutectic alloys with a narrow melting range can be used in a wave. For small to medium production runs, the displacement soldering machine should be tried (see Figure 5-17). This unit offers great promise for all nonstandard alloys; if necessary it can be adapted to mass production.

The majority of alloys can be made into powder and can be used in solder paste or cream form. This technique is convenient for all solder applications and will be discussed later in this Chapter (Sections 5-7 and 5-8).

5-6 SOLDERING OPTIONS FOR SURFACE-MOUNTING

There are two distinct methods of surface-mounting using solder. They apply to either or both sides of the printed circuit board and can be used together. They are:

- The direct flux and hot solder application. Examples include dip and wave soldering.
- The preplacement of solder and flux, followed by a reflow operation. Examples include the use of paste and vapor phase reflow (Fig. 5-1).

Figure 5-1 gives a general outline of the methods of application and the location of the surface-mounted components. For those who are familiar with both hybrid and printed circuit technology, there are only minor changes in surface-mounted device assembly and soldering. But for anyone who is new to surface-mounting, the changes are substantial. Let us classify the processing methods by mode of soldering.

1. *Molten solder application.* This method includes dip, drag, wave, and displacement soldering (see also Chapter 4). Heat and molten solder are applied simultaneously, directly to the joint area. For surface-mounted devices, this also means total immersion of the component in the hot solder. Obviously it affects the way components are made.

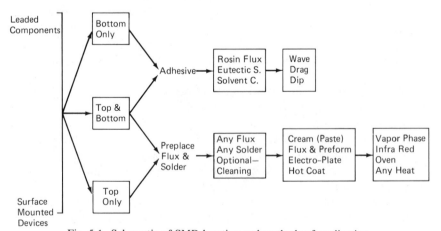

Fig. 5-1. Schematic of SMD location and methods of application.

The impact on assembly techniques focuses on a way of securing the component to the board prior to soldering. This is achieved with the use of adhesives that cement the device in place. The glue can be dispensed by mass screening, individual pressure extrusion, and other methods. Adhesive technology is a new discipline for most printed circuit assembly operations.

The components must be placed on the adhesive and secured by a cure cycle. Assembly can be done using mass placement techniques or as individual pick-and-place schemes. This is another area where a printed circuit assembler must master a new approach to equipment and precision.

The effect on soldering is also basic and results from the phenomenon of shading or skipping. When a board with surface-mounted devices is wave soldered, for instance, small gas pockets are created behind some of the components. These bubbles prevent the solder from covering adjacent pads and wetting them (see Fig. 5-18). As a result, many joints are skipped and require touchup. The gas is a combination of trapped air, flux fumes, and other volatiles. Most U.S. producers of wave equipment have developed a special double wave for this application (see Figs. 5-19 and 5-20). The first part of this system is turbulent, and the force of the solder displaces any trapped gases. The second part of the wave is conventional, eliminating bridging, icicling, and similar rejects. These waves are often tied together, and appear as one unit. A Japanese wave solder machine displaces the gas pockets by bubbling nitrogen through the wave. Another Japanese patent solves the problem by drilling vent holes in the pads.

The creation of bridges and icicles is also a function of board layout. As long as components can be placed in their preferred orientation, trouble is minimized [i.e., leadless chips, small outline transistors (SOT's), small outline integrated circuits (SOIC's), etc.]. This is not possible for devices that have leads in more than two directions (i.e., quads, leadless chip carriers, etc.). When such devices are soldered, some bridging is inevitable. Here the patented hot air knife in the Hollis GBS system offers relief (see Figs. 5-21 and 5-22).

Another area of concern is our inability to ensure total flux removal unless special precautions are taken. This may induce some people to return to rosin technology, where retained residues do not pose a problem. Relearning to use milder fluxes may prove traumatic to most such organizations.

2. *Preplace Flux and solder, then heat.* In these processes, we preplace discrete amounts of flux and solder to the joint area before heat triggers the melting and bond formation. This is different from the procedures in Item 1 above, where molten solder is directly supplied to the

joint, providing both heat and bonding solder. This group of processes is best understood by separating material preplacement from heating. It will be discussed in greater details in the next few section of this chapter.

5-7 PREPLACING SOLDER AND FLUX

This section deals primarily with the preplacement of solder and flux. The heating methods used for reflow are discussed in Section 5-11.

There are many automated mechanical and electronic assembly processes that require preplacement of the flux and solder in the joint area. Once they are in place, a heating operation activates the flux, melts the solder, and affects wetting. This procedure is common in semiconductor and hybrid assembly. It is used with soldering as well as with the brazing process.

The logistics of preplacement must be better understood. Let us consider the problems in assembling a large number of small joints simultaneously. The issues are as follows:

- How can the joining materials (flux and solder) be placed in the joint area?
- How can the quantity of joining materials be controlled to give uniform fillets?
- How can the joining materials and the surface-mounted component be held in place prior to heating?
- How can the joints be heated to melt and wet rapidly and simultaneously, in order to prevent heat damage?.

Fortunately, we do not have to invent new and unique methods, but can use established and proven techniques. As this industry expands, innovations tailored to surface-mounting will be developed. At present, the common preplacement methods in soldering are:

- The use of solder cream or paste.
- Pretinning with hot solder.
- Electroplating of solder.
- The application of solder preforms.

These methods are listed in order of preference and likelihood of use. Note that external flux must be added to the joint in the last three methods.

The application of solder cream or paste, like that of adhesive, require, the mastery of a new technology. Screens and metal masks are used to print the materials on the board in prespecified patterns. Such methods are totally alien to the present printed circuit assembler. So is the use of solder preforms, which are employed extensively in the semiconductor industry. Hot tinning and electroplating can be preformed by outside vendors. Unfortunately,these methods leave uncontrolled or skimpy quantities of solder, respectively.

1. *The use of solder creams and pastes.* This method is becoming very popular. It will therefore be covered in detail in a separate section (Section 5-10).

2. *Pretinning with hot solder.* The joint area can be directly coated with solder alloy, a method referred to as *tinning*. In this hot coating process, the surface is cleaned (degreased), fluxed, hot coated with molten solder, and cleaned. One or both parts to be joined may be treated, according to their special needs. Remember to reflux the joint area before final assembly. Make the joint by remelting the solder and fusing the parts together, a method also termed *reflow*.

The hot coating of solder by dipping is the least controllable operation in regard to the quantity of solder deposited. Surface geometry, flux type, solder temperature, dwell time, speed of immersion and rate of withdrawal affect coat thickness. Full control of these parameters helps keep the deposit uniform.

3. *Electroplating of solder.* Both pure tin and tin-lead solder can be applied by electroplating techniques in a controlled quantity. This method can be used, but it requires a relatively thick plating localized in the joint area.

The use of pure tin is diminishing because of the whisker growth phenomenon (see Section 7-11). Whiskers have caused field failures, and government specifications prohibit the use of tin after December 1982. Bismuth, antimony, silver, and other elements can be electroplated individually, but they cannot be codeposited with either tin or lead. This limits their use as a surface coating.

4. *The application of solder preforms.* Another simple way of preplacing solder and flux in a joint uses preforms. These are solder shapes manufactured beforehand to specific sizes. Upon melting, they give highly reproducible volumes of solder for joining. Preforms are available in the following forms:

- Spheres—produced from molten solder.
- Stampings—punched from foil.
- Wire forms—made by shaping wire or core solder.

There are no detailed catalogs for preforms, since they are made to order. The variations in size, alloy composition, and flux make it impractical to stock universal preforms. The selection depends on the method of application, joint configuration, and cost. More information on the design for the use of preforms can be found in the literature (Ref. 5-1).

To help locate preforms precisely in the joint, recesses can be incorporated in the parts. This is not always feasible on flat printed circuit boards. It is possible to hold down preforms with a viscous flux applied to the flat board and fuse them in place. They can also be placed as interconnected strips. During melting the solder climbs onto the metallic lands and separates easily, leaving no solder on nonmetallic surfaces. The use of preforms requires only *imagineering*, a combination of imagination and engineering (see Fig. 5-2).

The preforms are available as solid metal; external liquid flux must then be used. They can also be obtained flux coated, which requires no additional flux. Some wire forms and stampings are made of cored solder, with the flux inside.

The use of a small amount of flux in the final reflow or joining operation is vital to reliable results. Some attempts had been made to solder without flux, which would eliminate the need for final flux removal. This, however, proved to be an unreliable process with spotty results, because joint

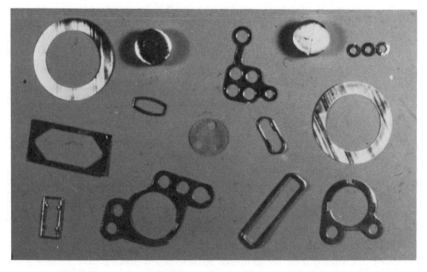

Fig. 5-2. A variety of solder preforms. (Courtesy of Alpha Metals)

quality suffered. If postsolder flux removal is not feasible, use a harmless nonactivated (type RA) or mildly activated rosin (type RMA) flux and leave the residues in place.

5-8 SOLDER PASTE AND CREAM

From its inception, hybrid technology was based on thick film materials, which were applied by screening. It made sense, therefore, to develop electronic-grade solder compositions in paste form that could be applied in the same way. These materials have been developed, together with sophisticated equipment suitable for mass production. *Paste* is the standard term, although rosin base materials are also referred to as *creams*.

These pastes consist of powdered solder suspended in a flux base and a suitable vehicle. Let us discuss them individually:

1. *Metallic powders.* Powdered solders must be *prealloyed*; it is not enough to mix tin powder and lead powder to make a paste. The tin-lead must be alloyed before the powder is made. Most solders can be made into powder, although many are not readily available. They are usually made from a melt by blowing or splashing on a spinning wheel, into air or controlled atmospheres. The metal powder must also be classified into particle size groupings. Pollution control makes these processes expensive.

3. *The flux system.* It consists of two components, which are blended together as follows:

- Flux base.
- Active ingredients.

The base gives the paste body. It may consist of rosin, resin, or a water soluble ingredient. The nature of the base determines the methods of cleaning. Together with the inert ingredients (vehicles, plasticizers, etc.), it gives the formulation its unique thermal resistance. The *thermal resistance* is the ability of a flux to withstand a combination of soldering time and temperature without breaking down or losing efficiency.

The active ingredients give the flux its chemical strength, apart from that of the base material. They may be solvent or water soluble. Their use follows the same rationale as liquid fluxes. The rosin family is even covered by the same government standards.

3. *Vehicle and plasticizers.* The rest of the formula is needed to give the material its consistency, suitable for screening (rheology), or other

properties depending on the method of application. *Rheology* is the deformation and non-Newtonian flow of liquids or pastes. It defines the behavior of the paste during shearing while being screened or extruded in the dispensing process.

This part of the paste is one of the most closely guarded trade secrets of this industry. Even small variations in composition can have dramatic effects on the behavior of the paste, as described below.

Solder pastes and creams are used in a large variety of applications. They must have different properties to suit each method of dispensing, handling, and heating. They are therefore available in many forms, such as the following:

- Nondrying compositions for ease of screening.
- Curable materials that can be dried on the work for ease of transport and handling.
- Tacky formulations that can hold components in place.
- Solvent cleanable materials.
- Water washable materials.
- Formulations in which the residues can be left on the work.

5-9 SOLDER PASTE MATERIAL AND QUALITY PROBLEMS

There are three common material and process problems worth discussion here. With planning and understanding, their effect can be minimized:

- The formation of solder balls.
- Voids and the entrapment of flux fumes.
- The shelf life of the paste.

1. *The formation of solder balls.* With the use of paste or cream solder ball formation is often unavoidable (see Section 7-15). These solder balls form when a quantity of paste is heated over an isolated nonmetallic surface. As the powdered solder melts, it coalesces into larger droplets that normally join the solder fillet. But in these isolated locations, the droplets cannot find their way back to the fillet, and thus form solder balls that stay embedded in the flux residue. There is no guarantee that these metallic particles will wash off with the flux, especially in tight spaces. (see Fig. 5-3).

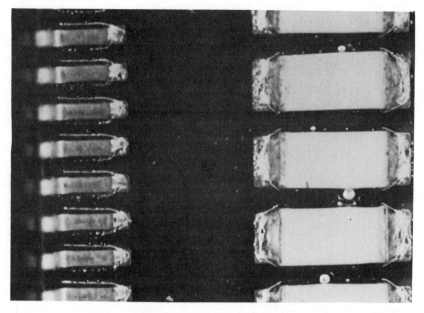

Fig. 5-3. Solder ball from paste that was not removed with flux residues.

Such isolated paste islands can be formed in several ways:

- Misregistration of screen pattern or dispensing equipment.
- Overflow of the spreading paste during a slow heating to melting. Remember, the flux melts much sooner than the solder and starts to spread over all surfaces.
- Oxidized and tarnished powder used in the paste. The flux here is unable to overcome this internal nonmetallic shell in time. The spreading flux eventually reacts with the powder, but the droplets that form can no longer coalesce and join the solder fillet. For more details, see item 3 below.
- A weak flux that cannot overcome the solderability of the work has the same effect as the oxidized powder described above. The solder balls are carried away from the joint before the surfaces can wet.
- Flux contained moisture or volatiles that were not driven off during preheating. This causes spitting during soldering, carrying solder balls to the surrounding areas.
- Excess paste applied.

To reduce this problem, there are several possible methods which should be tried:

- Apply the paste in a smaller but higher pattern than the metallic surface available.
- Use the right heating sequence. For example, heat from the bottom, melting layer after layer from the surface up. This reduces migration.
- Dry or cure the paste before soldering.
- Use only fresh materials with as strong a flux as possible.
- Use a controlled amount of cream; avoid an excess or an overrun.

2. *Voids and the entrapment of flux fumes.* When a paste or cream is used between horizontal flat surfaces, gases can become trapped. This cause a phenomenon commonly referred to as the *Swiss cheese effect.* The high ratio of flux to powder in this homogeneously mixed material is inevitable. (Remember that pastes are sold on a weight percent, not volume percent, basis). During heating the flux gives off fumes due to thermal decomposition. These gases become trapped between flat surfaces where there is no venting. On cooling, the metal solidifies with the cavity. Later the flux fumes condense to minute residues that are often not visible during cross sectioning. Thus, they give the appearance of voids. The voids formed should not be confused with freezing vacuoles (see Section 7-20).

The voids are found during failure analysis, cross sectioning, and X ray testing. They can be minimized by providing a passage for the fumes being generated. Slanted surfaces, round leads, and reflow techniques help. In many cases, the voids have been found to be harmless but they reduce the cross-section area.

3. *The shelf life of the paste.* The shelf life can be reduced by physical suspension problems and chemical deterioration.

The suspension of a powdered metal with a density several orders of magnitude larger than that of the flux paste is a problem. In the case of a eutectic tin-lead (density 8.4) -rosin (density 0.9) paste, the ratio is 1:9.33. The vendors try to load the paste with a high metal content to minimize the problem, but separation still occurs. In a good formulation, careful stirring will restore the mixture. However, this temporarily changes the rheology of the paste, and the freshly mixed material should be allowed to stand for several hours before it is tested or used.

This high specific gravity ratio is also responsible for the large flux content of creams. For the eutectic solder-rosin combination mentioned above, Table 5-1 presents a comparison of flux to metal ratios by weight and volume:

Table 5-1. Solder Paste Metal to Flux Ratio.
(A comparison between Weight Percent and Volume Percent[a])

WEIGHT PERCENT Wt %		VOLUME PERCENT V %	
METAL	FLUX	METAL	FLUX
75	25	24.32	75.68
80	20	30.00	70.00
85	15	37.78	62.22
90	10	49.10	50.90
95	5	67.06	32.94

[a] For Eutectic 63/37 Tin-Lead and Rosin.
Note: The ratio may change depending on the flux vehicle and the amount of air introduced into the paste during mixing.

In some cases, the flux reacts chemically with the metallic powder to form a white, crusty material. This is a chemical breakdown of the paste and cannot normally be reversed. In other cases the cream may simply dry out and harden, forming a solid crust on the jar. No amount of mixing will restore the smooth consistency of the paste, and hard solid particles will stay in the mix. These chunks may not prevent the use of the material for some applications, but in general their presence is not acceptable. If the viscosity of the paste is not critical, lift out and discard the hardened top portion.

Pastes and creams have been used in electronics for over 30 years. They provide a reliable method of bonding, provided the disadvantages are understood and controlled.

5-10 DISPENSING SOLDER PASTE AND CREAM

A major advantage of pastes and creams is their suitability for automatic dispensing and mass production. It is possible to preplace both the flux and solder simultaneously in these forms. The materials are located precisely where they are needed. In addition, it is possible to predetermine the quantity of solder per joint.

The major methods by which creams and pastes have been applied are:

- Screen or mask printing.
- Pressure dispensing.
- Dipping.

1. *Screen or mask printing.* The printing of paste materials is an established technology. Both contact and off contact printing are used. The type of screen determines the quantity of paste transferred. A metal mask (similar to a stencil) offers higher paste quantities. Such masks are usually combined with a stainless steel mesh for strength and accuracy of transfer. The metal mask is usually prepared by chemical milling.

The deposits from silk or metal screens are rather skimpy. The screens are prepared with photographic emulsions and are much lower in cost. They wear out faster than the masks and may stretch, giving poorer definition.

2. *Pressure dispensing.* Pastes can be extruded from an orifice like a needle or larger tube using air pressure. The materials are available and supplied in plastic tubes designed for such equipment. The quantity dispensed depends on the pressure and pulse duration. Hand held syringes are also available.

The continued impacts on the paste by the application pulse pressure may cause the material to separate. When this happens, the flux-to-metal ratio in the deposited paste may change slightly. This change is more pronounced where negative pressure is used to hold back the material for more accurate quantity control. The higher the temperature, the more likely the materials are to segregate in this fashion.

With the proliferation of surface-mounting applications, some automatic dispensing equipment using this method of extrusion has become available. The dispensing head is moved over the board or mounted on top of an numerically controlled table (XY-NC). The same equipment can deposit the adhesive needed to hold down surface-mounted components or to apply the solder cream.

3. *Dipping.* Dipping leads into the paste or cream is a simple but effective application method. Quantity control is relatively poor, and the cream may not be localized only in the joint area. It is a very inexpensive method, however, and easy to use. Low viscosity materials with a lower metal content are generally used.

5-11 METHODS OF HEATING FOR SOLDERING WITH PASTE

The soldering methods using paste can best be classified by the heating method used. The success of each method for a specific application depends on the sequence of heating and the overall heat transfer. One must consider the possible heat damage to the assemblies and the reliability of the solder bond formed.

In most mass soldering processes, localizing the heat on the joint area is impractical. Therefore, the entire board is heated to soldering temperatures. This benefits the assembly by reducing the danger of heat shock, but it also influences the heat stability of the board material. In general, excluding air (oxygen and humidity) from the work during the heat cycle helps to minimize chemical reactions like tarnishing.

Discrete localized heating is also possible for small scale production. This is achieved mostly by conduction, using heater bars and converted soldering irons. The use of convection by localized streams of hot air or liquid is also possible. Finally, radiation like infrared or low power laser radiation can be applied focused and diffused. The process is suitable for mass soldering as well as for spot heating.

The high labor content required for selective (localized) heating makes this method more expensive and suitable only for repair and replacement. With the aid of numerically controlled (NC) tables, it can be adapted to automation.

Whenever the flux and solder have been preplaced, both surfaces must reach the bonding temperature at the same time. Often nonuniform heating will cause the solder to wet one surface only. In essence, the solder runs away from the joint area. By the time the mating part reaches the correct temperature, no solder is available to bridge the gap. Often the flux also becomes inert on the surface that heats up last, preventing reliable joining (Fig. 5-4).

Besides wave soldering, there are many other methods of mass heating. In fact, any process that raises the assembly to wetting temperatures is a potential soldering method. By preplacing the flux and solder in the joint area, such soldering procedures can be simplified. For a list of 14 heating procedures for soldering see chapter 5 of Ref. 5-2.

The most likely heating methods for surface-mounting of components at present are:

- Vapor phase reflow.
- Ovens with air or inert gases.
- Hot liquid reflow.
- Radiation

1. *Vapor phase reflow*: This method is based on heat transfer from the condensing vapors of a boiling liquid to the work. The work absorbs the enthalpy of condensation, plus heat transfer by convection. This dual action makes the process very efficient and tends to heat an assembly uniformly. It is not possible to heat work above the boiling point of the liquid used, which is a major advantage. The high density of the vapor also excludes air from the process, which minimizes metal tarnishing and heat damage.

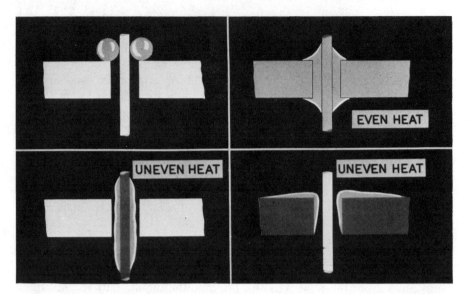

Fig. 5-4. Diagram showing uniform heating. Note that the solder spreads over the hot surfaces depicted here by darker gray. The sketch on the upper left depicts a cross section through a wire in a hole with a ring preform around the top. On the top right even heat caused uniform fillet formation. In the bottom the heat is uneven causing solder spread over the wire only (left), or the board (right).

A variety of nonflammable organic liquids can be used. They have a range of boiling points that makes soldering with different alloys possible. It is even feasible to make *piggyback* solder joints (in which one higher-temperature solder will not remelt while an alloy with a lower melting point is applied). At present, the cost of these liquids is relatively high. For details, see Table 5-2.

Table 5-2. Vapor Phase Soldering Fluids.

BOILING POINT (°C)	(°F)	DENSITY (g/ml)	SURFACE TENSION (dynes/cm)	TRADE CODE—NAME		SOURCE
174	345	1.88	16	FC-43	FLUORINERT	3M
215	419	2.03	19	FC-5311	FLUORINERT	3M[a]
215	419	2.01	18[b]	APF-215	MULTIFLOUR	Air Products
230	446	1.82	20	LS/230	GALDEN	Montedison
253	487	1.90	18	FC-71	FLUORINERT	3M
260	500	1.84	20	HS/260	GALDEN	Montedison

[a] —FC-70 and FC-5312 also by 3M, have same boiling point but different properties.
[b] —Estimated value.

The equipment is similar in principle and construction to a vapor degreaser. The liquid is kept boiling inside a sump, where the vapor is generated. The vapor rises to the work chamber, where the work is raised to the soldering temperature. In this process, the work is heated not only by temperature equalization, but also by the heat of vaporization which is released during condensation. As the vapor liquefies (recondenses), it drips back into the sump. Any flux or solids that are washed off fall into the boiling liquid, where most foreign materials solidify from the heat. These solids are removed by filtration, which requires the cooling of the liquid and is thus not continuous. The removal of active chemicals from the sump picked up from the work or generated by liquid breakdown may require treatment like neutralization.

The top of the work chamber contains cooling coils which prevent vapor loss. A large free board (top space) helps contain the drag-out of the vapors. These vapors are so heavy that they have little tendency to spill out. Any material losses must be minimized because of safety (toxic) considerations in the workplace and the high cost of the liquids used.

The density of the vapor also excludes air (oxygen) from the process. The high temperature prevents moisture from collecting inside the sump, unlike low boiling vapor degreasers. To prevent the loss of the expensive fluid, some equipment has a secondary lower cost vapor phase on top to contain the expensive ingredients. Heat instability in this second layer of vapor at the interface can cause corrosive breakdown. For equipment details, see Figs. 5-5 to 5-7B.

2. *Ovens with air or inert gases.* The flow of individual hot air streams, as well as hot air in a furnace, can be used in this convection method. The efficiency of heat transfer is relatively poor, requiring longer exposure times. As a result, gradual uniform heating is possible, giving internal conductivity a chance to equalize temperatures. In addition to convection these ovens also transfer heat by *infra-red* (IR) radiation. The wave length of the source determines color sensitivity, with the middle to far infrared range being blind to color differences. The ambient temperature inside the IR oven also has a large effect on heat transfer by convection, which helps overcome the shadowing effects.

There are well established methods for defining, controlling, and monitoring the heat profile in hot air ovens. Such equipment, along with specific heat profile development methods, has been in use by the semiconductor and hybrid industries for years. The method is considered very reliable and is worth investigating. Equipment using individual hot air streams is inexpensive and easy to use (see Sections 5-20 to 5-23 in Ref. 5-2).

A set of typical heat profiles for the use of solder creams are shown in Figs. 5-8 and 5-9. Note that in all cases we must pay attention to the heat

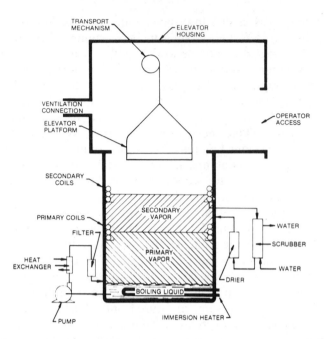

Fig. 5-5. Diagram of a dual-vapor batch Reflow unit, for vapor phase condensation soldering. (Courtesy of Dynapert-HTC)

transfer into the work. Conveyor speed determines heating time, while zone temperature determines temperature rise. These two factors, when equated to the work mass, make precise thermal excursions possible. However, we must also consider the properties of the fluxing media (in flux/preforms and pastes). Thus we must build in a volatile evaporation stage, which could be done beforehand in a separate piece of equipment like a batch oven. This shortens the dwell time in the conveyorized equipment. It is obviously also feasible to incorporate the flux outgassing stage in the inline process.

The basic principle of any heating operation is to create an adequate thermal cycle with a minimum of damage to the work. It is also true that thermal decomposition becomes more rapid as the temperature is raised. In the case of printed circuit board construction materials (resin), physical damage (warpage, etc.) starts once the T_g is exceeded. Chemical decomposition becomes pronounced at or above the melting point of solder. The same holds true for electronic components, depending on their construction. Thus an ideal heat profile strives to minimize the time above the

Fig. 5-6. View of a batch vapor phase reflow unit. (Courtesy of Dynapert-HTC)

solder melting point which is needed for wetting. This can be achieved by prolonging the lower heat soaking time, which poses less danger.

Gases other than air can be used in this type of equipment. These atmospheres should be dry because any water content increases the oxidation potential and negates some of the advantages of the method. Two types of materials are normally used in conjunction with furnaces: inert and active. Inert materials like nitrogen have an obvious advantage. For soldering, however, none of the standard reducing gases (e.g., hydrogen, forming gas, etc.) can replace fluxes. To be effective (in chemical reduction), high temperatures are needed, which are impractical for printed circuit boards and other plastic containing electronic assemblies.

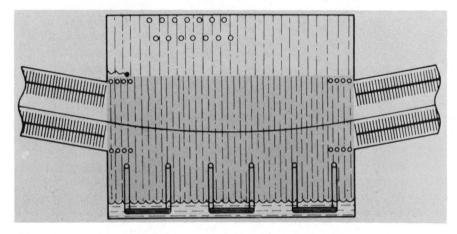

Fig. 5-7A. A three heater inline vapor phase reflow unit. (Courtesy of Hollis Automation)

3. *Hot liquid reflow*: The use of heat transfer media like oil for soldering has been in use for years. The term *reflow* here is historic, rather than descriptive. In some applications, oil is used only to remelt a previously deposited solder coating (a true reflow operation). In other cases a bond is created between two or more parts using preplaced solder and flux (a true soldering process which, however, is also called *reflow*).

A variety of reflow media are available, not all of which are oils. The name originated from the first processes, which used real oil. Today, many heating media in this category are not oil based, and some are even high-temperature detergents, which are also water soluble.

Fig. 5-7B. View of an inline vapor-phase reflow unit. (Courtesy of Dynapert-HTC)

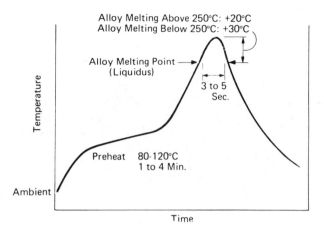

Fig. 5-8. Typical heat profiles for a single pass (one step) paste reflow soldering process. (Courtesy of Alpha Metals)

As in vapor phase and inert gas soldering, the heat transfer medium excludes the air. Judicious selection of the fluid can facilitate the wetting process by lowering cohesive forces and increasing interfacial tensions.

The equipment is relatively simple, and large solder pots can be used to heat the liquid. The parts are submerged in the pot until the solder joint is completed. The oil can also be applied in a hot stream or wave.

Patented processes have been developed which incorporate this princi-

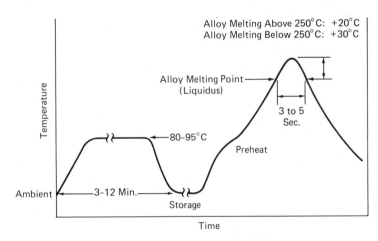

Fig. 5-9. Typical heat profiles for paste soldering process with a predrying cycle (two-step). (Courtesy of Alpha Metals)

Fig. 5-10. Special IR oven for surface mount soldering. The wave length of the IR source is carefully adjusted for uniform heating of surfaces with different emissivities. (Courtesy of Argus International)

ple into mass soldering processes like wave soldering. The ultimate configuration places the wave entirely under a reflow medium. With other types of equipment, a secondary oil wave is used to remelt the solder joints from the wave. This second oil wave reduces defects like icicles and bridges (Fig. 5-11).

4. *Radiation.* Both focused and unfocused radiation have been used to solder. The heat source configuration and distance from the work determine the heating profile. Heat intensity can be controlled by voltage regulation and exposure time.

Focused or slightly defocused infrared energy has been used for localized soldering as well as total heating. The heat is generated by a source like a quartz iodide tube located inside a reflector. For localized heating, an aperture or diaphragm helps to shield sensitive areas. Inline equipment is normally based on nonfocused strip sources and a conveyor.

The heat intensity can be very high, but it is easy to moderate. The heating rate depends on the part's emissivity and is a function of color, surface finish, and other factors. Here the dark color of solder creams and their high surface area help melt the solder rapidly while minimizing overall damage.

Uniform heating may be a problem when parts have very different shapes and masses. The speed of localized heating, however, is more attractive, when solder joints must be made without too much heat effect on the rest of the assembly. Careful prototype testing is needed to set up the equipment (Fig. 5-11).

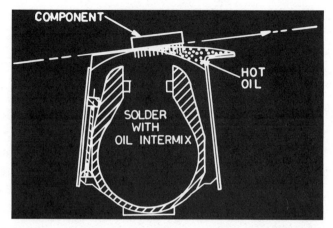

Fig. 5-11. Secondary oil reservoir behind solder wave to reduce rejects. This wave is especially useful in lead-wire tinning, as it eliminates flags (icicles) on heavier deposits. (Courtesy of Hollis Automation)

It is also possible to use other forms of radiant energy like the laser for soldering. The laser beam must be low enough in powerto avoid joint destruction. With careful aiming, very localized soldering without heat damage to the surrounding area has been reported.

5-12 LEADLESS CHIPS AND THEIR TERMINATIONS.

Leadless chips is a term that generally applies to passive components like resistors and capacitors. They are rectangular or round in cross section and may be bare or coated with an insulation (Fig. 5-12).

As the name implies these chips have no leads, but rather metalized terminations that can be classified as follows:

- A thick film material like fired silver or gold, sometimes with the addition of platinum or palladium.
- A precious metal loaded organic resonate.
- A mechanically attached metallic part.
- A surface electroplating like nickel or copper over a sensitized insulator.

1. *A thick film material.* Since the chip itself is made mostly by thick film techniques, fired conductors are part of the scheme. These are basically mixtures of low-temperature-melting glass frit (also called *flux* in the

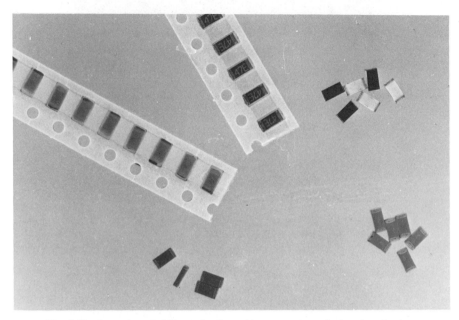

Fig. 5-12. A number of different leadless chips, some in typical reel packaging.

thick film industry) and metallic flakes. They are applied to the ceramic substrate as pastes, then dried, and fired. During this heating process, the glass melts and wets the surface of the device. This forms a bond that holds the thick film down while the metallic flakes form a conductive path.

A solder bond can form only on the metallic filler, and not to the glass. The amount of metal flakes that jot out of the surface is minute and finite. In addition, the metals normally used are easily soluble in solder. This gives causes to the leaching (scavenging) phenomenon, in which overexposure to molten solder removes the top film. As a result, the solder cannot adhere to the glassy layer underneath, and dewetts. Thus the solder may adhere to the termination initially and then pull back. This may happen on initial soldering or, more likely, during remelting in touchup or repair.

In order to retard scavenging, several steps can be taken (see also Ref. 5-3):

1. Load the solder with the same metal to reduce the solubility potential. For example, on silver we use 62Sn/36Pb/2Ag solder instead of 63Sn/37Pb.

2. Use a solder alloy that has less solubility potential for the thick film metal—low-tin high-lead solder for silver, for instance.
3. Lower the solder exposure time and temperature in order to slow solubility.
4. Use nonaggressive fluxes like pure rosin (water whiter type R) or mildly activated (RMA) formulations.
5. Apply the solder using a static process and heat only once. A dynamic wave application will wash off the metallized surface more readily.

These suggestions apply to bare thick film terminations only. If the film has a barrier plating of a nonsoluble metal like nickel, scavenging is not a problem (see item 4 below).

2. *A precious metal loaded organic resonate.* Instead of glassy thick films binders, organic resins with metallic fillers can be used. They require a much lower curing temperature than the glass firing temperature, which simplifies device manufacturing. However, they cannot withstand soldering temperatures as readily as glass before the organic binder starts to disintegrate.

The thick film resonate is usually covered by an electro-plated barrier coat, as described in item 4 below. Without it, soldering is very difficult and lacks bond reliability. Even with the plated cover coat, joint integrity is a function of soldering temperature and time. It is considered the least desirable chip termination.

3. *A mechanically attached metallic part or cap.* This termination normally has a round configuration. It is also referred to as a *metallized eLectrode face bonding* (MELF). As the manufacturing processes of these small leadless devices becomes more automated, this type of termination holds a lot of promise.

4. *A surface electroplating like nickel or copper over a sensitized insulator.* This technology is very similar to the additive process used for manufacturing printed circuit boards. The best plating is a material that is nonreactive with the solder, like nickel. This makes the most stable termination material combination.

In general, it is necessary to stabilize all of the terminations described above. In addition, it is advantageous to improve their solderability and prolong their shelf life. This can be achieved by electroless or electroplating techniques. Bear in mind, though, that these finishes must stabilize the system and improve solderability. They do not change the quality of the original bond to the substrate.

Let us see how the nature of the plated surface affects the soldering process. The metal coatings that can be deposited fall into three categories:

1. Barrier Plating, where the deposited metal is not soluble in solder to any measurable degree. Nickel is commonly used this way. It does not react with most solder alloys under standard exposure (time and temperature) conditions. Since nickel is a difficult metal to solder, it is seldom left exposed (see item 3 below).
2. Buildup plating, where a substantial layer of metal is deposited on the surface to prolong termination life during soldering. Copper and silver are good examples. Their presence on the termination increases film thickness and conductivity. At the same time, longer solder exposures are possible.
3. Solder coating, where solder deposits are applied for good solderability and prolonged shelf life. This category is also referred to as *pretinning* materials. The metals used are primarily fusible at soldering temperatures. Pure tin and tin-lead are good examples, although pure tin is less desirable because of whisker growth problems. These electrodeposits are often reflowed after plating, a process that involves remelting of the layer. The pretinned layer can obviously also be applied molten.

A word about gold plating and its solderability is in order. Gold has many undisputed advantages for mechanical contacts because of its unusually low breakdown voltage. For soldering, however, gold spells nothing but trouble. It forms very brittle intermetallic compounds and therefore should be avoided as a finish for soldering. In addition its very high price is an incentive for the application of only ultrathin coats. These coats are not sufficient to retain good shelf life for soldering. the government has taken a good attitude to gold in its specifications, it states that the gold must be removed prior to soldering.

5-13 SOLDERING LEADLESS CHIPS

There are no rules limiting the methods used to assemble and solder leadless chips. They are normally bonded to a printed circuit board in one of three ways:

- Wave soldering (bottom of board).
- Solder paste and reflow (top of board)
- Soldering of the top and bottom simultaneously in a single pass.

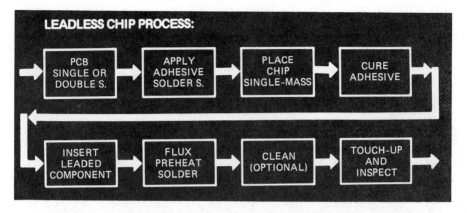

Fig. 5-13. Outline of Wave solder process for leadless chip components.

1. *Wave soldering.* The general process used in conjunction with wave soldering is outlined in Fig. 5-13. While variations are possible and will develop with time, this is the simplest procedure.

Let us review a typical assembly line for mounting chip components to the underside (solder side) of a board. Automatic equipment for inline and batch processing is available. Here is a step by step description;

a. Adhesive application. In order to wave solder the chip components, it is necessary to attach them to the board with an adhesive. They can be applied by a variety of methods including screening, dot transfer, pressure extrusion, and spotting.

The adhesive is applied to the solder side of the board, which travels through the wave. This method is suitable for single-sided, double-sided, and multilayer boards. The adhesive volume is quite critical; too little is obviously dangerous, while too much will bleed beyond the component outline or on the pads (see Fig. 5-14).

It may be possible to mount components higher off the board surface by placing and curing a spot of adhesive. A second drop can then be applied on top of the previous cured adhesive, yielding a higher spacing.

b. Chip placement. The chips are located on the adhesive applied as in (a) above. They can be handled individually or with mass placement techniques (see Section 5-17). These include vibrator feeders, magazine unloading, tape and reel systems, and others.

c. Adhesive curing. Once the leadless chip is placed in the adhesive dot, it must be cured. This secures the component in place during subsequent handling, assembly, and soldering operations. The curing method

Fig. 5-14. View of chips with glue spots that bled beyond the device outline.

depends on the adhesive chemistry. Heat, radiation (UV), air, and drying are some of the obvious methods.

d. Final assembly. After the leadless chip components have been se-cured to the surface, the other conventional components are added. Some of them may be inserted automatically while the rest are assembled manu-ally. Obviously, the adhesive used [in (a) above] must withstand the rigors of these operations.

e. Soldering. The finished assemblies are now fluxed, preheated, and soldered. Details of this process are given in Sections 5-14 and Chapter 4.

f. Cleaning. Flux removal and overall cleaning are feasible only when board design is correct. Flux may become trapped under the flat configu-rations of the chip. Flux selection thus becomes critical in relation to the adhesive dispensing system and placement techniques.

g. Quality control. Inspection and automatic testing are somewhat more cumbersome, but follow standard rules. A good definition of a high quality joint is required (see Section 8-15). Because of the unique configu-ration of the joints, it is necessary to follow the guidelines set up for hybrid circuits. Keep in mind the larger differences in the coefficient of thermal expansion between the materials in a printed circuit assembly.

h. Touchup and repair. The term for solder joint correction is *touchup*, while component replacement is called *repair*. See Chapter 9 for details.

Touchup for insufficient solder, excess solder, bridging, and the like is easy. It requires special irons and tools, used with the same general precautions needed in plated-through hole applications. The main point is to minimize heat and pressure during solder iron contact. Proper operator training is vital (Fig. 5-15).

The replacement and removal of components held down by an adhesive is more complex. It is a function of the adhesive used and its strength at various temperatures. The more popular resins can withstand wave solder temperatures for a short time, even though they soften. During replacement the heat of the soldering iron is higher, and this weakens the adhesive. When the resin is above T_g it is possible to force the chip off, usually

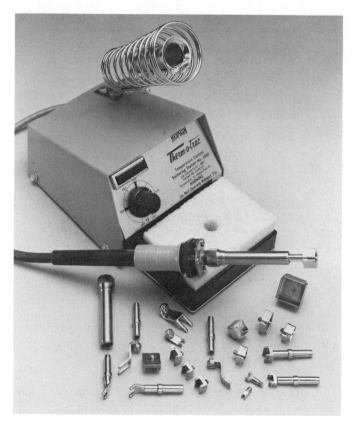

Fig. 5-15. Special iron and tips for chip removal. (Courtesy Hexacon Electric)

Fig. 5-16. Chip being removed with iron tip. (Copyright Manko Associates)

with a twisting motion or a push. Special tip configurations are available to remove these components where needed (Fig. 5-16).

Small-outline leaded components (see Section 5-15) can also be soldered in this fashion, provided they are not too wide. SOTs and SOICs are examples of surface-mounted components, glued on and wave soldered. The large solder fillet deposited on the lead negates compliant lead flexibility, and only narrow bodies will withstand the TCE mismatch.

2. *Solder paste and reflow (top of board).* The general process for chip soldering to the top side of the board is not limited to solder paste and reflow. Any other preplacement of flux and solder in the joint area (see Section 5-7), followed by heating (see Section 5-11), can achieve the same results. But manufacturing ease has made the paste-and-reflow popular.

Let us review a typical assembly line for mounting chips on the top (component side) of a board. Automatic equipment for inline and batch processing is available. Here is a step by step description:

a. Paste application. Using screen or mask printing (see Section 5-10), solder paste is applied simultaneously to all chip pads on the board. For best results precleaning the surface is advisable, it will improve cream registration and retention. Cleaning will also ensure solderability of the board, especially if pastes with mild fluxes are used.

Multiple application of an extruded paste spot will also serve the same purpose. The deposit through a needle however will always be round, and may not match the contour of the pads.

b. Chip placement. Next, the chip is located on the board over the pads (see Section 5-17). The solder paste used should be tacky and hold the chip in place. If there are adhesion problems, check to see if the chips are in need of precleaning.

Placement equipment is available for mass placement, or individual pick-and place. The accuracy of placement is not as critical as with adhesive usage. If the paste can hold the chip in place, solder self aligning properties will ensure repositioning during wetting (see item d below).

c. Paste drying. Before reflowing the paste, we must drive off the volatile ingredients in the formulation. Paste vendor instructions for time and temperature should be followed. The drying temperature is a function of the melting point of the flux base. In order to reduce solder ball formation, the temperature should not exceed the melting point of the flux resins.

The drying time is a function of the paste deposit height. The time needed to drive off the volatiles from a heavy deposit are much longer than the time needed for thin layer.

Once the paste is dry and cool, it can hold the chips in place for ease of handling. Some paste formulations make it possible to invert the board without losing the chips. Without proper fixturing, however, the board cannot be soldered in the upside-down position. When the chips are held by the dry paste below the board, the reheated flux base will melt long before the solder and the chips may shift or fall off.

d. Paste reflow. When the assembly is finally heated for reflow, it passes several important temperature levels. First the flux base melts and softens, and the flux activation system come into action. Then the solder powder melts, and the liquid coalesces into a molten pool. Finally, the wetting temperature is reached and the solder spreads over the surfaces to form a fillet.

During wetting, the important phenomenon of *self alignment* takes place. The chip actually floats on top of the molten solder, and the surface energies pull it into place. For further details see Section 5-17 and Fig. 5-27.

e. Solder solidification. The last step in the chip soldering process is the solidification of the liquid metal. It is important to prevent vibration from this part of the process, or *disturbed* fillets will result (see Section 7-5).

As in any industrial process, there are many variations possible for this paste-and-reflow process. The above outline, however, should serve to understand the process as a whole.

3. Soldering of the top and bottom simultaneously in a single pass. If it is desired to place chips both on the top and bottom of a board, this can be done in a single pass. One possible sequence of events is as follows:

   a. Attach the chips to the bottom of the board with adhesive and cure, in the standard process.
   b. Invert the board and place the chips on the top side with paste and dry, in the standard process.
   c. Add all conventional components if any, by insertion through the plated-through holes.
   d. Solder the bottom with a dual (chip) wave, in the standard process. Simultaneously, or immediately after the wave, add more heat from the top to reflow the cream.

In this combined process, the board is soldered from the bottom in the normal fashion. During this process, the temperature on the top of the board rises to allow the solder to wet up the plated-through hole. By supplying additional top heat at this stage, it is possible to reflow the cream to form the top joints.

The amount of additional heat required can be reduced by the use of low-melting solder paste. If the melting point of the solder used for the top fillet is low enough, additional heat is not required. This is actually a *piggy-back* solder joint, and the temperature differential will depend on the application.

The additional heat can be supplied by top radiation, with units similar to top preheaters. Hot air was demonstrated to give good results, and overcome radiation shadowing problems (Ref. 5-3).

5-14 MORE ABOUT WAVE, DRAG, AND DIP SOLDERING

Chapter 6 is devoted to the wave soldering process, and the reader is referred to it for more details. This section outlines the unique needs of chip components, and other surface-mounted devices (SO types, and chip carriers).

The molten solder application can take many forms, such as dip, drag, wave, and displacement soldering (Fig. 5-17). In these methods, heat and

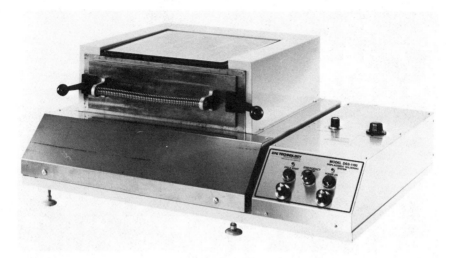

Fig. 5-17. A displacement solder machine, where vacuum is used to pull the solder up to the board, at controlled heights. (Courtesy EPE Technology)

molten solder are applied simultaneously directly to the joint area. For surface-mounted devices, this involves total immersion of the component in the hot solder. Such heating obviously affects the way components have to be made.

The basic problem in soldering low profile components mounted to the underside of the board is a phenomenon called *shading* or *skipping*. When components such as leadless chips are passed through a liquid solder reservoir, flux, fumes, and air are trapped underneath the printed circuit board. They collect behind the raised component profile, where the flowing solder cannot wash (displace) them away. These materials interfere with the contact between the liquid solder and the pad or adjacent areas, preventing wetting. As a result, the solder skips over the pads, and no fillet is formed.

When there is contact between a horizontal planar surface (the board) and a liquid (the solder), there are no escape routes for trapped gases. Some materials vent through adjacent holes; others remain trapped underneath.

In order to understand shading better, let us analyze how it occurs in the wave. Figure 5-18 shows a schematic of a board traveling through a wave. The area immediately behind each component is vulnerable to shading, because the laminar flow of the wave is interrupted. This is shown on the right side of the figure, where the wave is outlined. The

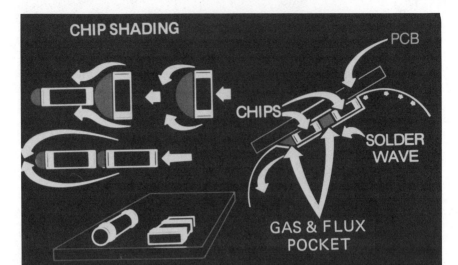

Fig. 5-18. Schematic of shading. Note that the proximity of devices plays a big role. (Copyright Manko Associates)

juxtaposition of the components to one another on the right helps explain why this occurs mostly in densely populated areas. The component outline (rectangular vs. round chip) also has a bearing.

In the contact between a horizontal planar surface and a liquid, there is an additional geometric factor—the pad *profile*—which causes skipping. Flat pads with little or no solder coating are often left unsoldered. This is especially true for solder-mask coated boards, where the resist is as high as, or higher than the solder coating. On the other hand, a large quantity of solder on the pad wets well. This effect is also influenced by pad size, with small pads showing more shading than bigger ones.

The trapped gasses are of varied origin:

- Air that was not flushed away.
- Flux volatiles not driven off during preheating.
- Flux thermal decomposition products formed in the wave.

Some flux residues can also be trapped in this fashion.

To avoid shading in the wave, several solutions have been offered. The most common are:

- A double wave configuration.
- A drag and wave combination.

- A nonfunctional vent hole in every pad.
- A gas bubble introduced at the base of the wave.
- High profile pads (above the solder mask).

1. *A double wave configuration.* Most U.S. producers of wave equipment have developed a special double wave for surface-mounting application (see Figs. 5-19 and 5-20). The first part of this pumping system is turbulent and has a multi-directional solder flow which displaces any trapped gases. The second part of the wave is conventional, eliminating bridging, icicling, and similar rejects. These waves are often tied together, and appear as one unit during pumping.

2. *A drag and wave combination.* The combination of drag and wave, or two separate waves, yields good results. The double dip operation provides a gas escape route in the gap formed in the equipment. This concept usually requires more dwell time in the molten bath.

3. *A nonfunctional vent hole in every pad.* There is a Japanese patent that covers nonfunctional holes drilled specifically to vent trapped gases.

Fig. 5-19A. The two pumps in a double-wave chip nozzle. Note that the first wave is turbulent and the second wave is smooth. (Courtesy of Electrovert Ltd.)

Fig. 5-19B. The two pumps used in a double-wave chip nozzle. (Courtesy of Hollis Automation)

This works well on low-density, single-sided boards but is less efficient as the aspect ratio grows. Functional holes like plated-via holes are excluded from the patent.

4. *A gas bubble introduced at the base of the wave.* Another wave soldering machine displaces the gas pockets by bubbling nitrogen through the wave. The gas introduced at the base of the wave displaces solder sideways when the bubble hits the board. This sideways pressure pushes solder into the shaded areas. Only inert gases (nitrogen) should be used to avoid excessive drossing.

5. *High profile pads (above the solder mask).* The pad may have to be modified. After plating reflow, an advantageous convex solder profile is formed. Thin solder coatings such as those left after solder leveling pose a problem. Pad solderability is obviously also a factor in shading.

The very high density possible with surface-mounted components creates an increased tendency to bridge. So do multileaded components like SOICs and "J" Leaded devices. As in conventional soldering, the creation of bridges and icicles is also a function of board layout. As long as conventional through-the-hole components can be placed in their prefer-

Fig. 5-20. A view of the double chip wave working in tandem. Note that there is only a slight separation between the waves. In this view the edge of a hot-air knife is visible at the end of the smooth final wave. (Courtesy of Hollis Automation)

Fig. 5-21. Bridging on SOIC.

Fig. 5-22. Closely mounted chips shorted by solder.

red orientation, trouble can be kept to a minimum. With the new surface-mounted devices (i.e., leadless chips, SOTs, SOICs, LCCs, etc.), this is not possible. When such devices are soldered, some amount of bridging is inevitable (see Figs. 5-21 and 5-22). Here the patented hot air knife in the Hollis GBS system offers relive (see Section 4-20).

Another area of concern in soldering comes from our inability to ensure total flux removal unless special precautions are taken (see Sections 6-18 and 6-19). There is now a trend to return to rosin technology, where retained residues do not pose a problem. Relearning to use milder fluxes may prove traumatic to many organizations. Many companies, however, are successfully using organic intermediate (organic acid) type fluxes and washing in water.

5-15 SOLDERING SURFACE-MOUNTED LEADED COMPONENTS

There are a variety of leaded components made specifically for surface-mounting. The leads are very compliant and are meant to compensate for any mechanical stresses by bending, thus absorbing the forces. The oldest known devices in this category are the flat packs that have been used by government and industry for years. These components have relatively

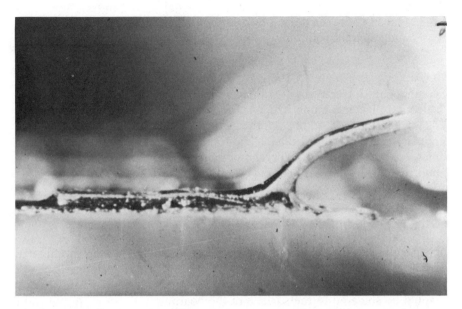

Fig. 5-23. A well-formed and soldered flat-pack lead. Note that the foot print of the flat pack was very long.

long compliant leads shaped to absorb the stresses of mismatched expansion. The leads on smaller and more recent devices are much shorter and are referred to as *compliant leads*, *"J"* - *leads*, or *gull wing leads*.

The basic solder joint derives its reliability from the heel of the lap joint (Fig. 5-23)

It would be desirable to bend the leads in such a manner that they would be parallel to the board. This is not practical, and thus we tend to form the lead with the toe up (in front of lead, away from the component). This gives us a good heel configuration (behind the lead, near the component). In this case, the solder should not rise more than half the height of the vertical lead to the knee. For further details, see Section 1-14.

On some of the smaller devices like transistors (SOTs) the lead may be so short and soft that it is not formed. The narrow width and small mass of these devices cause little stress. Thus lead shape is not critical. The leads of narrow SOICs can also be buried in solder without trouble, while wide-body devices will fail.

The attachment of leaded devices falls into two major categories according to the location on the board:

1. Surface-mounting at the top and bottom with the use of preplaced solder and flux (i.e., solder cream, etc.).

2. Placement of the device with adhesive to the solder side of the board and standard soldering techniques (double wave soldering, etc.).

The size and shape of the leads make them prone to bridging. Some components have leads extending in three or four directions, making selective placement impractical. The solutions to the bridging problem depend on the method of soldering. For wave applications, the use of the hot air knife eliminates bridging and removes excess solder from the knee of the bend. For paste and reflow applications, proper solder quantity control and the correct heat profile solve the problem.

5-16 THE LEADLESS CHIP CARRIER

The ultimate in packaging density, using surface-mounting, would be a leadless chip carrier. It can be best described as a planar multicontact surface acting as a connector for complex microcircuitry. Each contact would be soldered to the surface of the board.

The concept has been developed by the hybrid industry, where it works well. However, there are a number of problems that have prevented the use of this configuration on printed circuit boards:

- Mismatched expansion between device and board.
- Each solder joint is too small to withstand substantial stresses.
- The device has many blind, uninspectable areas.
- Touchup and repair are very complex.

As a result, the designs, which are more than a decade old, have found only limited use on special modified circuit boards. An additional technological breakthrough is needed to make them universally usable on circuit boards.

Soldering techniques are limited to preplacing the solder and flux and reflowing. For height control under the device, solder preforms are used. To ensure proper spacing all around, a high melting center like a copper ball can be used. As indicated in Chapter 1, uniformity of solder joints in this configuration is vital. Any joint that is stronger than required causes an uneven stress distribution and a more rapid failure of another solder fillet. Thus, it is not expected, that leadless chip carriers, can be soldered with a wave even when a hot air knife is used. Preplacing accurate quantities of solder on precisely designed circuit pads appears the only feasible method to date.

5-17 A WORD ABOUT SURFACE-MOUNTED COMPONENT PLACEMENT

Since there are no leads in a hole to locate the components on the board, surface-mount placement is quite different. In addition, the methods used for packaging and handling these devices are obviously new and unique. Since these parameters have only a slight impact on soldering, we will review the subject only in general terms, dwelling on solder related topics more fully.

Surface-mounted devices are bulk packaged in several ways. The most popular are:

• On reals.
• In magazines.
• As loose bulk.

Figures 5-24, and 5-25 show some of these packaging concepts. The one that has the greatest promise is the real, which is covered by EIA and ANSI specifications. We have actually taken an established concept, the 8 mm movie, and adapted it to our needs. The holes on one side help ensure accurate transport and positioning, while pockets in the tape itself hold the devices. Obviously, the width of the tape is not limited to the old film size, and technical needs dictate wider strips. Sophisticated developments of this concept include static-free materials, preorientation of device polarity, and so on.

There is less standardization in the field of magazine design. These are usually adaptations to individual equipment needs and specific vendors. Some adaptation of the *DIP stick* (a channel holding dual inline package like ICs), have been made for the small outline dual inline devices (SOICs).

The use of loose bulk devices is limited at present. Vibratory feeders seem to damage the components by abrasion, and it is also impossible to orient them by polarity.

The equipment used to transport, pick-and-place the devices, and so on, can be modular and simple or highly automated and complex. Figure 5-26 shows such an installation. From a soldering point of view, we are interested only in the end result, which is transferred to the soldering station.

The accuracy of the adhesive spot is a good example. In terms of location, it cannot mask any part of the pads to be soldered. In height, it determines the solder fillet size and washability. In addition, device placement is critical, since it is fixed by the curing of the adhesive. With too much misalignment, the solder cannot bridge the gap.

Fig. 5-24A. Closeup view of pick-and-place vacuum head, locating a component onto a printed-circuit board. (Courtesy of Universal Instrument Corp.)

Another factor that affects the accuracy of placement is the topography (profile) of the pad. By itself, pad contour is a function of board finish. A plated (not reflowed) or leveled board has an easy to use flat finish but may pose shelf life and solderability problems. A thicker fused solder coat has a rounded contour (solder meniscus) that makes precision placement difficult. The thick, hot solder coat has a long shelf life and good solderability.

Fortunately, the assembly process using paste and reflow is not as critical. During the soldering process, the device is free floating and is positioned by the wetting/surface tension forces. The molten solder has a great tendency to *self align*. In other words, the wetting and surface

Fig. 5-24B. Closeup view of reels and tape holding devices. This pick-and-place system has many component feeder types—tape, horizontal vibratory, gravity, etc. (Courtesy of Dynapert—EMHART Machine Group)

Fig. 5-25. View of tape feeders, "ski-slope" gravity feeders, and linear vibratory feeders mounted on each side of the unit. (Courtesy of Dynapert—EMHART Machine Group)

Fig. 5-26. Overall view of a complex flexible high-speed computer controlled surface-mounted device placement machine. (Courtesy of Universal Instrument Corp.)

energies of the solder/base-metal system can often move the device into its correct position.

To understand this important phenomenon better, let us review a simple experiment. Figure 5-27, shows a series of copper coupons in various stages of wetting and self alignment. Each copper coupon was punched from a 0.032 in. (0.8 mm) flat sheet and is 1 in. (25.4 mm) in diameter. These good sized planchets are then lightly fluxed on one side. A solder preform weighing 0.5 g is placed between the fluxed surfaces, as shown on the left side of Figure 5-27. A spot of paste of equivalent size would give the same effect, and prefluxing of the copper would not be needed. The copper-solder combination is then placed on a hot plate and allowed to melt. As the solder wets the two opposing surfaces, the process of self

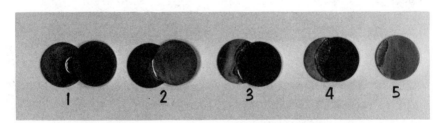

Fig. 5-27. The self-alignment phenomena. The parts on the left are before heating, the assemblies in the center were removed from the heat during the process. What appears on the right as a single disc is the end product of two solder aligned planchets.

alignment begins. The top copper piece is lifted and carried over to completely cover the bottom part. If one pulls the top piece back part of the way, it will quickly jump back when released (but only as long as flux is still present—when the exposed surface drosses over, self alignment stops). Similar demonstrations with leadless chip carriers have been shown to overcome up to a 24° rotational misalignment.

5-18 CHIP-ON-BOARD (COB) TECHNOLOGY

Another process, still in the development stage, is the chip-on-board mounting of active devices to the printed circuit board. *Integrated circuits* are manufactured on small silicon chips, and are then mounted inside of protective packages, with input/output (I/O) leads. The most popular package for IC chips is the dual inline (DIP) configuration. Inside the DIP device, the little chip is connected to the I/O leads with tiny wires (wire bonding). A much smaller adaptation of this technology is being used for the printed circuit industry. Bare IC chips are mounted directly onto the printed circuit board, and connected to the circuitry. A drop of organic sealer (glob) is used to protect the chip and bonds.

There are a number of COB processes that differ slightly from one another, as follows:

1. Chip-and-wire technology. In this process the chip is attached to the board by the user. Then the wires are bonded to the chip and the board. The wire bonding processes are standard in this industry (thermocompression, thermosonic, and ultrasonic).
2. Tape-automated-bonding (TAB) technology. In this method the chip is already mounted on a thin metallic conductor, and wire bonded to thin I/O leads. The conductors are made from a thin metal/plastic laminate, by photographic-imaging and chemical etching techniques. The plastic acts as a carrier tape, and has a similar configuration to photographic film. This makes it easy to handle and position, and prepunched widows in the plastic make the TAB bonding possible. A portion of the plastic backing is retained on the board. The I/O leads are bonded to the board by reflow soldering, conductive adhesives, or thermocompression bonding.

As yet, no one method has emerged as dominant for integrated circuit attachment. The board assembler is not always prepared to handle bare chips. In the TAB method, he has some measure of protection, but only SOICs and standard ICs are robust during mechanical handling.

REFERENCES

5-1. Howard H. Manko, "Design for Solder Preforms," *Materials in Design Engineering*, March 1964.

5-2. Howard H. Manko, *Solders and Soldering*, 2nd Ed. McGraw-Hill Book Co., New York, 1979.

5-3. Howard H. Manko, "Selecting Solder Alloys for Hybrid Bonding," *Insulation Circuits*, April 1977, pp 27-30.

5-4. M.F. Comoford and A.V. Sedrick, "Design Guidlines for Achieving High First-Time Solder Yields with Mixed Component Technology". *NEP/CON West Proceedings*, February 1986, pp797-806.

6
THE CLEANING PROCESS AND EQUIPMENT

6-0 INTRODUCTION

In the modern industrial environment, cleanliness serves two basic purposes—technical and cosmetic. Fortunately, the need for contamination-free assemblies was limited in the earlier generations of electronic designs. To realize this, one has only to review printed circuit assemblies that have been in service for a year or two. They pick up a lot of dirt and contamination, but continue to operate. A relatively small portion of today's printed circuit boards must be absolutely clean. It is estimated that less than 5% of the present conventional circuitry is sensitive to current leakage.

The trend towards miniaturization has brought with it the need for cleanliness. Systems such as hybrids (thick and thin films) and microelectronic devices have such high density that they cannot function without contamination control. They are produced in clean rooms, with purified air and operators wearing special clothing. As printed circuit board users move toward surface-mounting, and fine-line design, they too will have to learn to avoid the danger of *electronic dirt*. Thus the need for cleanliness is now shifting from critical assemblies to the lower end of the electronic product range. Segments of the industry which up to now have been immune to dirt problems will have to learn how to clean in this new environment and how to keep the work clean.

Cleaning has become an integral part of the assembly and soldering operation. It is triggered by the use of fluxes that may have to be removed. This operation also provides a convenient opportunity to remove all the contamination picked up during prior processes, storage, and handling.

Before discussing postsolder cleaning, we must define what dirt is and how the chemistry of cleaning works, and understand why cleaning is required. Then we will cover the processes of water washing and vapor degreasing. We will also discuss *what is clean in electronics?*. For additional reading material the reader is referred to Ref. 6-1, pp. 227-233, and Ref. 6-2.

In general, we must develop a philosophy of product cleanliness. A philosophy that considers the impact of every single manufacturing step on the cleanliness (reliability) of the final assembly. In addition, we must extend our concern to the contamination buildup during product use by the customer. This is often coupled with the need for cleaning during repair and field service.

Finally, we must be concerned with contamination control during manufacturing in such areas as storage, handling, and assembly. These, in turn, affect process yield by decreasing such parameters as solderability, coating adhesion, and other factors. The cleaning techniques described in this chapter are suitable for both of these purposes. Specific details are covered in other chapters (see Section 2-11).

6-1 DEFINITION OF ELECTRONIC DIRT

A typical definition of dirt starts: "Any foul or filthy substance, as mud, grime, dust, excrement, etc." (*The Random House Dictionary*, unabridged edition). The rest of the definition refers to earth and soil or to the social aspects of the term. This is a totally inadequate technical definition in dealing with electronics and the printed circuit board.

For our industry, we must therefore redefine this concept as follows:

Dirt—Any foreign substance adhering to the surface which is not an integral part of the material structure and which was not deliberately placed there for a specific function.

Examples of electronic dirt are dust, lint, metallic chips, air pollution sediment, dried or moist films of solution residues, and unwanted wax or oil layers.

In this book, we are not addressing ourselves to the aesthetic appearance of an assembly. Cosmetic beauty does not add to the technical functions of electronic equipment. In general, *cosmetics* implies hiding the true appearance of a surface by applying a false layer.

6-2 THE HAZARDS OF ELECTRONIC CONTAMINATION

Traditional low-density electronic hardware was generally cleaned for cosmetic rather than technical reasons. In the last decade, automatic test fixture (bed of nails) requirements rather than electronic or technical needs have forced more companies to remove flux residues. As a result, violations of the rules of cleanliness were not manifested as major failures, and the industry grew complacent. There seems to be a basic lack of appreciation as to why we clean.

In the electronics industry, the basic technical reasons for cleaning are:

1. To prevent electrical problems
2. To promote coating adhesion
3. To eliminate corrosion hazards
4. To preserve mechanical properties
5. To facilitate inspection (cosmetic)
6. For automatic testing with a bed of nails

These objectives, individually and in combination, make the need for contamination control of high density circuitry of paramount importance. Let us review them in more detail.

1. *Electrical problems* can be divided into several categories. The most obvious are the current leakage problems, which lower the insulation resistance of the assemblies. Here *ionic contamination* and *adsorbed organic materials* are the major culprits (see Refs. 6-3, 6-4, and 6-5). Obviously, conductive particulate matter such as metallic chips can cause the same problems.

Nonconductive contamination can also cause electrical problems. Such dirt films will interfere in make-and-break contacts, plugs, and edge connectors. Rosin and adhesive residues are typical examples.

2. *Coating adhesion* is another major factor. Organic coatings such as photo resists, solder masks, and conformal coatings depend on a contamination-free foundation for good bond strength. The danger of trapping conductive or corrosive materials under permanent coatings poses an obvious hazard.

In electroplating, interfacial contamination does not always prevent the adhesion of the deposit. Severe solderability problems, however, result from such entrapment when fusible or soluble metals are deposited (see Ref. 6-6). In addition, flaking of contact layers during use has been reported, and corrosion protection is substantially diminished by interfacial dirt.

3. *Corrosion hazards* can come from harsh processing solution residues, but this is not the only problem. A much more insidious danger comes from less aggressive materials whose potential is not recognized. The reason is obvious: whenever a potentially dangerous chemical like an etching solution, a plating electrolyte or a strong flux is applied, precautions are taken for their total removal. This is not true of less hazardous materials like human perspiration, improperly removed activated rosins,

and the like. These slow-acting chemicals, therefore, are the cause of major concern.

The danger of corrosion totally consuming small conductors or causing embrittlement of larger parts is one part of the problem. In addition, the corrosion products themselves are usually electrically conductive in the presence of moisture and can cause shorts or humidity-dependent current leakage (Fig. 6-1).

4. *Preservation of mechanical properties* is relates to the proper behavior of rotating surfaces (bearings), the fit of tightly mating parts, and the preservation of close tolerances. Here cleaning must often be selective and not affect the lubricants and sealers deliberately applied.

5. *Facilitating inspection and cosmetic reasons* should not be neglected. Contamination-covered surfaces defy proper visual evaluation. Many defects such as heat damage, delamination, and blistering can be observed only on clean surfaces. The impact of a sparkling clean assembly on the end user should also not be overlooked; it is often considered a sign of "good workmanship".

6. *Automatic testing requirements* dictate clean solder fillet surfaces so that the bed of nails makes repeatedly good contact. Rosin flux residues,

Fig. 6-1. A corrosion failure on a printed Circuit board, due to improper flux removal. (Copyright Manko Associates)

for example, can cause interference, and false failure. Periodic cleaning of the test fixtures is costly and time-consuming.

Flux location and the mode of soldering make a big difference. The automatic probe contacts the bottom (solder side) of the printed circuit. In wave soldering, gravity pulls any residual flux over this surface, since it is on the bottom. During hand soldering for touchup and repair, the board faces up. Thus gravity does not pull the flux over the fillet, and the flux ring stays on the board. Poor cleaning practices smear these residues over the solder joint, causing a problem (see Section 6-17). It may be advisable to consider leaving touchup flux on the board for reproducible testing.

With such sound reasons for contamination control, one might expect the industry to emphasize cleaning. However, the costs involved have slowed the adoption of contamination control. This includes more than the cost of the operation itself in terms of cleaning materials, equipment, and labor. The cost of compatible components and hardware, the restrictions in design and assembly, and the extra handling requirements must also be included.

In the long range, cleaning will have to become a normal part of electronic manufacturing. With the inevitable advent of more dense circuitry and surface-mounting, contamination control will be forced on the industry. The economics will be aided by automation that will replace much of the hand labor used. As computer-aided manufacturing becomes the norm, it will be easier to monitor accidental soil introduction.

6-3 ELECTRONIC DIRT CLASSIFICATION

The electronics industry is strictly interested in classifying dirt according to the damage it can cause. Some contaminants are electrically harmful; others are not. From this vantage point, we can divide soils into insulators and conductors, each affecting the assembly in a different manner. Insulators are detrimental only if their presence interferes with the passage of electrical current in devices which depend on contacts. Conductive dirt is damaging when it causes current passage across areas that should be insulating. This current leakage is the most serious effect of dirt in printed circuit boards.

Dirt poses an additional, and by no means unimportant, hazard; chemical corrosion. It is obvious that contaminants which chemically attack the surface must be scrupulously eliminated. Cleaning preserves the physical integrity of the assembly. Let us remember that chemical corrosion normally requires the presence of water or moisture. It is also greatly increased by the presence of dissimilar metals in intimate contact. Both of these conditions (moisture and dissimilar metals) are found on the board assembly while in use. Fortunately, the effect of corrosive materials is so obvious that gross contamination is seldom left on the boards.

6-4 CLEANLINESS LEVELS

It is unrealistic to expect electronic equipment to remain absolutely clean. Let us imagine what happens to a wave soldered assembly after we have washed or degreased it. With an average industrial process, it can be made clean at this point. Subsequent operations like inspection, touchup, repair, testing, intermediate storage, and final assembly are then performed. These are not carried out under clean room conditions by people who wear gloves. The assemblies thus become recontaminated by human handling, sedimentation from the air, and contact with work benches, storage bins, and so on. The level of cleanliness after wave soldering bears little resemblance to that existing at the final stage of assembly.

The exact level of contamination on a surface, unfortunately cannot be measured by any single piece of equipment. Nor is there a simple way to relate contamination levels to the reliability or operational life expectancy of electronic hardware. To date, no universal method of evaluating the degree of cleanliness needed in a particular electronic branch or industry has been established. There are too many parameters involved because dirt acts in several ways, as described earlier. In assessing the effect of dirt, each assembly is unique. Let us consider some of the major parameters:

1. Electrical current leakage depends on the quantity of ionizable material in the dirt found on the surface (see Section 8-11). In addition, it is affected by surface resistivity, which, in turn is affected by non-ionic organic compounds adsorbed into the surface. Both of these properties are sensitive to the humidity in the ambient environment. Furthermore, leakage is a function of the voltage gradient, conductor spacing, and circuit impedance.
2. Chemical attack (corrosion) depends on the chemical nature of the contamination. It is also a function of humidity and temperature. Galvanic corrosion is accelerated by the presence of dissimilar metal junctions.
3. Conductivity interference occurs when insulating dirt is deposited on contact surfaces. This depends on the quantity of the dirt, contact pressure, breakthrough voltage potential, and other factors.

All of the above reasons underline our inability to be specific about dirt levels. No two assemblies are alike in behavior. In addition, contamination is deposited haphazardly on surfaces. Thus no uniform rules are possible.

From a practical viewpoint, however, the question of how much cleanliness is needed for a specific product can be simplified. Based on cumbersome tests or on the equipment's expected use. For example, home appliances such as radios or television sets obviously require less stringent cleanliness levels than an airplane's navigational system or equipment for a radar-tracking station, upon which human lives depend. Theses contrasting examples help establish the fact that no one clear rule can be applied to all types of equipment.

It is not difficult, however, for a knowledgeable individual to establish general cleanliness guidelines for a specific type of equipment or an industry. The problem usually lies in maintaining and monitoring these standards, because small contamination levels are not visible to the naked eye. Such an invisible small quantity, however, may cause failure to an electronic assembly. We will therefore discuss how to obtain optimum cleanliness and allow the individual to compromise and apply these guidlines to his own situation.

There is one complex evaluation which can serve as a guide. It is possible to take an absolutely clean assembly and recontaminate it artificially with known levels of dirt. Several groups of samples are needed, representing ever increasing levels of known foreign materials. The test boards are then assembled into the final unit and placed in a humidity chamber. The units are powered up and operated under 90% relative humidity at 100°F for one week. This test shows not only current leakage but also the effect of corrosion. The latter can be detected by close visual inspection. This test also has limitations, since it assumes that the dirt is uniformly dispersed over the surface. In reality, dirt from a splash or fingerprint is very localized. Our test method, however, cannot detect or report the location or concentration of contamination in specific areas (see Section 8-11).

6-5 CLASSIFICATION OF CLEANING SOLUTIONS

One of the most universal and efficient ways of removing contamination from assemblies is by flushing them with a liquid. This removes loose particulate matter and dissolves adhering films. Nonsoluble particles can be removed by any liquid but may require the addition of mechanical scrubbing or agitation (ultrasonics, brushing, spraying, etc.). It is therefore easiest to discuss dirt in relation to its solubility in cleaning fluids.

Soluble dirt can be divided into two major categories, each related to the properties of the solvents. The solvents fall into two groups: The first forms true solutions and are collectively called *nonpolar solvents*. Here molecules of the solute (the dirt) are simply interspersed between molecules of the solvent. Examples include some alcohols, the ketones (i.e., acetone), the halogenated hydrocarbons (i.e., trichloroethene), and others. Typical nonpolar soluble dirts include oil, grease, wax, and rosin.

The second group is called *polar solvents*. In these solvents, the molecules, because of their electrical charge distribution, can separate the solute into ionic particles. A typical polar solvent is water, and a typical ionizable solute is table salt. Ions of sodium and chloride are formed upon the solution of table salt in water (Fig. 6-2).

The effect of ionization changes the conductivity of water dramatically. At one extreme is ion-free (deionized) water, which is practically nonconductive. As polar contamination builds up, the amount of ions formed is proportional to the increase in conductivity. At the other extreme are highly conductive solutions, also called *electrolytes*. Their presence on an electronic circuit is obviously disruptive. Ionic solutions are also corrosive to many metals and cause galvanic attack when dissimilar metals are in contact (Figs. 6-3 and 6-4).

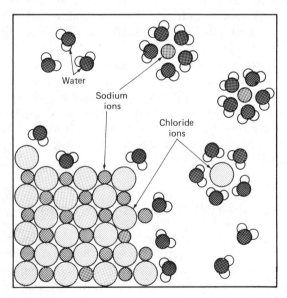

Water

Sodium
ions

Chloride
ions

Fig. 6-2. Simplified reaction of the polar molecules of water and table salt.

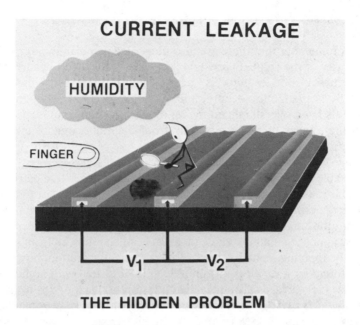

Fig. 6-3. Effect of finger print on a clean board in humid air.

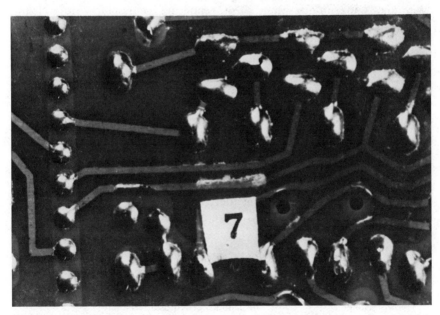

Fig. 6-4. A corrosion failure on a printed circuit board, due to improper flux removal. Moist flux residue caused electrical failure, in addition to the formation of green corrosion products. (Copyright Manko Associates Inc.)

Let us remember that there are also nonionizable polar solutions. Here the integrity of the solute is preserved in spite of the electrical activity of the solvent. A good example is sugar, which is soluble in polar water. Practically no ionization occurs, and the solution is not conductive. Although this is an oversimplified model of solvents and solutes, it demonstrates the danger of these materials from the standpoint of electronic equipment use.

Table 6-1 shows the conductivity of various moist dirts. Note that only the ionizable materials are conductive and cause danger to electrical circuits. The reason is simple: even small amounts of ionic dirt (like body salts in a fingerprint) can become conductive with the humidity in the air. Note that the same materials may also be chemically active when moist.

It may appear that all ionizable materials can be removed from surfaces by merely applying a polar solvent. Ionizable contamination, however, is seldom found by itself; it is usually mixed in or coated with nonionizable materials. If these are substances like oil, grease, wax, or rosin, they do not dissolve in water. In this case, we need a nonpolar solvent first, followed by a water wash. In other words, it is first necessary to remove all nonpolar dirt. This is normally called a *degreasing operation* and leaves all of the ionizable materials exposed. These can then be removed with a secondary polar water wash, rendering the assembly truly clean.

Few nonblended industrial solvents are both polar and nonpolar, and safe to humans (health) and factories (fire) at the same time. Unfortunately, the simple alcohols which fall into this category are flammable and not suitable for mass use. Isopropyl alcohol, however, is a good cleaner for hand soldering because it is used in small quantities at a time. If we try to achieve a double cleaning action with a single liquid, we must use cleaners which actually are blends of both polar and nonpolar solvents. These are erroneously called *bipolar* solvents, and are specifically designed for electronic assemblies.

The most desirable bipolar solvent must therefore be an azeotrope (sse azeotrope definition in Section 6-6) which boils at a constant temperature and distills both components together. The recovered bipolar solvent thus has the same cleaning potency as the original liquids.

Table 6-1. Electrical Conductivity of Moist Soils.

TYPE	PROPERTIES	EXAMPLE
Nonpolar	Insulators	Rosin, wax, oil
Polar nonionizable	Insulators	Sugar, some surfactants
Polar ionizable[a]	Conductors	Salt, most activators

[a] Often hygroscopic in nature.

6-6 SOME IMPORTANT PROPERTIES OF CLEANERS

The selection of the right solvent for an application is a complex process. There is a wide range of physical and chemical properties that require consideration. The following is a list of the more important ones, in no order of importance:

1. *Film drying characteristics*: This is a property of solvents that deals with two phenomena at the same time: the solvent ability to flush away dirt from the surface and its rate of vaporization. Unless the solvent can carry off the dirt without evaporating first, streaking occurs. In the process of draining from the surface, the solvent evaporates, leaving the dirt behind in irregular patterns.

This phenomenon is directly related to the evaporation rate or vapor pressure, which in turn is dependent on the boiling point of the solvent. The low-boiling solvents tend to streak much more than the higher-boiling ones.

2. *Soil capacity:* This is measured by an arbitrary scale which is referred to as the *kauri butyrate factor*. It refers to the quantity of this compound which a solvent can hold at room temperature. (Remember that solubility increases with temperature.) This measurement is accepted as a general scale for the efficiency of solvents (see Table 6-2). It is fortunate that this scale correlates well with the solubility of rosin in the common solvents.

3. *Compatibility with materials*: A good cleaner must be absolutely compatible with all the plastics and materials used in the assembly. The rate of attack of the various solvents on materials is the basic reason for the variety of cleaners used in electronics. The stronger solvents, capable of removing more dirt, often attack some material or coating on the work. This requires the use of more costly yet less efficient or aggressive cleaners. These are actually poorer cleaners, but safe to the assembly (see Table 6-2).

4. *Surface tension*: This is important only in regard to judging the solvent's ability to penetrate into crevices and displace other materials. In general, the cleaning action of a liquid depends on the process more than on surface tension. Such factors as agitation, fluid movement (flushing action), and temperature are more important and will shorten the cleaning time.

5. *Solvent stability*: This refers to chemical stability in use. For obvious reasons, the industry generally strives to use strictly nonflammable cleaners. The majority of solvents in this category are either chlorinated or fluorinated hydrocarbons. These materials form hydrochloric or hydrofluoric acid upon decomposition. Such acids are too aggressive for

Table 6-2. Nonflammable Cleaners.

SOLVENT	POLAR YES	POLAR NO	BOILING POINT °F	KAURI BUTANOL NO.	FLUX REMOVAL O.A.	FLUX REMOVAL R	FLUX REMOVAL R.A.	MATERIAL COST (GAL)	RECLAIM
113[a]		√	117.6	30	?			High	√
112[b]		√	199	70		√		Very high	√
Tri[c]		√	166	124		√		Low-medium	√
Perc[d]		√	249	90		√		Medium	√
Blend	√	√				√	√		√
H₂O	√		212	N/A	√			Very low	?
H₂O[e]	√	√	212+	N/A	√	√	√	Low	No?

[a] Trichlorotrifluoroethane;
[b] Tetrochlorodifluoroethane
[c] Trichloroethane
[d] Perchloroethylene
[e] With saponifier

printed circuit boards. Solvent breakdown is triggered by the presence of many industrial dirts like rosin or other organic acids, reactive metals (like zinc, and aluminum), and water. A good solvent, therefore, must be stabilized, and remain stable, even under the extreme temperature conditions that might be encountered in the heating section of equipment (as in vapor degreasing or distillation).

6. *Toxicity*: One misunderstood property of solvents is their safety. The average engineer places too much reliance on the *threshold limit values (TLV)* or *maximum allowance concentration (MAC)*. This is a suitable rating for the safety of liquids after they have evaporated in air. It is actually the quantity of solvent in parts per million that can be inhaled in an eight-hour work day without causing deleterious effects. This system alone, however, does not account for the volatility of the liquid. An additional critical factor is the speed with which solvents evaporate from a given surface or from the cleaning equipment to contaminate the air. Thus, a low-vapor-pressure, high-boiling liquid may be much safer, by several orders of magnitude, than high-vapor-pressure, low-boiling solvents because it takes much longer to reach a dangerous level in air. At present, there is no practical scale or method that includes this consideration. Common sense and caution are urged.

7. *Azeotropic composition*: An azeotrope by definition is: *a precise quantitative mixture of two or more liquids that has a modified boiling point*. The azeotrope boils at a lower temperature than either of its con-

stituents. The composition of the vapor is the same as that of the liquid, and therefore the ingredients cannot be separated by distillation.

This is an important property of solvent blends that contain polar and nonpolar ingredients. The azeotrope is the only type of mixture that should be used in a vapor degreaser (see Sections 6-14 to 6-16).

Matching the solvent to a particular job thus becomes a highly specialized field. Unfortunately, most people try to select cleaners by price, and not necessarily for their total economy.

6-7 CLEANER ECONOMY

The true worth of a cleaner cannot be measured by its cost per pound or per gallon. It must be measured by the number of units reliably cleaned in a system. Table 6-3 gives a comparison between inline cleaning systems and can serve as a guide. Let us briefly touch on some of the hidden cost that must be included:

1. *Evaporation loss* Because of the inherently volatile nature of cleaning solutions, there is an inevitable evaporation loss. Thus, a low vapor pressure and a high boiling point are desirable properties that reduce costs. To evaluate this condition, place equal quantities of various cleaners in identical beakers. Heat them to the process temperature and record their rate of evaporation.

2. *Cost of disposal or recovery*: The economy of a cleaner depends on the ability to recover the used solvent. If a material is distillable, it can be used over and over, provided its chemical stability is adequate. The importance of having the stabilizer and other protective additives distilled together with the azeotrope is self evident.

Residues and spent solvents must be disposed of in special ways that increase the cost of the cleaning process. They are normally picked up by specialists for a considerable fee and disposed of in a costly chemical burn operation. The cost may add up to 10% to the cost of the new solvent.

The costs of water and effluent are less prohibitive for the soldering industry. In most cases, the spent flux and cleaning solutions can be flushed down the sewer without treatment (see Ref. 6-7).

3. *Energy requirements*: In addition to solvent cost, one must consider the energy requirements of the system. This is not only an equipment consideration, but depends on the boiling point of the solvent. In vapor degreasing, the energy consumption of low boilers is usually higher than that of high boilers because the former require electrical refrigeration in the condensing coils. In water cleaning, the energy cost is split between the hot water used and the heat of drying.

4. *Equipment investment*: This is obviously a major contributor to the economy of a process. Since capital equipment is written off over a number of years, one must study the future of the cleaning system selected. Possible Occupational Safety and Health Administration (OSHA) and Environmental Protection Agency (EPA) restrictions (on a national or local level) may shorten the useful life of the investment. Check on impending legislation in your area for a guide to the projected longevity of the process.

5. *Labor content*: The labor involved in each cleaning process must be closely scrutinized (Table 6-3). The decisions here are equipment oriented for a given volume. One must select either batch processing (high labor content) or inline operations (minimal labor) according to their cost effectiveness.

Table 6-3 shows a comparison of inline cleaning costs for units of comparable size. These costs were developed for inline cleaning only and are based on a five-year depreciation of the equipment. In order to find a common base for this evaluation, the monthly trichloroethane material cost was used as a base of 100. In making comparisons then, the cost of the equipment and energy used were calculated and the total cost was established. Note that inline cleaning in all cases had the same labor content and maintenance costs, which were ignored. The low material cost of the water system is upset by the high energy required to operate it. This energy is needed to heat up both the cleaning solutions and the rinse water, as well as for drying.

Table 6-3. Comparative In-Line Cleaning Costs.
(Per Month)

SYSTEM	SOLVENT	FLUX	EQUIPMENT[a]	MATERIALS	ENERGY	TOTAL
Degreaser	Chloroethane only	Rosin	200	100	100	400
Degreaser	Freon TE	Rosin	270	235	100	600
Degreaser	Chloroethane azeotrope	Rosin	200	110	100	410
Washer	D.I. Water only	Organic acid	210	45	250	505
Washer	D.I. Water + Neutralizer	Organic acid	210	90	250	550
Washer	D.I. Water + Additive	Rosin (or) O.A.	210	130	250	590

[a] 5 Years depreciation.

6-8 WATER WASH SYSTEM

Water washing of printed circuit boards has become an acceptable method in recent years. It is suitable for rosin as well as organic intermediate fluxes. At first blush, water washing seems to be an inexpensive solution to the cleaning problem, but this conclusion is very misleading. While the equipment costs are roughly the same as those of solvent cleaning, the major difference lies in the energy required (see Table 6-3).

Water, with or without additives, is the most universal cleaner. It is not limited to any type of flux and offers the user flexibility. The exception is the *synthetic activated* (type SA) flux, which was designed for solvent cleaning only.

Water by itself is strictly a polar solvent and is incapable of removing polar soils such as rosin, oil, and grease. If the contaminants on the work, however, are known to be all water soluble, this becomes the least expensive washing method.

The cleaning range of water can be expanded by the addition of other chemicals. The electronic industry uses *saponifiers*, these are alkaline materials which react with materials such as rosin and oil to form a washable soap. Small amounts of saponifiers are usually added when water-soluble fluxes are used to remove accidental nonpolar dirt. Larger amounts of saponifiers are needed to remove rosin-type fluxes.

The principles of the water process are:

1. The solvency of the contamination.
2. The chemical conversion of soil to a washable form.

Here again, heat and agitation facilitate the cleaning process.

The equipment resembles dishwashers in the kitchen (batch) or cafeteria (inline). Metallic and polypropylene cabinets are used; plastic constructions have proved to be more energy efficient (Fig. 6-5). Design variations are found in the spray configuration, the volume of liquid pumped, and the location of air knives (Fig. 6-6).

The ability of the equipment to dry the work is of secondary importance. Only retained water is objectionable, since it may retain too much contamination. After cleaning and drying, the boards seek to reach equilibrium with the humidity in the room. This determines the overall moisture content of the work. Overdrying gives false electrical values until the moisture content is stabilized with the room humidity, and is wasteful of energy. One must remember that the cost of drying is an important part of cost effectiveness.

The quality of the water is also a matter of common sense (see Section

Fig. 6-5. Typical polypropylene water washer. (Courtesy of Hollis Automation)

6-9). Its purity is directly related to the cost of generation. Only in a small number of cases has the author found it necessary to use expensive deionized water. Here are two examples: In the first case, a high-impedance circuit had to be cleaned once again after touchup, inspection, and testing. It was then immediately sealed in an enclosure and kept clean. In the second case, the cleaning was done just prior to conformal coating. Here the deionized water was needed to prevent future mealing due to vesication (see Ref. 6-2).

In general, soft water is sufficient, and with proper processing the work will even pass the stringent ionic requirements of MIL-STD-28809 (see Ref. 6-8). Of course, the purity of soft water is not fixed; it depends on the mineral content of the water to be processed. If the water is in the range of half hard to hard (Table 6-4), it can be safely treated and used for most electronic assemblies. While all processed soft water contains ionic sodium (used in softening to replace the insoluble heavier metals), its overall mineral content is not harmful. This holds true only if no water is allowed to stagnate on the board or work; even deionized water will cause damage when left on the work too long.

The chemical composition of the water being used is important, but so is the makeup of the effluent. Fortunately, this can be easily controlled for the soldering industry (see Ref. 6-7). Little if any water treatment is needed by the user who has made a judicious selection of the soldering

Fig. 6-6. Multidirectional water spray in action. (Courtesy of Hollis Automation)

chemicals. Sometimes the quality of the used rinse water makes it very attractive to consider a close loop system. The rinse water is reprocessed and used again; this is especially important in areas where water is in short supply.

6-9 GRADES OF WATER

Let us consider first the water which is available for cleaning. This can be divided into the following categories:

- Tap water—untreated by the user.
- Soft water—treated.
- Deionized water—treated.

This classification is based on the contamination content of the water:

1. *Tap water*: The quality depends on the source of the water and may contain the following impurities:

a. *Organic matter* mostly from natural sources. This is prevalent in areas where the water comes from above-ground reservoirs. Decomposed vegetation, algae, and waste products from wild-life make up the bulk of this category. Unfortunately, industrial waste and raw sewage can also find their way into a water system. Many of these solids can be filtered out, and the balance is not considered harmful to electronic assemblies.

b. *Halogen ions*, often introduced by the water supply system for medical purposes. Chlorine is used to reduce the danger of bacterial contamination. Fluorides are introduced in some public systems for dental care. For most electronic work, these additives in the quantities added are seldom a problem.

c. *Metallic ions* from the minerals dissolved in underground water. These should be further divided into harmless and "hard" elements. The harmless elements, sometimes referred to as *soft* or *light* elements, include sodium, and potassium.

The hardness of the water is due to the metallic ions of the heavier elements. Elements such as calcium, magnesium, and iron are found in this category. They reduce the ability of the water to sustain dirt in solution because their salts are insoluble and precipitate out. Hard water also tends to coat the inside of heating elements and spray nozzles with mineral scale and organic scum. These interfere with the action of the equipment and requires substantial maintenance, usually referred to as *descaling*. Table 6-4 lists the grades of water relative to their hardness.

Table 6-4. Water Hardness.

PARTS PER MILLION (PPM)[a]	GRAIN PER GALLON (GPG)[a]	U.S. GEOLOGICAL CLASSIFICATION
0–60	0–3.5	Soft
60–120	3.5–7.0	Moderately hard
120–180	7.0–10.5	Hard
180–500	10.5–29.2	Very hard
500–1000	29.2–58.4	Not for human use
1000 and over	58.4 and over	Saline

[a] Equivalent of $CaCO_3$.

as clearer.

(Oregon tap water is soft.)

vith your local water company
t do so for all seasons of the
ange. On the average, melting
vater. You can also have your
t remember that one spot test
· treatment companies are an-
ι be biased.

2. *Soft water*, is the simplest and most economical form of treated water. In this process, the undesirable *heavy ions* are replaced by sodium ions. A selective ion exchange resin is first saturated by the user with sodium ions from rock salt. Then the water to be treated is passed through the resin bed. The undesirable ions have a greater affinity for the resin and replace the harmless sodium. The soft water leaving the unit is free of hard ions, but contains sodium ions instead. This grade of water is slightly conductive.

In general, this water is suitable for use in the majority of electronic applications. The presence of small amounts of sodium and the slight conductivity are unimportant. Products correctly washed with soft water can pass the most stringent requirements of cleanliness tests.

The cost of softening water is low as compared to the cost of maintenance that untreated water requires. It is substantially lower than the cost of deionized water described in item 3 below. It is usually cheaper to rent a softening unit with full service than to descale the equipment weekly, as required when hard tap water is used. The rent or buy decision depends on local conditions.

3. *Deionized water*, is the purest and most expansive grade. As the name implies, all foreign ions are removed in the process. Reverse osmosis and complex resin beds are used individually or in combination. The water is so clean that its solution potential is a problem. It can be handled only in special plastic or glass-lined pipes or containers. It is actually corrosive to most metal surfaces (including stainless steel), which dissolve in the water and recontaminates it.

Obviously, this water is suitable for use in all electronic applications. Its high price, however does not justify its use unless it is absolutely necessary. It is recommended for use under the following conditions:

a. The assemblies are to be coated immediately after cleaning with a conformal coating. In this case, any contamination left on the surface causes *vesication*, a process whereby slow permeation of moisture to the interface between the work and the coating sets up osmotic cells. These cells lift the coating off the work and may cause corrosion if they are ionic. They can also produce current leakage when present in a sensitive area. Upon drying these areas become visible, and the condition is referred to as *mealing*.

b. The assemblies are sensitive to contamination and will henceforth be processed under clean room conditions. Alternatively, they will be encapsulated or hermetically sealed immediately after cleaning.

c. External specifications, not under user control, impose the use of this grade of water.

In all of these cases, one must keep in mind what happens to the printed circuit assemblies after cleaning. If no precautions are taken, they become easily recontaminated during subsequent handling, processing, and storage.

6-10 WATER-BASED CLEANING PROCESSES

Having selected the appropriate grade of water for your use, it is time to decide on the process. Water based cleaning systems can be divided into four major categories:

1. *Water only*: This type is suitable for all applications where water-soluble dirts will be encountered. In soldering, organic intermediate flux, water soluble oil, and other such materials can be cleaned in such a system. This is by far the cheapest and simplest process. The quantity of water used depends on the quality of the water and the level of cleanliness desired. Table 6-3 gives the approximate cost of such a system using deionized water as the cleaning material.

2. *Neutralizer addition*: This process is little used because of its cost. It helps to minimize the volume of water used. An additive is introduced at the first washing station as a neutralizing agent. This additive interacts chemically with the organic flux residues and any other acidic materials. If properly formulated, it also facilitates the quick solution of many metallic salts. It forms many readily soluble complex metallic salts. These neutralizers are normally available with a chemical indicator that helps monitor the pH of the solution. As long as the cleaning solvent can neutralize more organic flux, the color indicator will not change.

3. *Surfactant addition*: Neither one of the systems described above can remove traces of nonpolar soils such as wax, oil, and grease. These are often accidentally deposited on surfaces in pre-solder operations. They are only partially removed by the mechanical impinging of hot water. Experience has shown that small amounts of such materials can be tolerated by many segments of the commercial electronic industry. If more thorough cleaning is desired, however, a cleaning agent must be intro-

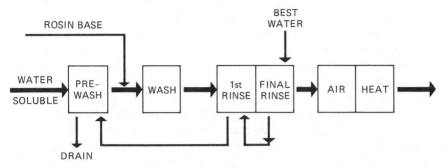

Fig. 6-7. Schematic of a water wash system. Note that the boards with rosin bypass the prerinse stage. Also, in all water-soluble systems, the wash stage may be repiped as another rinse.

duced into the water in the form of a nonfoaming surfactant (sometimes referred to as a detergent), which will help remove traces of these materials.

4. *Chemical (saponifier) addition*: Finally, it is possible to remove nonpolar dirts through a chemical reaction. Normally, an alkaline additive is placed in the washing station to interact with rosin and nonpolar materials. The reaction is one of saponification, which results in a soap-like material. These materials are easily water washable, if not entirely water soluble. This method can be recomended to all users for rosin and organic fluxes alike. For rosin, the quantities of saponifier are large. For water-soluble fluxes, only minute amounts of saponifier are needed.

Saponifying additives and reaction products, because of their surface tension, tend to foam. A nonsilicone foam suppresser is often added for trouble free use. Furthermore, the saponifiers that are available are heat sensitive. This limits the temperature of the water in the wash module. It would be desirable, of course, to heat the water above the rosin melting point for easy flux residue removal by a melt-wash combination (Fig 6-7).

6-11 EFFLUENT DISPOSAL AND THE EPA

One important characteristic of all washing additives is their effect on the effluent. They must meet all of the Environmental Protection Agency (EPA) requirements (see Ref. 6-7). Most solder chemical producers are aware of this requirement and will provide the right formulations. These must be harmless biodegradable compositions, and have the right pH, and so on (true for flux, oil, cleaner, etc.).

In the long run, the availability of water may become a serious problem. The unrestricted use of water for industrial purposes may lead to a water shortage even in areas where no drought exists. This, however, should not cause undue alarm in the selection of the water based cleaning processes. Reclamation of rinse water is entirely feasible and economical over the long run.

The wash water, in the first stages of the cleaning process, cannot be reclaimed economically . This is due to the presence of the soldering chemicals, the wash additive, and miscellaneous forms of dirt. Fortunately, the quantities are small and probably will continue to be discarded once a day. The rinse water, however, can be readily recirculated through a filtration unit and deionizing equipment. Unpublished studies by two corporations on the reclamation of their postsolder cleaning water are encouraging. They indicate that the cost of the close loop equipment and materials have a payback of between 28 and 36 months. Under those circumstances, the installation of a recirculation system is entirely feasible.

It is difficult, at this juncture to predict future developments in the cleaning industry. Automatic inline cleaning solvent systems and water wash systems are fairly similar in their total cost. Economic pressures, EPA and OSHA legislation, and similar outside pressures may push the industry in either direction. Without these external influences, it is anticipated that a balance between solvent and water washing will be reached.

6-12 SELECTING A CLEANING METHOD

For those who are selecting a cleaning method for the first time, this is a most confusing task. It is often resolved on an emotional rather than an engineering basis.

In reality, technical considerations and the true cost of the operation are the determining factors. The following is meant as a guide only; in the author's experience, conditions change from case to case. Normally, local industrial conditions and the unique needs of a product will determine the best choice.

The selection of the cleaning system can be divided into several independent steps:

1. Material selection.
2. Process determination.
3. Equipment identification.

Let us discuss them in more detail.

1. *The material selection*: Is a process that should be part of the product design but seldom is. It is usually left up to the manufacturing engineer, who must find the right compromise. The crucial issues here are:

    a. The optimum flux selection.
    b. The appropriate cleaning fluid.
    c. Any OSHA or EPA impact.

Included in these considerations are the effect of the soldering chemicals on the assembly.

2. *The process selection*: This hinges on the materials selected in step 1. Here the determining factors are:

    a. Assembly geometry.
    b. Work volume (inline or batch).
    c. Equipment and know how on hand.

Fortunately the number of alternatives is limited, making this determination easier.

3. *Equipment identification*: Is related to the purchase of new machinery and the choice of a vendor. This must be made on a total cost basis and include the following:

    a. Equipment and installation cost.
    b. Material usage and losses (evaporation, drag out, etc.).
    c. Spent material disposal expense.
    d. Labor and maintenance cost.
    e. Energy and space.

It is best to calculate the cost on a unit basis or a standard time period basis. The advice of a schooled cost analyst should be sought.

In spite of the higher initial cost of inline equipment, it can usually be justified because of the high labor costs in the United States. Inline equipment is not operator dependent, and with simple controls will give more uniform reliability. In addition, the amount of solvent and/or water used per unit of work is much smaller, and material losses can be held to a minimum.

The controversy about *solvent versus water* cleaning will continue, and will depend on many external conditions (safety, environment, etc.). The author has no particular preference. In addition, careful calculations with a variety of clients have shown repeatedly that there is no major cost issue in the removal of a specific rosin flux with either system. There is a dramatic cost reduction of 30-40% in switching from the rosin family to water soluble fluxes.

6-13 SOLVENT CLEANING

This is a much more traditional method of cleaning in electronics than water-based processes. The solvents and the vapor degreasers have existed from the inception of this industry.

Only nonflammable organic solvents are used in the electronic industry for obvious reasons. Since the fluorinated and chlorinated solvents are nonpolar by themselves, they are blended with polar alcohols (Ref. 6-3). Preferably, these should be azeotropic mixtures, which are still nonflammable.

The solvents are limited to the removal of nonpolar rosin- and resin-based fluxes. These contain activator systems which, by themselves and after thermal breakdown, are mostly polar in nature. The solvents are not used with water-soluble intermediate organic fluxes, erroneously also referred to as "organic acid fluxes."

Solvent cleaning is based on three factors:

1. The solvency of the contamination.
2. The purification of the vapor phase.
3. The condensation of clean solvent on cooler work.

All of these principles working in tandem make solvent cleaning an economical process. The reuse of the same solvent is made possible by either external or internal distillation or the use of the vapor phase.

The most common solvent cleaner is the vapor degreaser (see Sections 6-14, 6-15, and 6-16). Both batch and inline equipment are available. Vapor degreasers as such have limited mechanical agitation in the dip and spray cycles. Cleaning equipment using forceful solvent sprays offers distinct advantages in cleaning speed and efficiency. These are coupled with a distillation cycle for solvent recovery which provides recycled solvents at an elevated temperature to improve dirt removal. These degreasers also use a vapor phase cycle to help dry the finished work; the drying cycle also limits solvent loss through drag out.

Other mechanized methods, like cold cleaning with rotating brushes, are available. This process is normally used only for bottom cleaning. It is suitable for printed circuit assemblies with *nonwettable* components or solvent-sensitive plastics on the top. The equipment consists of a number of counter-current cascading tanks mounted in series. Brushes are partially submerged in the solvent and rotate in the opposite direction to board travel. The work is mounted on a conveyer that keeps it out of the solvent but in contact with the brushes. In this fashion, the boards are bottom cleaned by the brushes as they pass from contaminated to ever cleaner solvent tanks (the counter-current principle). The dirty solvent is removed continuously and sent to a still for recovery. The fresh solvent from the still is introduced at the end of the line, where it contacts the work just before it leaves the unit. As the solvent becomes contaminated, it cascades forward, to be removed for recovery at the end.

Unfortunately, ultrasonic cleaning through cavitation with solvents or water systems is banned by government specifications. It is also frowned on by segments of the semiconductor and microelectronic industries. The reason is the fear that cavitation may cause resonance inside the devices. This, in turn, would weaken or destroy some of the wire bonds and interconnects. As a result, this excellent mechanical scrubbing and cleaning process has not become popular in the printed circuit industry.

Both EPA and OSHA regulations have been threatening the uncontrolled use of organic solvents. The technology is available today, however, to control the escape of these chemicals into the work environment. Its use should satisfy OSHA and avoid personnel health hazards.

The problems with the EPA are much more serious. The unfortunate criminal disposal of these solvents by illegal means has introduced traces of chlorinated solvents into the drinking water of several communities. This is a matter of great concern to management, and many companies shy away from the use of solvents for this reason. There are, however, safe and legal disposal methods which should be available through the solvent supplier.

The selection of the right solvent is made on the basis of the data in sections 6-5 to 6-7. It is a compromise between material properties and their attack on the plastics of the assembly. The cost of the operation is also a consideration. Obviously, only nonflammable solvents should be used, and they must be azeotropes containing both polar and nonpolar components.

Solvent cleaning can be done in several ways, including cold dipping, brushing, and, of course, vapor degreasing. In all of these processes, the operators must not expose their skin to the solvent for prolonged periods. The solvent removes the oil from the skin, which, in turn, causes serious problems. Some people are also selectively allergic to solvents. It is therefore prudent to eliminate large-scale hand cleaning from production unless protective gloves are used.

6-14 VAPOR DEGREASING PRINCIPLES

Vapor degreasing is based on the following principles:

1. Only the solvent boils off; the dirt stays behind in the sump. Thus the vapor phase is always pure.
2. The clean vapor condenses on the cooler surfaces of the work. This causes a continuous flushing action with clean solvent that removes dirt. The contaminated solvent drips back into the equipment.
3. The temperature of the work rises with vapor condensation and eventually reaches the same level. When the board temperature matches the boiling point of the cleaner, all rinsing or cleaning stops.
4. A cold dip or spray cycle cools the surface for additional multiple cleaning and vapor condensation cycles.
5. Hot work by itself does not drag any solvent out of the equipment facilitating solvent economy.
6. With azeotropic solvents, polar and nonpolar components are present in the vapor phase in the same ratio. If the liquid is nonflammable, the vapors stay the same. In addition, the stabilizers, acid acceptors, and other additives keep the solvent stable and reusable for reasonable periods.

The biggest advantage of vapor degreasing is the multiple reuse of the same solvent. This helps the material economy and makes the use of expensive materials possible.

These principles are used in both batch and inline processing. It is important to understand the individual steps in the process. That way, we can take advantage of the full potential of the equipment. Let us describe a batch cleaner first.

6-15 BATCH VAPOR DEGREASING

The stand alone batch cleaning unit consists of several sections and may also be connected to a still (Fig. 6-8). Let us discuss these separately.

1. *A sump*: This is where the solvent is boiled by steam or electric heat. This is where the soil accumulates, and unless the unit is connected to a still, it must be drained periodically.

Never immerse the work in the dirty boiling sump, or dirt will bake on (see Section 7-25). A metal structure may be used as a platform to prevent the work from reaching the boiling liquid.

Immersion may be acceptable if the sump is connected to a continuous

Fig. 6-8. A batch vapor degreaser, showing upright boards being sprayed with clean distillate during the last rinse and cooling cycle prior to final vapor cleaning. (Courtesy of Hollis Automation Inc.)

still. In this case, the unit's recovery rate must be compatible with the rate of dirt buildup.

Solvent maintenance is vital because the buildup of dirt increases the solvent boiling point. This may bring the organic compounds close to their breakdown temperature. At this point, chlorinated and fluorinated compounds give off dangerous hydrochloric and hydrofluoric acids. These can attack the work and dull the solder joints. Therefore, the temperature rise in the unit should be monitored and kept below 5-7°F above the boiling point. It is also possible to measure the acidity of the solvent with special indicator paper or by water extraction. In general, the pH should be kept above 4. Consult your solvent manufacturer's literature for details on your specific solvent. In some cases, suspending baking soda in the sump helps control its acidity.

2. *The vapor zone*: This extends over the boiling sump and the cold rinse section described in item 3 below. It is the major work area of the degreaser. It is in this section that the hot, rising vapor condenses on the cooler work. This is the area that gave the method its name.

The vapor is heavier than air and tends to stay as low as possible. But with a constantly boiling sump, it would build up enough volume to overflow the equipment. To prevent this, a cooling coil is mounted on top of this section. It serves not only to contain the solvent, but also to generate cold pure distillate that is channeled into the hidden cold reservoir in item 4 below. The cooling coil is located not at the top, but somewhat lower. The area above the coil is referred to as the *free board* and serves to minimize solvent loss.

The vapor level should be maintained at the lower to middle part of the coil. In addition, no drafts or air currents should be allowed over the free board, to eliminate undue solvent loss. Keep the unit covered when not in operation.

The exposed top section of the cooling coil condenses water from the air on humid days. This water drips into the unit with the solvent and can extract desirable additives or some of the alcohol components. While this is true for solvents that boil below 212°F (100°C), a good piece of equipment will have a water separator or a dryer build in. If water gets into the cold reservoir, it will float on top of the heavier solvent, in what looks like oil patches on water.

3. *The cool rinse reservoir*: Located next to the boiling sump is a collection reservoir holding cool solvent. This material is slightly contaminated from the work. It is used to rinse and cool the work after its initial vapor exposure. It is thus acceptable to dip the work into this section.

The liquid in this section is continuously replenished from the overflow of the hidden tank (item 4) or the cold solvent spray (item 5). It, in turn, overflows back into the boiling sump. This section requires little maintenance and should be kept free of debris.

4. *The hidden cold reservoir*: The condensing solvent from the cooling coils is collected and channeled to this section. On the way, the solvent passes the water separator, or a dryer, where condensed water from the humidity in the air is removed. These dryers need monitoring and maintenance.

The cold solvent that is collected here is very pure as a result of the distillation process. It is thus suitable for direct contact with the clean work. It is applied through the pump and spray wand described in item 5 below.

Any overflow is used to replenish the rinse section (item 3 above). If the solvent must be changed completely, this section can be pumped out with the spray wand. Under normal sump cleanout, this section is not emptied, since it contains no dirt.

5. *The spray wand*: A pump takes the clean distillate from the hidden reservoir and moves it to a hand held wand. A spray nozzle forces the solvent at an increased pressure toward the work. The cold solvent gives the work another clean rinse while cooling it once more.

The pump is usually foot actuated, while the wand is hand held. Thus operator technique in applying the spray is important. It is also possible to mount the work on a hoist and pass it automatically between permanently mounted spray nozzles. The cycle is also triggered and timed automatically.

The proper use of the batch degreaser is operator dependent. While a hoist can be set to compensate for this factor, it is still important to learn the five steps of good cleaning:

Step 1. Lower the work into the vapor phase above the sump. The work must be cooler than the solvent boiling point, or no cleaning will take place. In addition, white residues may form (see Section 7-25).

Hold the work in this stage until all dripping from it stops. From 90-120 seconds is usually a good time.

If the unit is set up for direct dipping in the sump, do not stay more than 45 seconds. Remember, you can do this only if you have a hookup with a continuous still and adequate solvent recovery output.

Step 2. Dip the work into the cool reservoir section for a rinse and cooling of the work. This removes some dirt and prepares (cools) the boards for a second vapor exposure. Remember that the cleaner in this section is slightly contaminated.

Stay in the cool solvent for 30-45 seconds, agitating lightly. Withdraw it slowly, letting the cleaner drip back. Do not shake the basket.

Step 3. Suspend the work in the vapor above the sump. This is a repetition of step 1, but with a cleaner work-load.

Hold the work in this stage until all dripping stops. Usually 90-120 seconds is a good time.

Step 4. Suspend the work over the cool reservoir and spray it with the wand. This gives it a final clean liquid rinse and cools it once more.

This should definitely not be done above the sump because the cold distillate slows boiling and vapor generation.

Spray over all surfaces for about 20-40 seconds. The wand should not touch the work.

Step 5. Leave the work in the vapor phase for a last condensation warmup cycle. This is a repetition of steps 1 and 3.

Hold the work in this stage until all dripping from the work stops. Usually 90—120 seconds is a good time. Raise the work slowly from the degreaser, it should be clean, hot, and dry.

There are several additional items to watch:

1. Make sure that the basket or fixture is not too large a heat-sink and lets the vapor freely through. Lower the empty basket into the vapor phase and note the time it takes to stop dripping. Any time in excess of 45-60 seconds is not matched to your vapor output.

2. Rack your work so that it stands as vertically as is feasible. Draining of dirty solvent is essential for good cleaning. Horizontal printed circuit boards are not properly degreased unless they are lowered one at a time.

3. Fast removal of work *drags out* solvent, which is expensive and bad for employee health. If you can smell the solvent in the area, check for problems.

4. Segregate the unit for flux removal only. Restrict any unauthorized contamination which will upset the cycle.

Sometimes an ultrasonic transducer is located in the vapor degreaser. Ultrasonic cavitation is an excellent way of accelerating cleaning but is prohibited by government and many industry specifications. It is suspected that the ultrasonic energy may harm microelectronic devices.

6-16 INLINE SOLVENT CLEANING AND VAPOR DEGREASING

The inline cleaner is basically the same kind of unit as the batch vapor degreaser. The work is loaded onto a belt and transferred through a series of cleaning stages. The boards travel horizontally or on a slight angle, one at a time, and are cleaned.

While older units simulate the batch vapor degreaser, more recent equipment has added forceful liquid sprays. Vapors are still being used to purify the solvent, heat the work to accelerate cleaning, and dry the work. The real cleaning action, however, is provided by high volume sprays.

The process is much less operator dependent than the batch units and can be computerized. Solvent losses are significantly lower than those in

Fig. 6-9. An inline solvent degreaser with vapor drying. The cleaning is achieved by forceful spray of hot liquid. (Courtesy of Hollis Automation)

batch cleaning. This solvent savings often justifies the replacement with inline equipment. See Fig. 6-9 for details.

The reduced labor content of the process results from the automatic loading and unloading of the unit. To achieve this, the cleaner must be located behind the wave solder output and an unloading conveyor. Unfortunately, this advantage is lost when pallets, fixtures, or stiffeners are used. These must be removed prior to cleaning for obvious reasons.

Maintenance of inline equipment is simplified when an external still is incorporated. Automatic level controls keep the solvent in balance with losses by drag-out and evaporation. Activated charcoal beds can be used to purge the vapors from the exhaust air. These beds, in turn, are regenerated with steam.

6-17 LOCALIZED HAND CLEANING

A number of hand soldering operations may be required after wave soldering and cleaning. These include touchup and repair, installation of nonwet components, and final wiring. These operations are normally performed with core solder that leaves a small residue behind. Most of the core solder used here is of the rosin family. It is possible to leave these residues in place, especially with rosin flux. However, industry frowns on this practice for two reasons. First, the flux interferes with the probes (bed of nails) in the automatic testing of the assemblies. Second, there is some unjustified fear of problems and, even more important, a desire for a pleasing cosmetic appearance.

The normal practice is to apply cleaning fluid to the joint with a brush and to wipe the residue around (Fig. 6-10). This is not really a cleaning process, since the flux is not rinsed off the surface. There is no place for the flux to travel, and it stays on the board. The contamination from the flux, finger-prints, and other sources is just spread around in a thinner (less visible) layer that covers a larger area. This is especially dangerous when activated rosin (type RA) fluxes are used. The fused rosin residue formed from such core solder during heating is relatively safe. However, when the residue is diluted with solvent and spread over a larger area, some of the protective properties of the rosin are negated. The activator and ionic dirt are more likely to become a current leakage hazard.

The practice of brushing liquid cleaner on a freshly made solder joint has merit if total removal is contemplated. In a good cleaning sequence, it loosens the flux, and cools the joint at the same time. The operator should

Fig. 6-10. Common ineffective method of removing touchup flux residues. The dissolved flux is only thinned and spread over a larger surface.

be encouraged to do this as soon as feasible after soldering (see Chapter 9). Flux that has aged on the joint for several hours is difficult or impossible to dislodge even in a vapor degreaser. After the flux has been loosened, however, there is no rush for full and final removal. The solvent, when applied within seconds, will softens the residue, and easy final removal can be done even hours later. This final cleaning must be thorough, using equipment such as a vapor degreaser or a bottom scrubber.

Cleaning of the entire printed circuit board or assembly is not always technically feasible. Often circumstances rule out total immersion because of sensitive components (switches, relays, pots, etc.) or subassemblies which will retain liquids. Sometimes the size or weight of the assembly prevents total liquid submersion. The cleaning device may be too remote from the repair station. This is especially true of field services and equipment at the customer's location.

In such cases, in situ localized cleaning is required. Adequate spot cleaning typically cannot be achieved by brush and fluid alone. The important element of flushing with a clean solution is missing. As indicated earlier, the flux residues should not merely be spread in a thin layer.

Proper spot cleaning, however, can be achieved with the use of the device shown in Fig. 6-11. Using this hand-held unit, the operator can achieve total dirt removal in a localized area. The principle of this unit is simple: the cleaning liquid is released through the hand-held handle under the control of a small valve. The brush helps dissolve the dirt, and the contaminated liquid is removed by suction actuated by a foot paddle (Fig. 6-12). Repeated operations of liquid flushing and vacuum suction help remove the last traces of dirt.

The proper use of this in situ spot cleaner is as follows;

1. Apply cleaning solution to the spot only.
2. Brush to speed up the cleaning process.
3. Remove the dirty solution by suction.
4. Repeat steps 1 through 3, flushing until all dirt is removed.

By repeatedly flooding with fresh solvent and removing it, very clean work can be obtained. In most applications, three to four cycles are sufficient.

The cleaning device works with both water and solvent systems. The water system is simple, and applies to both rosin and organic intermediate fluxes. In the case of rosin, a small amount of saponifier is metered out from a needle-topped plastic bottle directly onto the flux residue. The brush is then used to work the saponifier into the flux; after about 5 seconds, water is added. An additional scrubbing helps react all of the

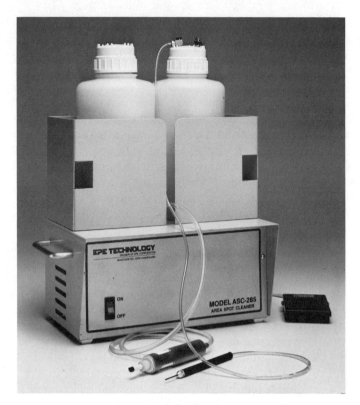

Fig. 6-11. General view of ASC unit for total *in situ* cleaning. (ASC—Area Spot Cleaner). The container on the left holds the cleaning liquid which is dispensed through the brush to the work. The used "dirty" liquid is vacuumed off the board and channeled into the collection chamber on the right. (Courtesy of EPE Technology)

rosin, which can then be removed by suction. This process is followed by several rinses until the dried board (by the vacuum) does not show the typical gloss of rosin residues. This is usually a sign that the board is clean.

The process is much simpler if only water-soluble dirt is to be cleaned. In this case no additive is needed, but the number of flushes is increased to ensure proper cleaning. Both sides of the circuit board must be cleaned unless flux residues are known to be limited to one side only.

A series of tests were conducted by the author to check on the capabilities of the hand-held in situ cleaner (Figs. 6-11 and 6-12). Excessive quantities of flux were placed under large, flat components and then heated to soldering temperatures for up to 10 seconds. The ability of the spot cleaner to remove the trapped flux was evaluated with an Alpha Iono-

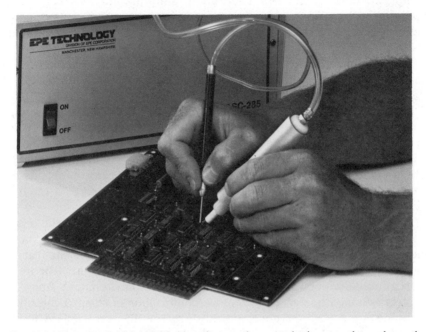

Fig. 6-12. Closeup of ASC hand-held tool, note the control trigger used to release the cleaning liquid is located on the brush. A food paddle, not visible, actuates the vacuum. In the "flushing" mode (under components, etc.) the cleaner is released on one side, while the used liquid is removed on the opposite end. (Courtesy of EPE Technology).

graph. The removal of both rosin and organic fluxes was adequate to meet MIL-P-28809 levels.

Water-based flux in core solder is also available. Most of these flux residues are harmless for a few hours after soldering. This is apparently due to the total drying of the residue during the heat of soldering. Small water washers for hand soldering are also available. These could be batch equipment, like a kitchen washer, or inline.

There is one drawback to water soluble core solder fluxes. They tend to blacken the tips of soldering irons faster than rosin and cause their dewetting. This calls for an increased maintenance program and offline retinning of tips.

6-18 CLEANING OF COMPONENTS AND DEVICES

Total removal of dirt, whether harmful or benign, is a known technology in most industrial fields. What makes electronic cleaning unique is the complexity of the assemblies. There are no uniform conditions that would make the process standard for the industry. Let us look at various components to illustrate this point. Printed circuit boards are covered in the next section.

The variety of devices used in this industry range from hermetically sealed units to unprotected components. A partial list is presented in Table 6-5.

Let us consider the washing damage to the component itself as the only criterion. Here any detrimental effect of the cleaning solution and its residues must be investigated. This obviously includes the components of the dirt that are either suspended or dissolved in the cleaning liquid. These components must be studied, since they may cause problems of their own. Nothing must attack any external or internal parts of the devices.

Even if the trapped contamination poses no direct danger to the function of the device, it may still cause trouble. Retained chemicals may bleed out with time and humidity. Penetration of tinning fluxes into plastic integrated circuit packages has recently created such a failure. The integrated circuits kept on functioning even in a humid environment. Approximately two month later, however, the combination of trapped chlorides and moisture caused the development of rust on the ferrous alloy leads and short circuits on the board. In another case, it took four to nine month of humidity exposure in the field for trapped chlorides to destroy the active device.

Only hermetically sealed units are always safe, especially if any leakers have been weeded out. In real life, however, there is always the possibility that some of the devices in the batch are leaking. Therefore, we must include this category with the next group of encapsulated components.

The process of encapsulation normally involves interfaces between metals and plastic materials. Even under the best adhesion conditions, these dissimilar material boundaries are subject to microcracks under

Table 6-5. Washability of Components.

TYPE	SAFE TO CLEAN	EXAMPLE
Hermetically sealed	Yes	TO-5 transistor
Encapsulated	Yes[a]	Plastic ICs
Encased only	Possibly[b]	Trim pots
Totally exposed	Mostly[b]	Coils

[a] With a proper wash temperature profile.
[b] If they can be flushed and do not trap liquids.

aging conditions. In addition to mechanical abuse, such as lead wire bending, there are stresses due to a large mismatch of the TCE. The capillary spaces that are formed (usually at the interface) may admit dangerous solutions which can cause the device to fail or bleed out with time and humidity.

There is a way to control the penetration of solutions into capillary spaces, microcracks, and leakage passage-ways. It is based on the *process temperature profile*, and relies on simple physics. The principle is based on the expansion and contraction, with temperature changes, of the air in the potential leakage gap.

When air is heated it expands, and in a blind capillary it will be expelled. Under these conditions, no dangerous contaminants will be admitted. However, during cooling, air contracts, and creates suction that can fill a microcrack. If a dangerous process solution is present during cooling, it will fill this space and be trapped there. Remember that a blind crevice, once contaminated, cannot be flushed out or washed by external rinsing. Even open capillaries cannot always be cleaned in the normal cycles used in the electronics industry.

In reviewing Table 6-6, we see that there are two major areas of concern. The first is the fluxing operation: excess flux must not reach any potential trouble spots. Judicious flux application can usually control this danger, which does not occur until the transfer stage. The period after soldering and before cleaning (the transfer stage), is the only cooling cycle in which excess flux can be pulled into a capillary.

Table 6-6. The Temperature Profile in Wave Soldering and Cleaning.

OPERATION	TEMPERATURE CHANGE	VOLUME CHANGE	DANGER	COMMENTS
Fluxing	None	None	Yes	If flux reaches a potential entrapment spot, it will pose problems later.
Preheating	Up	Positive	None	
Soldering	Up	Positive	None	
Transfer	Down	Negative	Yes	From excess fluxing at an entrapment spot (see Fluxing).
Cleaning	None[a]	None	None[b]	
Drying	Up	Positive	None	

[a] It is very important that the work be no hotter than the cleaning liquid.
[b] Only if cleaner and work temperatures match.

If the temperature of the work matches that of the first cleaning stage or is cooler, no danger exists. If there is any doubt, or if some components may be especially vulnerable, make sure that the first rinse water has the best purity level available (not the worst). The heat profile inside the washer must also be positive (going up), or at least uniform.

If we keep these simple relationships in mind during soldering and cleaning, we can keep leakers and encapsulated component from absorbing liquids. We must maintain a positive temperature profile in which no component cooling is allowed during the critical stages. This guarantees the exclusion of soldering chemicals (oil, flux, etc.) and washing liquids (water or solvent base).

Encased and exposed components can be washed, provided they do not trap liquids in recesses or the enclosure (Figs. 6-13 and 6-14). It is advisable to look for *drain passages* when determining the washability of a design. In this category, the undesirable removal of lubrication is an important consideration. Judicious selection of the cleaning medium is the only practical approach.

Fig. 6-13. Typical encased and exposed components, many of which have been washed for years without apparent trouble.

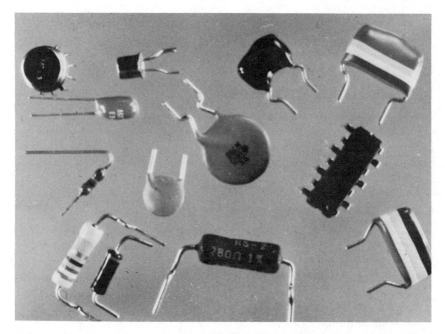

Fig. 6-14. Only two of these components (the metal cap transistors) are truly hermetically sealed. The others are only encapsulated, but definitely washable.

6-19 PRINTED CIRCUIT CLEANING

Good cleaning conditions include far more than component compatibility with the process. The design of the assembly is just as important. Let us take a printed circuit board as an example.

The most difficult area to clean on the printed circuit board is underneath the components and subassemblies. These areas are partially shielded from the cleaning fluid flow, and thus cannot be readily flushed. This problem is aggravated by the geometry of the gap created between the component and the board. This space may retain liquids or may be entirely blocked. In either case, total residue removal is impeded.

The cleanability of this critical gap depends on a variety of factors as follows:

1. Gap height and geometry.
2. The spacing of adjacent hardware.

3. Contamination and cleaner chemistry.
4. Cleaning solution surface tension or flow.
5. Equipment parameters.

These vary from case to case, and no generalities can be made. Let us consider them individually:

1. *Gap height and geometry*: There is a definite relationship between the component's height above the board and the ability to clean. Unfortunately, this is not an independent function, but let us discuss one specific case.

A study was made of cleaning an organic water soluble flux from a printed circuit board using an inline washer with the following stages:

- A prerinse
- An air knife
- A 0.5% saponifier wash (145°F–63°C)
- An air knife
- A soft water rinse (160°F–71°C)
- An air knife
- A final soft water rinse (175°F–80°C)
- Final air knives
- Flash heat dry

In this investigation, a modified industrial inline washer with standard spray nozzles was used. Soft water was available, and proved adequate. The standard ionic contamination test was used to establish cleanliness, using MIL-STD-28809 levels (Ref. 6-8).

A relationship between component height (gap) and component width was established through this test. Table 6-7 gives the results.

Table 6-7. Cleanability of a Gap.

CLEANING HEIGHT		DEVICE WIDTH[a]	
(in.)	(mm)	(in.)	(mm)
0.002″	0.05	0.006″	0.15
0.003″	0.076	0.010″	0.254
0.004″	0.10	0.025″	0.635
0.005″	0.127	0.250″	6.35
0.006″	0.15	1.50″	38.1
0.007″	0.178	Over	

[a] Not obstructed in cleaning direction.

Note that component width, not component length, was critical here. The smaller dimensions were simulated to provide data on round components, busbars, and shields. Extreme caution was also used to provide clear excess to the spaces in the critical direction, the width of the component—front and back. The front is obviously where the cleaning liquid enters. The component rear must be free to avoid liquid backup. These tests should be taken only as a guide; each organization must asses its own cleaning conditions. A similar test is planned for surface-mounted components with adhesive anchorage.

2. *Spacing of adjacent hardware*: This is also an important issue. If the cleaning steps are impeded by the contours of the surrounding hardware, good soil removal is obviously difficult. Liquid penetration alone is not sufficient; circulation and replacement are also vital. In this case air knives help to clear out the previous solution, thus enabling a purer liquid to remove more contamination (Fig. 6-15).

3. *Contamination and cleaner chemistry*: These factors are of major importance. We must make sure that all of the materials making up the dirt dissolve in the cleaning liquid. The key phrase here is *all the materials making up the dirt*, not the flux used. In the study quoted above, a 0.5% saponifier solution was used even though the flux was totally water soluble. The intent was to remove any accidental dirt, like finger-prints,

Fig. 6-15. An aerodynamically designed air knife used to blow off retained rinse water from component side of board. (Courtesy of Hollis Automation)

grease, or pollutants, which is not water soluble. Many of these types of contaminants are carried into the process on the components and boards, which can be received already contaminated. The saponifier used also tends to neutralize any residual organic acids.

Only recently has solvent-based inline equipment been made available that has good cleaner circulation and does not rely only on vapor condensation and dipping. This is important given the renewed interest in rosin based fluxes for surface-mounting of components.

4. *The surface tension of the cleaning solution*: This is obviously a critical factor when the process is based on a static cleaning mechanism. However, with today's inline cleaning equipment, the disadvantages of surface tension combinations are compensated for by dynamic cleaning methods. From a practical point of view, solvent efficiency and economy are much more important.

5. *Equipment parameters*: All cleaning is based on the use of progressively purer solutions, and the equipment determines the volume and dynamics of the process. Such parameters as spray direction, volume, pressure, and distance from the work are critical. They offset the importance of the cleaning solution's surface tension and the length of the equipment.

Cleaning of densely populated printed circuit boards is a daily process for many organizations. With proper care, both water and solvent systems can be used. The cleanliness of the end product, however, depends on many postsoldering and cleaning steps. For an overall view of the entire line, see the next section.

6-20 WHEN TO CLEAN AND A CLEANLINESS PHILOSOPHY

The electronics industry has yet to learn to *think clean*. We must look to the semiconductor and hybrid industries, although our needs are not as stringent. As stated earlier in this chapter, the need for cleanliness is growing as assemblies become miniaturized. The new fine-line printed circuit boards and the surface-mounted devices are catapulting us into an era where cleanliness will be a must. Thus an all-encompassing *cleanliness philosophy* will be needed.

Total cleanliness of the assembly is just as important as cleanliness of the subassemblies after each manufacturing step. As an example, let us review the needs of an average electronic product containing printed circuits. The critical areas where the board and components can become contaminated in the facility or at the user's location are:

1. Incoming materials and receiving inspection
2. Storage and handling
3. Assembly
4. Wave soldering*
5. Touchup* and add-on
6. Inspection and testing
7. Final assembly
8. In service
9. Possible field repair

Remember that dirty parts carry contamination into the process. This contamination may not be compatible with the cleaning processes in production. Let us review each step and discuss the pitfalls involved.

1. *Incoming materials and receiving inspection*: The cleanliness of incoming materials is related directly to their solderability and shelf life. Dirt left on by the vendor and contamination introduced by incoming procedures must be considered. Unless the parts are used on receipt, any contamination will affect their long range quality. Consideration should be given not only to surface contamination but also to inner layers such as beneath the surface of the solder mask (Ref. 6-3).

2. *Storage and handling*: In house storage must continue to preserve the integrity of the supplies. Detrimental chemicals such as sulfur fumes should be excluded, or solderability will suffer. The parts must also stay covered to prevent films of sundry organic materials and dust from settling and hardening on the surfaces.

The same holds true during handling operations. Kitting, transportation, and intermediate storage. These steps are also vulnerable to contamination buildup.

3. *Assembly*: During assembly the work is exposed to equipment and human handling. Lubricants, perspiration, and a whole range of contaminants can build up. While the dwell time of this dirt on the surface may be too short to affect solderability, these materials may not be compatible with the flux cleaning system.

4. *Wave soldering*: The wave soldering operation is the easiest to control, since we have jurisdiction over the soldering chemicals. These can be easily matched with the postsolder cleaning process.

Industrial and government specifications regarding cleanliness levels after soldering are available. However, as the diagram in Fig. 6-16 shows,

---

* Today only step 4 and part of step 5. receive serious attention.

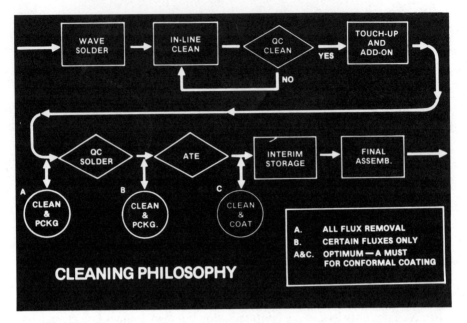

Fig. 6-16. Diagram showing the alternatives for final cleaning in flow chart form.

this is not the final production step. It is an important evaluation only in reference to process control, to make sure that the soldering and cleaning operations are performing properly.

5. *Touchup and add-on*: Touchup by itself can still be coupled with a postsolder cleaning process. The repaired work can be channeled back to the wave solder cleaning unit.

Often touchup is combined with the addition of nonwettable devices in a so called *add-on* operation. The problem here is that these components cannot go through the original post-solder wash, since they are sensitive to the process. Therefore, they cannot be cleaned after hand soldering unless the hand-held unit described in Section 6-17 is used.

As a result, flux selection for this process is critical. Two alternatives exist:

a. The use of a safe but easily washable formula.
b. The application of a flux that can be left on the board (RMA or R).

After the addition of nonwashable devices, the use of gloves or finger cots should be considered. It is recognized that they encumber the operator and increase costs. (The gloves get dirty and must be frequently replaced.) They may, however, provide the simplest solution to the problem. This is especially true for steps 6 to 9.

6. *Inspection and testing*: Inspection and testing are seldom believed to contribute contamination. However, unless they are performed under clean room conditions, they will add to the dirt buildup on the assemblies.

7. *Final assembly*: Final assembly involves very little soldering and is usually mechanical. What makes this step unique in most cases is our inability to clean the finished product.

8. *In-service*: In-service use of the equipment poses a series of problems. The manufacturer has no control over the environment in which the end user will operate. Sealing the unit, however, is not always feasible. Electronic devices generate heat, and convection or forced air cooling is often used to maintain a reasonable internal temperature. As a result, there is a continuous buildup of dirt on the surfaces, as depicted in Fig. 6-17.

The use of conformal coatings, sealing sensitive portions of the assembly, and filtering of cooling air are just a few possible solutions. They must be coupled with a design plan to conduct unwanted heat out of the critical areas.

Fig. 6-17. Dirt accumulated on a printed circuit board in normal household use.

9. *Possible field repair*: Possible field repair creates another potential danger point. Customer service is often performed with portable equipment in the users location. Careful training and indoctrination are needed to achieve contamination control at this stage.

REFERENCES

6-1. Howard H. Manko, *Solders and Soldering*, 2nd ed., McGraw Hill Book Co. New York, 1979.

6-2. Carl J. Tauscher, *The Contamination of Printed Wiring Boards and Assemblies*, Omega Scientific Services, Bothell, WA., 1776.

6-3. Howard H. Manko, "What Is Clean in Electronic Assemblies," *Insulation/Circuits*, July 1982, pp 31-33.

6-4. F.M. Zado, *NEP/CON Proceedings*, 1979, pp. 346-354.

6-5. J. Brous, *Electronic Packaging & Production*, July 1981, pp. 79-87.

6-6. Howard H. Manko, "Solderability a Prerequisite to Tin-Lead Plating", *Plating*, June 1967.

6-7. S. Bernstein and J.M. Green, "Mass Soldering and Aqueous Cleaning Process Operated in Compliance with Water Pollution Regulations," *Electronic Packaging & Production*, March 1981.

6-8. MIL-P-28809, "Printed Wiring Assembly", 1975.

7
TROUBLE-SHOOTING THE PRINTED CIRCUIT

7-0 INTRODUCTION

The origin of fillet imperfections that require touchup, is a complex subject. One must differentiate between design-oriented failures, solderability problems, soldering chemical inefficiencies, and process (equipment) trouble (Ref. 7-1). In addition, workmanship standards and inspection criteria are often the cause of unnecessary touchup (Refs. 7-2, 7-3, & 7-4). A discussion of what requires touchup in any particular organization or industry is beyond the scope of this book (Ref. 7-5). For details on setting up your own workmanship and quality standards, see Section 10-2.

It must be noted that many so called rejects may be considered perfectly adequate solder joints. The reader is referred to Chapter 1 for a design guide. It offers a method to calculate established joints and to generate fillet criteria. Such design reviews show that many accepted inspection practices are erroneous, accentuating cosmetics rather than technical quality considerations. This causes sizable yet unjustified costs for unnecessary touchup throughout the industry. Remember that touchup does not always result in improved reliability (see Chapters 8 and 9).

In the framework of this discussion, we will consider only those imperfections which are directly connected with the soldering (wave, reflow, etc.) process. It is assumed that board design, material selection, and presolder processing are all within acceptable limits, and therefore did not contribute to the failure. The comments are intended for use with established production lines where background information is available. They can also serve to optimize a new process and investigate its faults.

Specific problems associated with the soldering of printed circuits are given below, together with some suggested solutions. While many problems in this industry fall into repetitive patterns, the variations are endless. It would be presumptuous on the part of any author to prescribe fixed solutions. The following discussion is based on years of experience, but the reader must still solve his own problems.

7-1 TROUBLE-SHOOTING OUTLINE AND THE RESTORATION OF SOLDERABILITY

When a problem occurs, the first check must focus on the basics of the process. Let us remember that there are three major categories of causes:

1. *Material problems*: These include soldering chemicals like flux, oil, solder, and cleaning materials. As well as printed circuit surface coatings like protective coatings, temporary or permanent solder mask, markings, and the like.
2. *Lack of Solderability*: This refers to all surfaces, components (including surface-mounted devices), and printed circuit boards. The quality of the plated-through hole, if one is used, must also be considered.
3. *Process parameters deviations*: Deviations in equipment settings, maintenance, and external conditions. Temperature, conveyor speed and angle, depth of immersion, and similar parameters are directly machine oriented. Drafts, air pressure drops, voltage variations, and other such external conditions must also be included in the analysis.

Each problem is unique and no generalizations can be made. The following is a recommended logical sequence of checks that should help locate the source of the problem:

Step 1. The machine parameters are least likely to change; therefore, check them first. Use independent instruments for this purpose. For example, use an external thermometer to check temperatures and verify the accuracy of the machine's dials. Work from the list of optimum conditions which was prepared when the equipment was peaked or installed.

NOTE: *Under no circumstances should machine settings be adjusted to overcome a temporary problem: this will lead to more trouble.*

Step 2. Next, check all soldering chemicals. Remember that flux density, clarity, and color can be augmented with an ionic content check. Monitoring of solder purity should be an ongoing program, but a spot check will help. Refer to Chapter 8 for in-process quality suggestions and to Chapter 3 for a discussion of materials.

Step 3. The solderability of the leads or printed circuit boards is the most likely trouble area. Study the pattern of the problem to isolate any possible offenders. The probability of having more than a few groups of leads fail at the same time is small. Repeated problems on the same component can be corrected by obvious means.

Solderability restoration is a function of the surface materials (see Sections 2-5, 2-6, and 2-7). Try degreasing to remove dirt first. If the problem persists, you may have to use special chemicals or return the component to the vendor. See Chapters 8 and 10 for further details on solderability management.

Remember that plated surfaces need more specific treatment. Often it is the interface between the plating and the base metal that causes the problem. In retinning such surfaces, a double dip is mandatory.

The solderability of printed circuit boards can be restored by chemical means, which are quite drastic. A more recent process—air or liquid leveling—can achieve the same results with less chemical damage, although additional heat exposure is necessary. You can always try to clean the board first. Often degreasing alone will help. Sometimes water washing with a saponifier is required.

Step 4. Check the quality of the plated-through hole for signs of plating or hole generation (punching or drilling) defects. This is best achieved with magnification and back-lighting. The hole must be clean and smooth, with no visible nodules or cracks. Plating thickness should also be checked; see Section 8-9 for details.

The problems resulting from bad holes are not limited to poor filling or solder rise. Blow holes, cracked barrels, and trapped chemicals are also part of this problem.

7-2 POOR WETTING AND NONWETTING

Soldering depends on wetting for its properties. The quality of the solder joint can be equated with the perfection of wetting. Basically, solder only wets bare metallic surfaces that are not covered by tarnish (oxides, sulfides, etc.) or other nonmetallic contaminants (dirt, organic coatings, etc.). Imperfection in wetting can thus be translated into imperfections on the surface.

Poor wetting or nonwetting must be separated from dewetting, by the mechanism through which the phenomenon occurs. *Nonwetting* is a condition where the solder coat is not continuous, and well-defined areas of base metal show through. This problem is especially easy to see in the case of copper (Fig. 7-1). The reddish base metal is plainly visible through the tin-lead coating. Notice also that the outline of the base metal is normally rounded, this is due to the cohesive force of the molten solder.

Fig. 7-1. Nonwetting on copper surface. Note the rounded outline of the exposed copper areas, which is due to the cohesive forces of the molten solder. (Copyright Manko Associates)

Dewetting on the other hand, occurs after wetting had taken place (Fig. 7-2). The metallurgical inter-reactions at the wetted interface change the energy conditions there continuously. As the system cools, wetting energies decrease as cohesive forces increase. The liquid solder pulls back and solidifies as beads with a large dihedral angle. The surface left behind has minute quantities of tin and or lead reaction products with the base metal that look like tinned areas. Under very high power magnification, there are indications that dewetting is usually accompanied by many minute areas of nonwetting, and vice versa. In this book, we will handle the two conditions separately. For details on dewetting, see the next section.

Both poor wetting and nonwetting are intolerable. They reduce dramatically the strength and ductility of the solder joint. They also diminish electrical and heat conductivity through the fillet. Unfortunately, there is no known mathematical method for calculating the degree of imperfection created by poor wetting. Nor is there a meaningful measurement to indicate the percentage of coverage. Thus, by default, all the defects in this category are rejected.

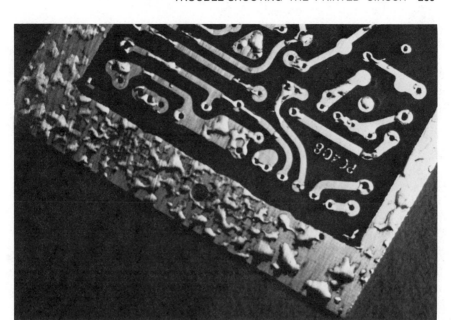

Fig. 7-2. Dewetted condition on surface. Note that the solder pulled back into beads and solidified. With any external force (leveling, slinging, shaking, etc.) the solder beads would be removed, since adhesion is very poor.

Stated differently, poor wetting normally results from the inability of the flux to prepare the surface for good solder adhesion. Less frequently, soldering conditions like time or temperature are not conducive to wetting.

Defects in this category are normally caused by one or more of the following conditions:

1. *Foreign contaminants*: These can be present on either the component or the board, and include such materials as oil, grease, paint, wax, and the like. These materials are normally referred to as "dirt" and are removed by suitable cleaning methods as follows:

- The more traditional solvents in vapor degreasers have been used successfully for this purpose. Make sure, however, that a suitable electronic-grade formula is used. Some commercial chlorinated solvents tend to leave harmful residues behind.
- More recently, with the advent of water wash systems for flux removal, the same equipment has been successfully used for the same purpose.

- Some foreign contaminants originate from the solder mask (resist). Sometimes they are due to poor screening techniques or misregistration; in other cases, the resist bleeds during curing. Cured plastic coatings are not easy to remove without damaging the rest of the board. Consult your vendor and his suppliers for details. Be careful if abrasion is used; it is a popular technique but may embed abrasive particles in the surface. For more details, see item 2 below.

2. *Embedded particles*: Foreign materials embedded in the surface are another widely documented cause of wetting defects. When using pumice or scouring powder on a soft metallic surface, it is easy to implant the hard material into the top layers. These nonmetallic particles obviously do not wet with solder. Nor are they removed by the fluxes used in the printed circuit board industry. The same applies to brushes and woven materials that have the abrasive incorporated in the structure.

- The only good remedy consists of etching the surface layers to remove the embedded particles by removing the underlying metallic layers. The etching materials are strong chemicals that require careful control and good rinsing.

3. *Silicone oils*: Silicone oils, although a foreign contaminant like those in item 1 above, are treated separately because of their unique nature. Silicone compounds are used as lubricants because of their tenacious adherence to surfaces even in very thin films. There is no known solvent that can remove silicone contaminants once introduced, hence their incompatibility with the soldering system, where they are considered a poison to solder wetting.

Sources of silicone contamination are numerous. One persistent source is the plastic bag, which is often made with silicone as a mold release. Today a whole range of safe plastic packaging materials are available. Another common source of silicone is the heat transfer paste often applied before soldering. Even if the silicone is not used near the soldering station, it has a tendency to spread throughout the process and eventually creep there.

- There are presently no known cleaning processes for most silicone oils. Thus the solution is diligent housekeeping and strict policing to minimize the problem.

4. *Heavy tarnish layers*: Layers that cannot be penetrated by the production flux are another source of wetting defects. Tarnish is normally

formed on all metallic surfaces, and a reasonable amount can be over-come by flux. The problem comes from excessive amounts formed during improper storage and/or previous manufacturing steps like burn-in.

Some simple solutions to this problem are listed, in no special order:

- Usually a stronger flux can overcome the wetting problem. Its use, however, may not be feasible for the entire assembly on the production line. Flux strength may be limited by specifications or technical reasons. In this case, one of the alternative corrective actions listed below must be taken.
- Pretin the offending parts with a strong flux and wash prior to assembly. Halide-free fluxes have been developed for this tinning process. This is acceptable even under strict government specifications.
- Treat the tarnished surfaces with a descaling solution, usually an etch. Rinse to remove all traces of chemicals and solder soon thereafter, before additional detrimental tarnish is formed during prolonged storage. Remember that the standard production flux can penetrate normal tarnish layers.
- An improper fluxing process can also cause poor wetting conditions. Spotty or nonuniform flux application can be rectified by equipment adjustment. Contaminated flux can cause the same problem.
- Insufficient time or temperature in preheating prevent the flux from reaching full activity levels. An increase in either time or temperature, will improve the situation.

5. *Solder alloy*: Solder too may not have the right combination of conditions (time, temperature, etc.) to wet properly. Remember that the melting point of the solder is not a good soldering temperature. A gradient of 100–150°F (55–80°C) above the melting point is required for good wetting.

To check on the solder try:

- Check solder composition and purity by analytical methods.
- Establish all soldering parameters with an external instrument (temperature, time, etc.).

7-3 DEWETTING

This defect is also an unacceptable quality condition, very similar to the poor wetting described in Section 7-2. In this condition, solder wets the surface initially, but after a period of time pulls back from the base metal interface (Fig. 7-2). The cohesive forces of the solder pull the liquid into droplet shape. The surface from which the molten solder recedes is no longer just base metal. During the wetting process, certain metallurgical reactions take place. Alloying and intermetallic compound growth may occur depending on the metallic system. This would be the case with copper, silver, and gold, for instance. Thus the surface left behind has a tinned look but is really a nonsolderable thin alloy and intermetallic compound layer. The reasons for dewetting are many. It is usually due to a large number of occlusions imbeded in the surface or the presence of contamination layers. The reasons for dewetting of plated surfaces are discussed below.

Dewetting in printed circuit boards is often associated with tin-lead electroplated deposits. Here the problem is not with the top surface, but with the base-metal to the plating interface. This is the condition depicted in Fig. 7-3. A similar mechanism can be observed in other plated metals (Ref. 7-6). Because the cohesive forces are responsible for solder pull-back, it is extremely important to make sure that all solder fillets are shallow. Excess solder will mask this condition because of the hydrostatic pressure of the heavy solder.

Fig. 7-3. Solder dewetting from a tin lead-plated printed circuit board. Note that the phenomena is affected by solder quantity where large amounts of the molten metal do not permit beading. The hydrostatic pressure is larger than the cohesive force, which can lead to misleading interpretation. In this case, the solder joint on the top right failed in service, separating from the board. (Copyright Manko Associates)

Resoldering of dewetted surfaces does not help. The contamination that interfered with the wetting process is covered with the beaded solder in many areas, preventing flux action.

Some suggestions for corrective steps are as follows:

- Strip the solder off the surface and clean underneath before re-soldering is effective. This is accomplished by:

1. Chemical stripping with solutions that do not attack the copper or board materials.
2. Leveling the board (melting the solder and stripping it with an external forces like a hot air knife or oil stream) and resoldering it with an aggressive flux. Two passes may be needed for full success.
3. For component leads or terminations, leveling is impractical, but multiple dipping or passes through the wave are effective.

7-4 WETTING PROBLEMS IN SURFACE MOUNTING

There are two problems associated with wetting and surface-mounting. The first stems from the small size of the solder pads on the printed circuit. The second deals with the interpretation of solderability on glass fired terminations. In all other respects, surface mounting is no different from regular components when it comes to wetting.

1. The problems associated with the small size of the pads are as follows:

a. The small size of the pad defies proper wetting interpretation if too much solder is deposited. This is a problem with preplaced solder and flux applications, for example.

b. There is a problem with covering all the small pads with solder by dipping methods, a phenomenon also called *skipping* or *shading* (see Sections 5-14 and 7-28). This may be the result of a relatively thick layer of solder mask surrounding the pad and trapping gases. It may also be a factor of the wave.

The solutions here depend on the identification of the problem source. Try the following:

- Inspect adjacent larger areas for solderability or remove any excess solder before inspection. Use a solder removal tool (wick, suction, etc.) or dip the board in a hot medium (solder, oil, etc.) and level manually. Leveling a board during testing can be done by slinging (like lowering the mercury in a thermometer) or acceleration (slapping the board or hand against a work bench). Be sure to hand level while the solder is still hot enough, and allow the residual solder to solidify without vibrations. Use high magnification (10-30X) for the interpretation of any restricted areas.

- If no large testable areas are available on the raw board, design a special test pattern on its edge. This portion is broken off for this and other tests, and will have the same ability to wet. Use standard solderability test methods for evaluation.
- See also the solutions to the skipping and shading problems discussed in Section 7-28.

2. Problems of interpretation of solder wetting on silver-fired surfaces (thick film terminations, like those on leadless chips) are due to the unique nature of the material. Thin films of solder are very rough in appearance and seem to "dewet". They are in reality well-wetted surfaces, only their appearance is misleading. Here a distinction must be made between bare surfaces in the as-fired condition and barrier plated (nickel) and tinned terminations. In the first instance, this problem prevails; in the second, it is not normally found.

If you have this problem, try the following:

- Use a destructive evaluation of the solder joint formed. Solder a representative sample of devices to a test board with an actual production pad pattern. Then pry the components off and examine the failure location. Wetting is good if:

    1. The joint failed in the solder
    2. The failure was between the fired film and the substrate
    3. The pad was pulled off

    Wetting was poor if the solder separated from the surface in question. This may also be a problem of silver scavenging (leaching), as described in Sections 5-5 and 7-24.
- Evaluate the components in a modified wetting balance (meniscograph). Slow immersion speeds and expanded sensitivity ranges are claimed to yield acceptable results. The author had no experience in this area at the time of writing.

7-5 DISTURBED (COLD) SOLDER JOINTS

This condition is defined as a fractured, uneven solder joint. It occurs when components move in relation to the printed circuit board during the

Fig. 7-4. Disturbed solder joint formed due to component movement during solidification. (Copyright Manko Associates)

freezing stages of joint formation (Fig. 7-4). The motion disturbs the orderly growth of the solder alloy grains, weakening the structure. In the worst case there are many micro-fissures and a fractured joint.

This defect forms with vibrations that can originate from several sources:

- The conveyor belt.
- Bearings, pumps, etc., that are not balanced.
- The ventilation or exhaust fan.
- Solder that is still molten at the conveyor exit.
- The touchup operator.

The condition is corrected by ensuring smooth transfer until the solder has entirely solidified. Reflowing the solder joint will correct this fault on touchup.

Signs of component movement and disturbed solder during freezing should always indicate doubt about the integrity of the joint. This problem requires a statement in the quality manual such as: *The solder fillet (for Sn63, Sn60, Sn62, etc.) surface must be smooth and ripple free.*

7-6 INCOMPLETE FILLETS, UNFILLED HOLES, AND POOR SOLDER RISE

Incomplete fillets are given a variety of names throughout the industry, ranging from "blow holes" and "pin holes" to "drop-outs" and "empties". In this book, we will discuss these problems separately, coin our own terms, and define them for clarity. Thus an *incomplete fillet* is one that either does not wick all the way through or does not wet all the surfaces provided for the joint. We must divide these into two categories according to the spatial direction of the fault:

1. Unfilled holes. The solder did not fill the hole all the way around (360°). This happens in single-sided, double-sided, and multilayer boards. (Fig. 7-5).
2. Poor solder rise. The solder does not rise to the top of the board. This happens only in boards with plated-through holes (Fig. 7-6).

If a new board design is soldered for the first time, the problem is much more complex. A thermal analysis is probably needed to understand the

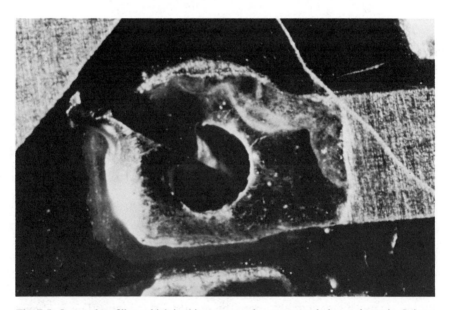

Fig. 7-5. Incomplete fillet, which in this case was due to a poor hole-to-wire ratio. It is an acceptable joint for this commercial application. (Copyright Manko Associates)

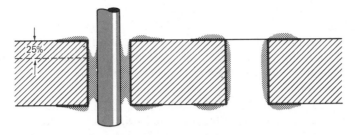

Fig. 7-6. Schematic of solder rise in a plated-through hole. In most cases 25–30% sag is permitted, provided there are signs of good wetting on the top of the board. Unused (via holes) do not have to be full, provided they too show signs of good wetting all around.

problem. When an established production line suddenly shows these problems, remedial action can follow the outline given below.

This discussion is predicated on the assumption that the machine and material were checked for such parameters as temperature, speed, flux, solder, and so on. If no deviation is found, try to identify the problem from the following list:

Unfilled Holes: The most common causes are:

- Solderability of components or the printed circuit board.
- Cleanliness of the plated-through hole.
- Solder resist misregistered on the hole.
- Large hole-to-wire ratio.
- Uneven wave or conveyor vibrations.
- Poor (warped) fixturing.

Poor Solder Rise: Check:

- Solderability of the component or printed circuit board.
- Cracks or nodules in the plating of the hole.
- Contamination in the hole.
- Misregistered solder resist.
- Flux starvation.
- Overheated flux.
- Weak knee (see Section 7-29).

If any of these problems recur only on specific lots of components or boards, check to see how they differ dimensionally from one another. They may also have material- or process-oriented differences.

Problems that are repeated on specific parts of the board are probably heat balance (design) oriented. The thermal requirements of the joints may be different from those of the rest of the board (see Section 1-19) and cause this fault. In this respect, remember that multilayer boards require special preheat, often even from the top. They are more vulnerable to this category of rejects.

The question of whether all holes must be filled and the issue of the height of the solder rise are subject to calculation and covered in Sections 1-5 to 1-10. In many cases, design considerations may make touchup of these faults unnecessary. This should be reflected in your own quality standards (see Section 8-2).

Such organizations as the Institute for Interconnecting and Packaging Electronic Circuits (IPC) (Ref. 7-7), and many government specifications permit a 25% sag (Fig. 7-6). In other words, the solder can drop back down 25% of the thickness of the board, provided that there is evidence of wetting on the top of the fillet. This is an indication that the solder wet but could not be retained on the component side of the board. In the case of multilayer boards the solder must always rise to the top layer for good internal layer interconnections.

7-7 EXCESS SOLDER

We define *excess solder* deposited on a joint when it obscures the fillet contours (Figs. 7-7A and 7-7B). This excess, in turn, can hide wetting deficiencies on the lead wire or board. It may also conceal the absence of a wire in the joint. Thus both board and lead wire contours must show. Finally, excess solder does not add to joint strength or conductivity and is wasteful.

Each organization should have its own workmanship standards that define the maximum quantity of solder per joint. If your company has none or you have an outdated specification, it is easy to generate a new one (Section 8-2). In general, however, the fillet surfaces, top and bottom, should be concave, and shallow.

The usual causes of excessive solder are:

- Incorrect depth of wave immersion.
- Preheat or solder is too cool.
- Flux starvation.
- Marginal solderability of component or board.
- Incorrect wave exit angle or speed.
- Insufficient oil in the intermix.
- Contaminated solder.

Fig. 7-7A. Excessive solder joint in center. Note that the contour of the printed circuit pad is not visible. (Copyright Manko Associates)

Fig. 7-7B. Excessive solder joint in center, here neither the pin nor the printed circuit pad contours are visible. (Copyright Manko Associates)

Corrective measures can eliminate the problem, and it may be desirable to resolder work showing this reject. For an effective rework method without heat damage, let the boards relax and regain resin strength for 4-6 hours. Only then should the boards be sent through the adjusted process for a second resoldering operation.

Excess solder joints must be considered a possible reliability hazard. They can readily mask poor wetting conditions and similar defects. They may also obscure lead outlines and assembly faults. There is no justification for permitting them to be used.

7-8 ICICLING

The term *icicle* is very descriptive of the defect (Fig. 7-8). The solder freezes in the process of draining from the wetted surface. The resulting icicles are usually conical, with a sharp point. They occur in wave soldering, dipping, and hand soldering. They can be found on printed circuit traces or pads, on the component lead wire, and even on individually dipped parts.

We must seek the causes in the methods of soldering. They are:

1. *Hand soldering*: Icicles that form when a soldering iron is used normally appear as little flags at the point of tip contact. In this case there is a thermal balance problem of too little heat.

The solution, however, does not call for a hotter iron, but rather a bigger heat supply (more calories) at the same lower temperature. This is

Fig. 7-8. Undesirable icicles. The cause here was poor drainage of a large land area. (Copyright Manko Associates)

Fig. 7-9. This icicle includes an adjacent lead wire. It is caused by a large ground plane and long leads. Redesign of the board was needed to eliminate a high probability that this fault would occur on the same board location. (Copyright Manko Associates)

normally obtained by using an iron with a bigger heat capacity, or a tip with a larger working contact area. A clean tip, adequate flux, and good soldering techniques also help (see Chapter 8, of Ref. 7-8.

2. *Wave soldering and dipping*: The causes of icicles with automated soldering methods are more complex. In addition to thermal balance, there are the effects of solderability, design, and equipment. Let us look at some specific factors:

a. Thermal balance

- Insufficient temperature output of equipment.
- A large ground plane acting as a heat sink.
- A heavy component draining heat from the joint.

b. Solderability

- Dewetting of the cooling surface.
- Poor solderability.
- Flux skips or starvation.

c. Design

- Wrong hole-to-wire ratio.
- Large, empty plated-through hole.
- Poor drainage on large lands.

d. Equipment

- Board too deep in wave.
- Rough wave exit.
- Dross on dip pot or in wave.

The cure for icicles depends on the source. Thermal and equipment problems are easily overcome by resoldering using corrective measures. Design problems must be touched up and redesign initiated. Solderability problems, however, must be handled in a special way (see Section 7-1).

The impact on reliability is also important. It signals problems in soldering that should be corrected. In addition, the spikes formed may become a problem in high-voltage circuits. They can also cause shorts with adjacent boards. Finally, they pose a hazard to the employees who handle the boards. During touchup, hot removal of the icicle (with a soldering iron) is preferred to mechanical cutting.

7-9 BRIDGING

In *bridging*, we have another condition caused by excess solder, where adjacent metallic conductors are shorted out (Fig. 7-10). This condition must be separated into problems of exposed conductors (bridging), and those that occur under the solder resist (see Section 7-10).

Bridging is usually caused by the following problems in design, materials, or equipment:

Design

- No drainage on conductors that converge.
- Exposed circuit runs which make sharp bends.
- Conductors too closely spaced in the direction perpendicular to the wave.
- Component leads bent erratically or too closely spaced.

Material

- Solder slivers (over-etching).
- Copper traces (under-etching).

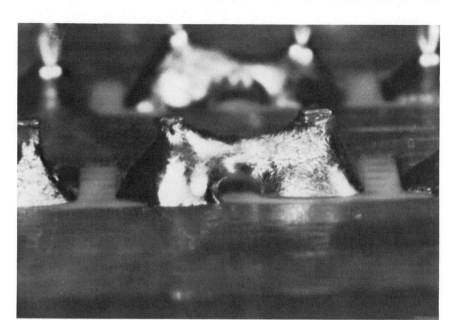

Fig. 7-10. This bridge was repeatedly formed in the same location (note second bridge in background). Investigation showed this part of the wide board sagged during soldering, which brought these IC legs too deep into the wave. (Copyright Manko Associates)

- Poor solderability.
- Flux starvation.
- Contaminated solder.

Equipment

- Low heat balance.
- Pumping of dross (Maintenance problem).
- Too deep in the wave.

While the corrective measures depend on the exact cause of bridging, touchup is needed for their removal. Use some external flux and follow the instructions given in Chapter 9 of this book.

7-10 SOLDER AND COMPONENT SHORT CIRCUITS

The term *short circuit* is usually abbreviated to "short" and refers to all total short circuits in an electrical system. Shorts are normally found under the solder resist (mask) or components. They should not be confused with bridges (see Section 7-9). A total short is an absolute failure; it can be identified and corrected easily. It is readily detected with automatic test equipment. When the short circuit is intermittent and activated by thermal expansion, vibration, or shock, it is more difficult to locate.

This failure occurs in several locations and should be viewed accordingly:

1. *Shorts under the solder mask (resist)*: These are usually the result of too much solder plating on the metallic lands of the board. Even if the solder is reflowed later, the quantity rather than the treatment is important. Remember that this solder will melt during wave application, mostly on the bottom side. As it melts, the internal stresses of the mask or resist cause shifting of the solder. This is visible as a crinkling of the solder under the resist. During this shifting, molten solder can be squeezed in between adjacent metallic conductors and cause shorts (also see Fig. 7-39). This has been observed more frequently on closely spaced fine-line boards for surface-mounting.

Some solutions to this problem are as follows:

- Less plating before reflow or solder leveling are effective in reducing the problem. Using one of these means, we can reduce the amount of solder under the mask.
- Larger line spacing would also help but is seldom feasible.
- Solder resist over bare copper circuitry (SMOBC), as in additive boards, offers the best and most elegant solution.

Repair involves the removal of the resist, which is mostly mechanical. Special care must be exercised not to damage the board. The solder can then be removed molten in one of the standard ways (see Chapter 9). Recoating the surface with resist should not be necessary.

2. *Shorts under mounted components*: This can be a design problem or a process oriented failure.

Design problems

- Exposed traces located too close to the top of a solder fillet.
- A metallic component body or lead wire placed too close to an exposed board land.
- Two components or bare lead wires touching (Fig. 7-11).

Process-oriented problems

- A pulsating wave.
- Boards that are immersed too deep in solder.

Fig. 7-11. Short created by two component leads touching. (Copyright Manko Associates)

- Outgasing during soldering (Section 7-17).
- Solder ball formation from cream or wave (see Sections 7-14 and 7-15).

Shorts underneath components are difficult to detect for obvious reasons. Presently, only x-ray techniques,which are slow and expensive, can be used. Once located, the short requires component removal to repair.

7-11 ELECTROMIGRATION, AND TIN WHISKERS SHORTS

Some short circuits occur by direct metallic growth on the surface or through the bulk of the printed circuit board material. These fall into two categories:

1. *Electromigration*: This is a phenomenon in which metallic filaments grow from the cathode toward the anode, induced by a direct current (DC) potential. Since the process is based on ionic movement, it requires the presence of a moisture film. It can occur on very clean surfaces, but the presence of foreign ionic contaminants accelerates the process. Of the metals used in printed circuit boards, silver, tin, and copper are most prone to electromigration.

The structure of the filament formed resembles a tree, with multiple branches developing just before contact is established. The amount of metallic conductor that is available to bridge adjacent metallic board lands is minute, and small currents will burn it off.

Electromigration can also take place through the thickness of the board material. This is usually accompanied by high voltages between internal layers of multilayer boards, where each layer is inherently thin. No failures through double-sided boards 0.032 in. (0.8mm) and up in thickness have been reported.

Electromigration can be demonstrated in the laboratory in the following manner. Select two closely adjacent conductors on a clean, bare copper board and trace two convenient contact areas. You may wish to solder wires to these contacts for future use. Clean the board carefully and make sure that there is no ionic contamination. Place the specimen under a microscope or high magnification for observation. Place a drop of pure water (deionized if available) on the test area and connect the contacts to a direct current source of 10 V. The process with plain copper may take about five minutes.

2. *Metallic whiskers*: These have also caused failures in printed circuit boards. When pure thin metallic coatings come under internal stress, this energy may be released by the formation of a whisker on the surface. The whiskers are thin metallic fibers that seem to grow out of the surface, hence the name. Zinc and tin are known to form such whiskers readily. Component failures have been documented from tin whiskers on integrated circuit packages coated with bright tin. This is the reason that pure tin plating is not a preferred coating.

Any thinly coated lead wire comes under stress during the assembly process. Sometimes the heat of soldering will reflow or anneal the coating. The danger, however, is that the coating will stay under stress, and a whisker will short out the circuit with time.

The whiskers can carry substantial amounts of current, and they tolerate enough voltage to damage the devices before they burn off. There are no test methods known today to predict the probability of whisker growth. It is best, therefore, to avoid the use of materials like pure tin.

7-12 CURRENT LEAKAGE

This is a condition closely associated with surface contamination. It involves the slow passage of trickle currents across insulating surfaces. It is usually a function of humidity and is very difficult to locate.

Three major categories are involved as follows:

1. *Ionic materials*: These are operative only in the presence of humidity. They can be detected by instrumentation and extraction techniques (see Section 8-11). For further details on ionizable contamination, see Chapter 6.

Once identified, the remedy is simple. Thorough cleaning in a polar solution will eliminate the problem. A close check on the presence of corrosion products formed before detection is suggested. Details on cleaning are given in Chapter 6.

2. *Nonionic materials*: These are absorbed or adsorbed into the surface. Their effect is also humidity sensitive. They originate from many processing solutions, including the flux, and can be avoided by judicious selection. Unfortunately, they are not easily detectable and require surface insulation testing under humidity. They cannot be identified by the same instrument used for ionic detection.

The cure depends on the vapor pressure of the material and/or its boiling point. These materials can often evaporate on their own with time or can be driven off by baking.

3. *Metallic and other conductive materials*: These include solder smears and graphite from pencil marks (Fig. 7-12) and result from poor

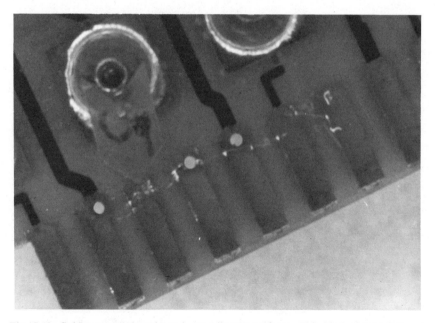

Fig. 7-12. Solder smear that shorted out adjacent surfaces. This kind of total failure is unusual, generally only surface resistance is lowered. (Copyright Manko Associates)

manufacturing practices. The soft solder is easily transferred to the insulation when boards rub against one another in stacking, transportation, and so on. The conductive material seldom shorts totally, but it lowers the insulation resistance.

Special care must be taken to avoid such contamination under the solder mask, where it cannot be cleaned off later. The metallic particles may originate from dirty scouring brushes and physical rubbing of board surfaces against one another. To protect both the supplier and the user, a standard insulation resistance pattern on the board is recommended. Before applying the solder mask, the fabricator can place the board for 8 hours in humidity. An insulation resistance check will indicate whether the board lot is clean. The user can use the same test with a 24 hour humidity soak. The longer time is needed to penetrate through the mask material.

There are no easy cleaning methods to remove such residues. Solder does not dissolve in organic cleaners or water, and graphite pencil marks do not come off by solution. Suitable chemicals may be too aggressive for the board. Mechanical removal with an eraser can damage the surface and spread conductive materials even further. Each case requires careful study. Care and prevention are the only true solutions to avoid this problem.

In general, these problems are also highly design oriented. Only high-impedance boards or high voltage areas suffer from small amounts of leakage. For further details on cleanliness levels, see Section 6-4.

7-13 WEBBING

Solder droplets and strings adhering to the insulator surface between the metallic lands are called *solder webbing* (Fig.7-13). Webbing is found on the bottom only, and should not be confused with solder balls found on top (see Sections 7-14 & 7-15).

There are four general causes of webbing:

1. Improper curing of the resin in the laminate or the solder resist, whichever is on the surface. This results in a surface that softens from the heat of soldering. Sometimes an additional prebake of the boards, either prior to component insertion or soldering, alleviates this situation. The advice of the resin or resist manufacturer should be sought on such a post-cure process.
2. A rough surface resulting from excessive scouring, solution attack, improper application of solder resist (under-curing, wrong formation, air bubbles in screening application),and so on.

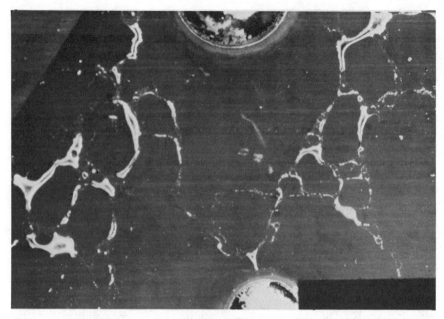

Fig. 7-13. Solder webbing. The solder mask was not cured enough, and solder adhered to the bottom of the board. (Copyright Manko Associates)

3. Contact between solder dross (metal and metallic oxide) and a printed circuit board that is starved for flux (due to a low solid content flux). This is usually an equipment and/or maintenance fault that is easily corrected.
4. Flux starvation. Check fluxer for deposited quantity and/or flux density.

Some corrective measures are suggested above, while others are implied in the cause. The problem of reducing the defects in a lot on hand is separate from corrective measures for the future and from the removal of webbing on existing boards. Try the following suggestions:

1. *Reducing webbing in a defective lot on hand*: Webbing has often been reduced with the use of a heavier flux or the application of a soldering oil to the surface. It appears that more viscous rosin fluxes help to minimize the problem. This is easy to understand in terms of heat transfer. The rosin is a poor heat conductor and provides a protective cushion to the board. Increasing flux density may prove a temporary solution, but it does not solve the basic problem.

Also, investigate the possibility of an additional plastic curing cycle, as in item 1 in the preceding list.

2. *Preventing future problems*: Once the source of the problems has been identified, purchasing and engineering specifications must be amended. Incoming inspection procedures should be generated, and above all, the vendors must be contacted.

Soldering a bare board on receipt, will reveal potential webbing problems. The board may be floated on a solder pot (for 10 seconds at 500°F) or passed over a wave in the production line.

3. *Removing webbing from boards*: This is a difficult, time-consuming, and costly operation. The bottom of the boards must be brushed very gently with a stiff bristle brush and a suitable solvent. Aggressive scouring or the use of metallic brushes can cause a serious lowering of insulation resistance. The solvent should be mild and should contain a polar ingredient to prevent static damage problems. It should also be capable of removing any traces of flux hidden behind the webbing. After flushing the boards with fresh solvent, the remaining metallic particles and dross must be removed with a tweezer.

It is obvious that solder webbing cannot be tolerated on the printed circuit boards. The danger is compounded by the possibility that the metallic particles would eventually become loose. They would thus be in a position to create short circuits in other parts of the equipment.

7-14 SOLDER BALLS FROM WAVE SOLDERING

Unlike solder webbing (Section 7-13), solder balls are found on top of the board (Fig. 7-14). Such solder droplets tend to have a spherical shape due to the cohesive force of the solder. They normally adhere to the surface because of the dried flux. Sometimes they are also embedded in plastic films like the solder mask or the legend inks, which soften when the molten solder touches them.

The mechanism that propels the solder to the surface is very similar to that which generates blow holes (see Section 7-16). The timing of gas generation is different, however; in this case, a large amount of volatiles react faster. The gas is thus generated earlier, while the top of the fillet is still molten. Hence the solder balls form on the top of the board, rather than form cavities on the bottom. For details on the generation of gaseous materials and their sources, see Section 7-17.

Most solder balls are caused by non-dried flux volatiles (thinner) or excess water in the flux. The sudden contact with the molten solder creates volumes of gas and causes solder eruptions. This catapults the solder in the form of balls over the top.

Fig. 7-14. Round solder balls on top of a board. The cause in this case was determined to be water (from factory supplied compressed air) in the foam flux. Note top fillet blowhole. (Copyright Manko Associates)

Many flux formulations have deliberate small amounts of water in their makeup, but this does not cause the problem. If solder balls suddenly occur, one of the following causes should be investigated:

1. Insufficient drying in preheat.
2. The presence of excess water in the formula.
3. A bad plated-through hole (see Section 7-17).
4. High humidity period in the plant.

The excessive amounts of moisture and water can originate from:

- Full drums or cans stored outdoor exposed to rain and snow, where water can accumulate on top due to the stacking flange. Temperature changes tend to pull the water through loosely closed bungs. Cover such storage areas or store the containers on their side. Always tighten all bungs, which may have loosened in shipping (see Sect. 2-4).

- Condensation in partially used containers kept in an uncontrolled environment. When flux is transferred to the machine in safety cans, the air gets into the partially filled drum. The moisture it contains will condense on the walls and contaminate the flux. Special driers are available to prevent this problem. It can also be minimized by closing the drums tightly and keeping them in a dry area.
- A mix-up in containers or a change in flux formulation. Changing flux vendors may also initiate this problem.
- Water in the foam flux from the shop air. The water and oil droplets carried by the compressed air must be removed by a trap and/or filter. Routine maintenance of these devices is vital.
- Wet parts entering the process, from a previous step. This occurs infrequently and is considered a poor manufacturing practice.
- Water carried back by the conveyor to the fluxer from the finger cleaner in water-soluble systems. Make sure that the air knife provided removes excess water.

Repair procedures fall into the same class as webbing, but the brushing is more difficult. The presence of the components makes the removal of solder balls from the top surface very difficult. Care must also be taken to find the balls under components, where they may be hidden (see Section 7-10).

Solder balls may dislodge at any time, creating a serious reliability hazard. Prevention is the only reliable way to eliminate this fault.

7-15 SOLDER BALLS AND OTHER PASTES PROBLEMS IN SURFACE-MOUNTING

Solder balls are one of the three problems associated with the use of creams and pastes. They are caused by inherent materials and process problems. We will discuss only those problems which appear from time to time. For further details on new assemblies, see Section 5-9. The failures are:

- The formation of solder balls.
- Spattering of the paste during reflow.
- Dirty solder residues from paste.

1. *Formation of solder balls.*: In order to understand the sporadic generation of solder balls, we must review the reasons for their formation. In general, solder balls form when a quantity of paste is heated over an

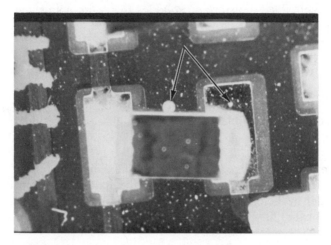

Fig. 7-15. Solder ball from paste which was not removed during cleaning. The ball is held by small amounts of flux residue. (Copyright Manko Associates)

isolated nonmetallic surface. As the powdered solder melts in the flux matrix, it coalesces into larger droplets. These small solder pools cannot wet the nonmetallic surface on which they are floating, and stay isolated. Unless they touch one another, they remain separated, and upon freezing form solder balls that stay embedded in the flux residue. There is no guarantee that they will wash off with the flux, especially in tight spaces (Figs. 5-3 and 7-15).

Such isolated paste islands can be formed in several ways. Each suggests its own remedies:

- Misregistration of a screen pattern or dispensing equipment.
- Application of excess paste, which runs over the edge of the conductor surface.
- Slow heating, which allows the cream to spread (sag) and migrate to nonmetallic surroundings.
- Poor quality cream.
- A poorly solderable surface, making flux too weak.
- Flux spattering (see item 2 below).

2. *Spattering of the paste during reflow*: Spattering of the flux during reflow can carry solder cream and/or balls to the surrounding areas. This usually stems from the following sources; the solutions are obvious:

- Flux contained moisture or volatiles that were not driven off during preheating.
- A change in the heating profile during soldering.
- A dirt layer on surface under the solder cream.

3. *Dirty solder residues from paste*: Dirty solder residues from the cream or paste are not a problem if they are cleaned off. However, they are an indication of problems with paste quality (see Section 5-9). This is normally due to:

- Tarnished or oxidized powder solder used in paste.
- Flux not strong enough to reduce the oxide skin from the solder powder in the cream.
- Overheated flux residue (not a paste quality issue)

7-16 THE BLOW HOLE DEFINED

There is a major category of faults normally referred to as "blow holes." In this chapter, we will define them very specifically. Some of the faults which are not covered by our definition are discussed in Sections 7-6 & 7-20 and Refs. 7-9 and 7-10.

The term *blow hole* is erroneously used as an all inclusive name for several imperfections in the solder fillet. We will define this defect as a cavity extending into the solder fillet which has erupted on the bottom side, having a solid, well formed fillet top. This condition is depicted in Figs. 7-16, 7-17, and 7-18. The imperfection extends deep into the plated-through hole, and usually the bottom is not fully visible, which makes it different than a pin hole (see Section 7-20). In most cases, the lip of the crater shows signs of violent eruption. The appearance of the opening, however, is a function of solder temperature at the time of fault formation. If the solder is still molten and hot enough, the circumference of the hole smooths out again due to the cohesive forces.

Small amounts of harmless gas are always being generated during soldering operations. They result from the process and from the relatively high temperature during solder contact. Most fluxes break down by heat and generate smoke; the laminate contains small amounts of moisture that escapes. Such small amounts of gases normally vent through the molten solder before the top solidifies and are not trapped.

But a true blow hole is the result of a larger and more violent gas generation from sources described in Section 7-17. In addition, during blow hole formation, the solder fillet top must have solidified while gas was still being generated inside (Figs. 4-12 and 7-17). Since the solidified

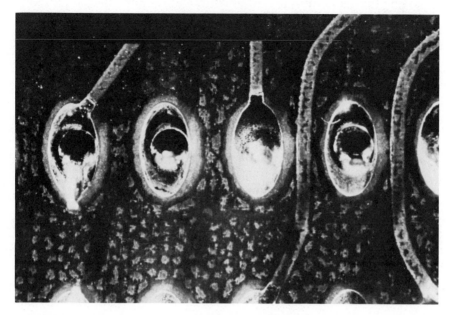

Fig. 7-16. Bottom view of blow holes. Note that bottom is not really visible.

top cap no longer affords an escape route to the evolving gases, they must escape through the molten bottom. This creates a violent eruption, especially since the gases still expand due to the high temperature of soldering.

These defects, which are truly blow holes can seldom be recognized from the top of the joint. In addition, not every joint that contains the evolving gas has the volume to burst out (Fig. 7-19). In this case, a large internal gas pocket is trapped, weakening the joint without being visible.

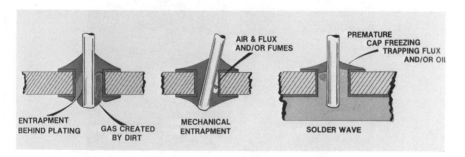

Fig. 7-17. Schematic showing blow hole mechanism. Note that the top of the fillet is already solid, while the bottom is still molten.

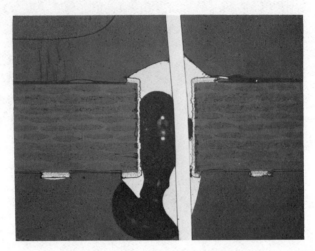

Fig. 7-18. Cross section through a blow hole. Note well-formed top of fillet. (Courtesy Alpha Metals)

Formation of the visible blow hole, then, depends on several ingredients which are:

1. A frozen top cap.
2. A source of gaseous material inside the fillet.

Fig. 7-19. Cross section through an adjacent fillet to Fig. 7-18. Note that the gas volume was not enough to blow out, and caused entrapment. (Courtesy Alpha Metals)

3. Enough volume of gas to blow through the bottom after the fillet has left the wave and before the bottom has solidified.

When the gas being generated does not have enough volume to blow out, *entrapment* occurs. A gaseous pocket trapped inside the fillet which, under similar circumstances, may have created a blow hole (Figs. 7-19 and 7-20). Therefore, it is likely that blow holes observed on a board are accompanied by several entrapments hidden in the adjacent fillets. The impact of this observation is threefold:

1. We must study the cause of the blow holes before we can eliminate them in future production (see Section 7-17).
2. We must study the results of repair and touchup, based on an understanding of the true physical properties of the joint (see Section 7-18).
3. Touchup of visible blow holes alone does not correct entrapment problems.

For specific cures and methods of repair, see Sections 7-18 and 7-19.

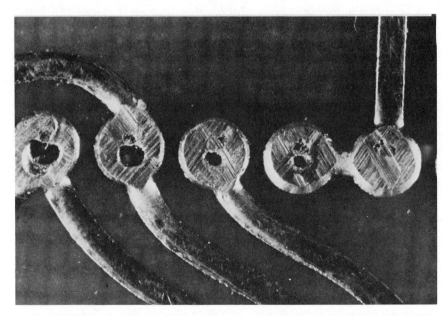

Fig. 7-20. Horizontal cut into a row of holes where only one blow hole was evident. Note that gas pockets existed in all adjacent fillets. (Copyright Manko Associates)

7-17 SOURCES OF GASEOUS MATERIALS

The sources of gaseous materials that create blow holes and entrapment in solder fillets are many. Some of the more common ones are listed in Table 7-1. While the term "blow Hole" implies a source of the gaseous material, the column on sources is more explicit. Proposed short- and long-range remedies are also listed. We will not consider the effect of the weakened structure of the fillet in general. This is differs in each case and requires specific analysis of design and usage parameters (Sections 1-4 through 1-8).

Let us discuss the cause of blow holes in order of decreasing frequency.

• Contamination in the plated-through hole or on the lead.
• Printed circuit board volatiles and absorbed moisture.
• Poor hole quality.
• Plating thickness and ductility, coverage continuity, and plating or fusing chemical residues.
• Flux volatiles.
• Mechanical blockage
• Premature freezing.

1. *Contamination in the plated-through hole or on the lead*: This is the result of poor handling and storage of both components and printed circuit boards. While a small amount of such contamination is unavoidable, the quantities picked up are often sizable. They can cause not only blow holes but also solderability problems.

Better storage and housekeeping techniques and assembly procedures help reduce the condition (see Sections 2-3 and 2-4). Sometimes an inline presolder cleaning process is the only way to reduce on accumulated dirt (see Section 2-11). The dangers from this category of blow holes or entrapment depend on the nature and source of the contamination (see Table 7-1).

2. *Printed circuit board volatiles and absorbed moisture*: Printed circuit volatiles (mostly in multilayer boards) are normally the result of inadequate printed circuit board processing steps. They include resin formulation, lamination techniques, curing time and temperature, and similar operations.

Absorbed moisture in the boards is another frequent source of blow holes (found in any type of board with a plated-through hole). The printed circuit laminates have a specific moisture content, which is a function of the resin used, and the quality of the laminate. The water content is

Table 7-1. Sources for Blow Holes and Entrapment[a].

TYPE	PROBABLE CAUSE	SUGGESTED REMEDY		POSSIBLE DANGER TO JOINT FROM RESIDUES (EXCLUDING WEAK STRUCTURE)
		SHORT RANGE	LONG RANGE	
Contamination	Poor handling and storage	Pre-solder cleaning	Improve house-keeping	Depends on corrosivity
Absorbed moisture	Humid environment	Pre-solder bake	Store in low humidity	None
PCB volatiles	Inadequate PCB processing	Pre-solder bake	Control PCB process or incoming inspection	None
Flux volatiles	Wrong flux or low preheat	Increase preheat	Change flux or increase preheat	Depends on flux residue (none for volatiles only)
Mechanical	Physical blockage	Depends on geometry	Design change	None
Plating continuity or process chemicals	Poor drilling or processing	Pre-solder bake and post-solder clean	Control PCB process or incoming inspection	Trapped electrolyte may bleed and cause corrosion
Premature freezing	Heat balance in wave process	Increase preheat or wave temperature. Slow down conveyor	Make sure top freezes in proper sequence	Trapped rosin or oil no problems recorded. Organic flux may cause corrosion

[a] In order of decreasing frequency.

299

however, a function of the relative humidity in the air. Thus, in humid periods, we have more blow hole problems than in the dry seasons. The quality and thickness of the copper plating in the hole also controls this effect (see item 4 below).

The short range solution for boards on hand is a prebake operation prior to soldering (see Section 2-12). This has generally become standard for multilayer boards. It may also be necessary to preclean the assemblies prior to baking. Dirt left on the surface (as in item 1 above) can harden during bake out and cause solderability problems.

For a long term solution, the problem should be discussed with the raw material manufacturer or board fabricator. No long-range reliability impact (in the form of additional deterioration) has been associated with this source of blow holes. It is frequently accompanied by measling and delamination.

3. *Poor hole quality*: Poor hole quality before and after plating is a common source of blow holes. A smooth, uniform hole is required for more than blow hole problems. Two aspects of hole perfection pertain directly to our discussion. First, there is the problem of loose fibers in the hole, which may be covered by plating to form a nodule. During lead wire insertion such projections are broken off, and plating continuity is disrupted. The second quality problem deals with cracks which also cause plating discontinuities. Any break in the plating of the through hole must be considered a potential source of a blow hole.

The mechanism is simple. The small amounts of volatiles and moisture inherent in the board generate gases, with no plating barrier to keep them back.

There is no way to correct the plating in the hole. Baking out the volatiles is the only solution (see Section 2-12).

4. *Plating thickness and ductility, coverage continuity, and plating or fusing chemical residues*: As the heading implies, plating parameters in the hole also affect quality. This is especially noticeable when plating solutions are trapped behind the barrel walls of the plated-through hole. The encapsulation of materials is caused by poor plating techniques and/ or damage done during the hole punching or drilling operation described in item 3 above. Unless hole generation is even, crevices are formed inside the laminate which retain solutions.

Plating thickness is also a factor because a thin wall is more readily ruptured (see Figs. 7-21 and 7-22). Copper thicknes of $0.001 \pm 0.0002''$ $(0.025 \pm 0.005$ mm$)$ in the wall are reported to be sufficient for this purpose. This is also enough to prevent small amounts of natural moisture from creating a blow hole.

As with plating discontinuities, there is no cure for existing boards.

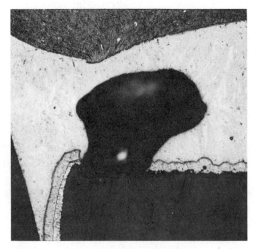

Fig. 7-21. Cross section of a poorly plated through-hole. Note that the copper was ruptured during blow hole formation. (Courtesy Alpha Metals)

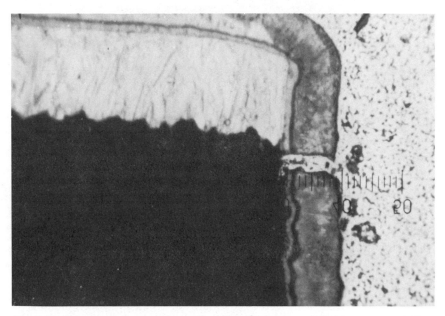

Fig. 7-22. Board from the same lot as Fig. 7-24, which was prebaked to remove volatiles. Note that the flux and solder mended a barrel crack. (Courtesy Alpha Metals)

Prebaking (see Section 2-12) provides only a partial solution. It is questionable whether drying of the trapped plating chemicals in a prebake is reliable. In addition, care must be taken after soldering to wash off any corrosive plating chemicals which might have come to the surface. This subject will be discussed further later in this section.

Only good process control and inspection by the fabricator can prevent this problem. Incoming inspection can help detect such imperfections before they reach the assembly line.

5. *Flux Volatiles*: These are the result of poor flux selection and/or inadequate preheating. They may also result from water accumulation in the flux (see Section 7-14).

One may easily change the fluxing material so that it can dry with existing preheat conditions. Additional preheating, if available, can also eliminate the residual flux volatiles.

The trapped flux volatiles are dangerous only if they contain corrosive materials from the flux.

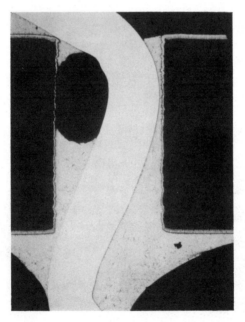

Fig. 7-23. Cross section through a solder fillet showing gases trapped under bent lead. (Courtesy Alpha Metals)

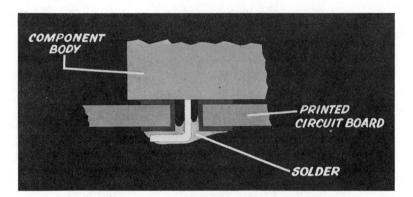

Fig. 7-24A. Diagram of a no venting condition which can lead to entrapment.

6. *Mechanical blockage*: There are physical blockages located at the top of the solder fillet that cause blow holes. They can originate from closely seated housings, connectors, or components. This condition is aggravated by insulation that runs on the lead to form a meniscus of nonmetallic materials (Fig. 7-23A). A special bend in such lead wires can solve the problem (Fig. 7-23B).

Mechanical blockage is also created by crimped component leads forced into the side of the plated-through hole (Fig. 7-24A and 7-24B). The blockage prevents the natural release of gaseous materials through the top side and causes entrapment.

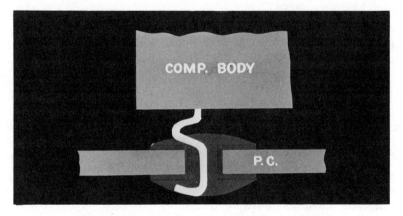

Fig. 7-24B. Diagram of good venting using a lead bend.

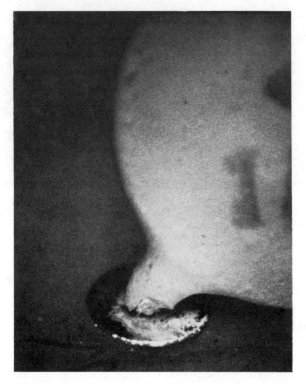

Fig. 7-25A. Top view of capacitor meniscus in plated-through hole. (see also Fig. 7-25B.)

This problem is normally corrected by proper design or assembly techniques. Its effect on reliability is strictly a function of joint design factors and the flux used, because there is no known chemical entrapment other than flux and or oil fumes.

7. *Premature Freezing*: This is the only wave solder oriented process problem. It is due to improper heat balance in the wave application.

It is simple to readjust the heat balance so that premature freezing will not occur. One or more of the following methods can be used:

- Increasing preheat.
- Adding top heat.
- Raising the solder temperature.
- Slowing down the conveyor speed.

Fig. 7-25B. Bottom view of same capacitor meniscus as in Fig. 2-25A. Note that part of the insulation meniscus is visible in the incomplete fillet.

The rule of thumb here is: *Make sure that the top of the fillet solidifies no sooner than 0.75–1.0 in. (20–25 mm) of travel time after leaving the wave.*

The danger from this type of heat balance entrapment depends on the nature of the soldering chemicals being used. Rosin fluxes, petrochemical hydrocarbon oils, and plain wax are not the source of any known corrosion mechanism or documented problems. Organic intermediate flux residues or water soluble oils may pose potential corrosion hazards. None have been reported since 1963, when they were first used in electronics. The nature of other chemical hazards must be identified before a comment can be made.

There is no danger of corrosion if the chemicals are truly sealed inside the fillet. The solder that encapsulates them acts as an effective barrier. The danger comes when the corrosive materials are located in a cavity that is connected to the outside world, like a blow hole or crack. We can hardly hope that these cavities will be cleaned during the washing cycle. Therefore, repair decision must be based on the type of chemicals involved. Again, rosin-based fluxes, with their specific chemistry pose little or no hazard; nor do the waxes and petrochemical based oils. Obviously, one must view organic acid fluxes, water-soluble oils, and plating chemicals with alarm. Their interaction with humidity may cause harmful materials to bleed out from the fillet cavity to the board surface.

We can make no generalizations about the plating chemicals which may have been retained behind the barrel. If the heat of soldering does not rupture the copper plating in the hole, no real danger is anticipated. However, if the chemicals become exposed due to a blow hole, a corrosion and conductivity hazard exists.

7-18 BLOW HOLE CURE

Finding a cure for blow holes requires a series of well-defined steps. Obviously the remedy is a function of the cause, which must be identified. The procedures outlined below can double for the diagnostic study and the full cure. Details hinge on the equipment on hand and the sensitivity of the board assembly.

Step 1. Contamination in the hole or on the lead wires can be tested for and removed by an appropriate cleaning process. We must usually treat the fully assembled printed circuit board because it is at this level of assembly that we detect the blow holes.

If all the components are held in place by a crimp, use the best process available in house. Send them through a water washer with saponifier, degrease them in vapor, cold wash in a solvent, and so on.

If the components are not mechanically secure, dip the whole assembly slowly by hand. Equip the operator with gloves and other required safety equipment. Have the boards slowly lowered horizontally into the solvent. This makes the cleaning fluid percolate through the holes and around the component leads. The board should then be soaked for 30–60 seconds and withdrawn slowly in the same manner.

Step 2. Dry the boards to remove all cleaning fluids from step 1. This step should not be confused with the prebake in step 4. Cold solvent-cleaned boards need be exposed to air for only 10–15 minutes for final evaporation. The work emerges dry from the vapor degreaser. Water cleaned boards must be dried more thoroughly with the use of a bake. In most cases, room temperature drying, after water washing, takes over 6-8 hours, and then only in low relative humidity (less than 60%).

Step 3. Wave solder the cleaned and dried board using the standard process. If the blow holes have disappeared, contamination was the reason of the problem, and the solution has been identified. If the problem persists, the cleaning process may not have been strong enough, but more likely cleanliness alone is not the problem. Try step 4.

Step 4. Prebake a cleaned board to remove all volatiles. Use the process described in Section 2-12 and use the maximum time for each temperature during the diagnostic stage. If the remedy works, you may want to investigate the shortest time that gives satisfactory results.

Step 5. Wave solder the baked boards to establish freedom from blow holes. Remember that the assembly starts to reabsorb moisture as soon as it comes down to room temperature. The rule of thumb is: *Solder the boards within 10–20 minutes from the time they reach 25°F above room temperature·*

If this works poor plated-through hole quality is part of the problem. It may be enough to bake the boards without cleaning. However, this is a dangerous procedure if there is also dirt on the surfaces. Such contamination may dry out and harden, making the assembly unsolderable. Check the need for precleaning carefully before side stepping this cumbersome cleaning process.

Step 6. If baking is the remedy, you must still establish the reason for the volatiles. Metallurgical cross sectioning of the board provides one good diagnostic tool. You can also inspect the hole with back-lighting and measure the thickness of the copper plating by conductivity (see Section 8-9).

Step 7. If cleaning and baking do not help, check to see if there is mechanical blockage. Relieve the back pressure by providing air and fume vents for each plated-through hole.

7-19 BLOW HOLE REPAIR PROCEDURES

As stated earlier, blow holes are formed when the fillet top cap is solidified and gas is generated. This implies that there is a sufficient volume of gas to break through the bottom. Experience has shown, however, that when blow holes are formed, many fillets end up with large internal cavities that cannot reach the surface (Figs. 7-19 and 7-20). They are hidden because the gas volume was too small to escape. As a result, there is a grave reliability hazard whenever the solder joint is marginal in design or quality. Some over designed joints may not be threatened by a hidden gas pocket, but this condition is not permitted in multilayer boards. It is safe to say that whenever there are repetitive blow holes visible on the board, there are just as many if not more hidden cavities.

Remember that blow holes are difficult or impossible to clean out in the normal printed circuit process. Some corrosive materials may be lodged inside the hole and/or the ruptured barrel (see Table 7-1). The repair of blow holes is, therefore, always associated with thorough cleaning. Some of the released chemicals may not clean in the same way as the flux residue.

As a result, the touchup of the board is problematic. On over-designed assemblies, fixing visible blow holes eliminates the washing hazard. But in most cases, resoldering only the visible defects is not enough. The entire board must be resoldered to fill in hollow fillets. This, unfortunately, cannot be done by the wave, since the same gas-generating mechanism may still be operative. Thus the first rule of thumb for production is:

When more than an occasional blow hole is found after wave soldering, stop additional production of this batch. Find a remedy before you resume production.

Using this rationale, we avoid the need for blow hole repair and hope to produce only reliable boards. The second rule of thumb applies to the repair and touchup of soldered assemblies where blow holes were detected. It states:

Touch up an occasional blow hole only. If more than one fault is found per board, suspect the presence of many hidden cavities. Notify quality assurance or the material review board. If repair is authorized, touchup 100% of all joints on the suspect board.

Touchup follows the procedures outlined in Chapter 9, except that liquid flux must always be used. Inspect the flux that comes to the surface for the presence of foreign materials. For example, a change of color in rosin fluxes from a light amber is an indication of foreign materials. Thoroughly wash all flux residues from the surfaces, since they may contain corrosive or conductive materials. Consult Table 7-1 for details.

7-20 PIN HOLES, FREEZING LINES, ALLIGATOR SKIN, AND OTHER SOLIDIFICATION IMPERFECTIONS

Pin holes, freezing lines, alligator skin, and other freezing imperfections are surface blemishes only. As long as they are shallow and the bottom is fully visible (washable), there is no reason to reject them. They affect the surface only; they do not weaken the fillet nor decrease its current carrying capacity. In addition, they are easily washable and thus pose no quality hazard. Some companies even include minor scratches and depressions in this category.

We classify this category as solidification imperfections because of their mode of formation. The solidification of any metallurgical cast structure from liquid metal starts at the cooler walls. As the metal solidifies and cools, it shrinks in volume. If one surface freezes last, it will be irregular due to this shrinkage, and blemishes called *pin holes* or *freeze lines* will develop. In some cases, the outline of grains become visible through the shrinking solder and causes the condition referred to as *Alligator Skin*. Here the surface of the fillet is crisscrossed by the freeze lines in a pattern similar to alligator skin.

If all outside surfaces solidify before the center of the metal, the result is *coring* or *piping*. After the outside shell has turned solid, freezing proceeds to the center and the hotter bulk. Here cooling and shrinking give rise to the well-known *casting cavity* formation. No regular casting is free of this defect unless unusual precautions are taken. Some freezing vacuoles also appear as small, round voids in the metallurgical cross section of the joint (Fig. 7-26).

These imperfections stem from the shrinkage of the metal volume during the phase change from liquid to solid. At the same time, shrinkage occurs during cooling, according to the TCE. Since the outer shell has solidified first, internal cavities are formed. Most spherical or near-spherical voids found during metallurgical cross sectioning originate this way (Fig. 7-26). These voids are free of chemicals and sealed from the outside. The vacuoles are a clue to the sequence of fillet freezing, indicating which part froze last.

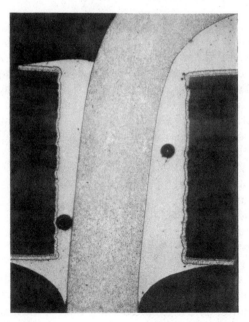

Fig. 7-26. Typical harmless freezing vacuoles in a solder fillet. Their presence indicates the area that froze last. (Courtesy Alpha Metals)

When one outside surface does not freeze with the rest of the skin, the pattern is different. In an open circular mold, the top surface usually stays molten longer. It solidifies in concentric rings, from the outside to the center, at roughly the same rate as the walls. As a result, the shrinking metal leaves a central *core* cavity that extends to the surface. With shallower molds like solder bars, shrinking causes a trough to form on the surface.

This is what happens in the average solder fillet on a board. Normally, the cooler top cap freezes first. The bottom of the joint solidifies next, and the center is last. As a result, we find small vacuoles in the structure of solder joints. This solidification pattern is always needed for blow hole generation.

As stated earlier, if the bottom freezes last due to special circumstances, small shallow cavities called *pin holes* form on the surface. They can also create shrinking patterns that are erroneously referred to as *stress lines*. When the freezing pattern creates depressions between individual metallic grains, we call this phenomenon *alligator skin*. None of these defects are considered a reliability hazard.

7-21 GRAINY OR DULL SOLDER JOINTS

Here we discuss a rough solder surface with small, gritty projections protruding through the top, or a none shiny surface that shows no signs of chemical attack. In either case, the solder joint is not as smooth and uniform as it should be. Both of these conditions are unique and should not be confused with disturbed joints (Section 7-5), freezing imperfections (Section 7-20), or high-temperature alloy joints (Section 7-22). Surface residues and surface attack are also excluded (see Section 7-25).

Some of the common reasons for these faults are listed in order of importance:

- Freezing patterns of high purity eutectic tin-lead.
- Impurities in the solder.
- Intermetallic compounds.
- Dross mixed in solder.
- Insufficient-heat fillets.

1. *Freezing patterns of high purity eutectic tin-lead*: The freezing pattern of a high purity eutectic (63/37) alloy (See Fig. 7-27) may be the cause. Here dendrite growth is fast, and the bulk of the metal freezes last. Thus the tips of the dendrites project through the surface especially on the edge of the fillet near the bottom.

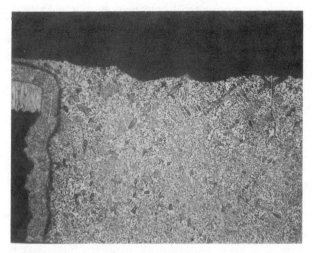

Fig. 7-27. Cross section of a high purity solder joint showing the dendrites projecting through the surface. (Courtesy Alpha Metals)

This condition appears with both hand and wave soldering. If you are sure the metal is pure, there is no reason for rejection or touchup.

2. *Impurities in the solder*: Solder impurities cause both grainy and dull solder joints. We can subdivide this category into two conditions:

a. Small amounts of metallic impurities such as aluminium, cadmium, and zinc, cause the solder to appear dull. These metallic contaminants may have been in the incoming solder bar or picked up from the work.

It is impossible to generalize on the effect of these impurities on quality. The reader is referred to Chapter 3 in Ref. 7-8.

b. Larger amounts of copper and gold impurities will cause the solder to appear dull. These metals are normally picked up from the work and must stay within given limits (see Section 3-13). This condition is normally not acceptable, and the solder must be changed. The rule of thumb is:

The solder fillet must have the same level of luster as the parent metal used.

In other words, if the solder is inherently shiny, like Sn63 or Sn60, the fillet must have the same luster. See Section 7-22 for a discussion of high temperature dull solder alloys and joints.

Units that have been soldered with contaminated solder must be evaluated. In critical applications, they can be resoldered to dilute the contamination level in each fillet.

3. *Intermetallic compounds*: When the contamination content exceeds the allowed limits, intermetallic compounds may form. These are crystals that float in the metal, and some rise to the top. They too impart a rough appearance to the joint. Intermetallic compounds of copper, gold, silver, and iron can be found in electronic applications. They all give the solder a gritty and often dull appearance. The first three impurities are usually picked up from the work. Iron intermetallic compounds, also referred to as *hard heads*, may come in with the bar. In the case of drastic overheating, iron is also picked up from equipment walls.

The level of impurity here far exceeds the permissible limits outlined in Section 3-13. Therefore, the condition should not be accepted, since the brittle compounds can initiate joint failure in critical stress planes.

4. *Dross mixed in solder*: Dross mixed in with the solder, causes the same problem. The source is normally poor equipment maintenance, low solder levels, or large dross deposits left on the surface.

In cases where dross is trapped only inside the fillet there is no outside indication of joint quality. Dross is often manifested as poor wicking to the top of the board and a lumpy solder appearance on other joints. Beside causing a weakened structure, which is poorer than with chemical entrapment, it poses no problems to the assembly.

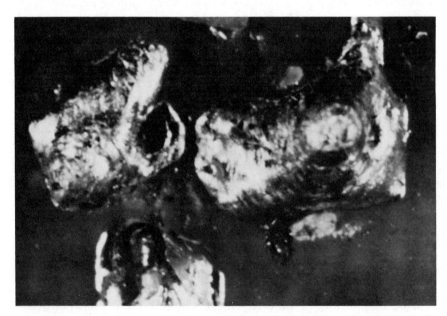

Fig. 7-28. A typical "insufficient heat" solder joint. Note the chalky and rough surface appearance. (Copyright Manko Associates)

5. *Insufficient-heat fillets*: An *insufficient heat joint* also has this appearance. It is usually chalky, and has a puddled on look (Fig. 7-28). *Puddling* is a term usually used to describe layers of solder on a joint that do not seem to flow together. It is formed when the joint temperature stays below the wetting temperature of the alloy or when flux starvation exists.

The remedy here is also the best diagnostic test. Just add a small amount of flux and remelt the solder in the joint. An insufficient heat joint will be transformed to a shiny fillet.

For identification of the source, metallographic cross sectioning is used. An evaluation must be made in each case to assess the weakening of the joints in question.

7-22 APPEARANCE OF HIGH- AND LOW-TEMPERATURE SOLDER JOINTS

Only sporadic use of high- and low-temperature solder alloys has found its way into the wave soldering of printed circuits. These alloys are used for applications where the standard tin-lead solders have the wrong melting point. For details on alloys with different temperatures see Chapter 3, in Ref. 7-8.

While most laminates can be used for higher-temperature alloys, the polyimide boards have the best heat resistance. Special high-temperature wave solder modules are needed for this application. Obviously, lower-temperature solders offer no real problems in relation to the laminate. They can be used to minimize the problems associated with the glass transition temperature (T_g) of laminate resins.

The surface-mounting hybrid industry has used, different melting point solders for years. The ceramic substrates used in hybrid technology can withstand higher temperatures with ease. This makes it possible to do *step soldering* (piggy-back soldering). In this process, a number of joints are made in succession with solders of decreasing temperature without the remelting of previous joints.

Now the use of high- and low-temperature solders is being extended to surface-mounting on printed circuit boards. This brings with it an interesting problem. The inherent appearance of high-temperature alloys is dull and often grainy. This lack of luster is due to the high lead content found in most of these alloys. Low-temperature alloys usually contain bismuth, tin, lead, and other elements. Both high- and low-temperature alloys have unique freezing characteristics that often result in very uneven surfaces (see Sections 7-20 and 7-21). This graininess, which is due to the freezing pattern looks very different from the appearance of the standard tin-lead solder. Untrained inspection and quality personnel have trouble quantifying the appearance of these joints.

We must therefore define the quality of all solder joints as follows:

The surface appearance of a solder fillet must be identical to that of the parent alloy. If there is a deviation from this appearance, a problem exists.

In other words, if the parent alloy is dull or grainy, the joint should be the same; if the solder alloy is normally smooth and shiny (like Sn63), the fillet must have the same appearance. In the latter case, we must apply a similar rule of thumb to that described in Section 7-21.

Under no circumstances should high- or low-temperature joints be rejected or reworked because of their appearance. As long as they look like the parent alloy, they are acceptable. It is suggested that each organization generate its own quality standards for such solders (see Section 8-2).

Also, remember to dedicate a special iron and tip for all touchup and repair with these alloys. The tip must be tinned with the same alloy to avoid contamination and melting point changes.

7-23 CRACKS IN SOLDER JOINTS

In this section we will discuss solder joints in general; in the next section, we will cover problems unique to surface-mounting. The reader is advised to read both consecutively, since there will be no duplication.

The term *cracked* is used to describe failure in three distinct areas of the solder joint. According to the location of the failure as follows:

a. At the component-solder interface.
b. In the solder joint itself.
c. At the solder-printed circuit interface.

Most failures are compound fractures, which include two or more of the above locations.

In the average printed circuit board, the likelihood of solder failure is very small. In properly designed single- and double-sided boards, the solder fillet has more than adequate strength. Conservatively, the strength of the solder is 2-3 times greater than that needed for single-sided boards and 6-10 times that needed for double-sided and multilayer boards. You can calculate your own designs from the data given in Chapter 1. This observation does not hold for surface-mounted components.

Overall joint strength obviously depends on the degree of fillet perfection. This, in turn, is a function of the materials, design, and process. It obviously also depends on the specific conditions in the use cycle. Remember that in printed circuits the solder is the softest but not the weakest link in the assembly.

Failure analysis follows well-defined metallurgical practices. Careful cross sectioning through the affected area provides a clue to the true nature of the defect. This is coupled with a stress analysis of the assembly to reveal the nature of the problem. Most cracks in solder joints are initiated by a localized stress that exceeds the material strength at that point. These locations are called *stress concentration* areas. The forces are generated by such causes as vibrations, shock, thermal excursions, and other events. They are complicated by the variations in moduli of elasticity, mismatches in coefficients of expansion, and the like between the boards, components, and solder. For further details on board design, see Chapter 1.

For practical reasons, let us discuss the sources of problems by category of origin:

1. *Solder oriented problems*: These problems normally stem from contamination in the metallurgical system. Intermetallic compounds, when concentrated in a critical area, weaken the joint because of their natural brittleness. This is especially true of gold and, to a lesser degree, of copper. The failure is a geometry and metallurgical problem. It is aggravated by:

- Contaminated solder in the wave.
- Excess soldering time or temperature.
- Unnecessary touchup or remelting.
- Elevated operating temperature of joints in equipment.

The solutions to these problems are self evident. If, for some reason, the use of less heat is not feasible, use a barrier plate such as nickel (see Section 1-12). Tin-lead solder and nickel do not form appreciable intermetallic compound layers at soldering temperatures or below. A nickel flash can thus prevent such failure.

2. *Dross-oriented cracks*: Dross and oxide layers mixed into the solder in the fillet cause mechanical weakening of the lattice and failure. This is normally the result of poor maintenance or housekeeping.

Check the skimmer in dip and drag equipment. Make sure that the level of solder is correct in dry waves, and skim dross regularly. This problem has not been reported in waves that use oil.

3. *Solderability-oriented problems*: Marginal solderability of component leads is the most common cause of cracks in joints. Solderability problems in printed circuit boards are less common and are easier to detect.

The term *marginal solderability* needs elaboration. There is no problem in determining solderability at either end of the quality spectrum. Good wetting, on the one hand, and nonwetting or dewetting are easy to discern. Between these extremes there is confusion, since most people use only visual criteria. There is no confusion in this respect when solderability is checked with a meniscograph. The problem stems from the fact that there are levels of wetting imperfections that allow the test surface to hold a sheath of solder without good wetting. This solder coating gives the misleading visual results. Under stress, such joints fail by cracking at the interface. A double dip in solder will often show these imperfections. When there is still doubt, solder the surface in question to another metal, then pull them apart. Look for the area of failure; with marginal wetting, there will be a split at the surface.

It is easiest to demonstrate the problem with a case history. A batch of boards had one heavy component with marginal solderability. The plated-through hole was of high quality and drew the solder up to the top. Cursory final inspection did not detect this fault, since the solder seemed to form a good fillet. The boards also passed electrical testing with flying colors. The joint would probably have continued to function, were it not for stresses that arose in use. Vibrations caused the heavy component to move and stress the fillet. After field service for 2–11 month, a number of these connections failed (Fig. 7-29A and 7-29B).

Analysis showed that the component lead had marginal solderability as measured on a meniscograph. A stress analysis revealed that the component was exposed to vibration from a cooling fan mounted nearby. The weight of the component, which was larger than average, caused excessive stresses under vibration, and the joint failed. Failure was intermittent at first and later became an open. Identical boards that were not yet installed were tested, and recorded a joint strength below 1 oz of pull. Several boards on the shelf showed fine cracks beginning at the same location. It is apparent that this failure mode is time and stress dependent (the duration and magnitude of stress). It is not always traceable because it is intermittent before catastrophic failure occurs.

The lesson is simple: solderability must be checked prior to assembly and soldering. Such tests as the wetting balance check are more precise than the visual check.

4. *Design oriented problems:*—are also responsible for cracks in solder joints. These fall in to four categories:

a. Weak structural design, where the inherent strength of the solder is not sufficient. This applies particularly to surface-mounted components like leadless chip carriers.

Fig. 7-29A. Cracked solder joint in field service due to marginal solderability of the lead wire. Note the location of the crack. (Copyright Manko Associates)

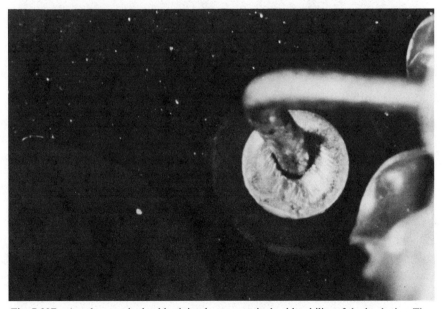

Fig. 7-29B. Another cracked solder joint due to marginal solderability of the lead wire. The solder shows "flow lines" indicating partial wetting. (Copyright Manko Associates)

b. Designs where the board undergoes excessive movement due to vibrations, usually caused by poor chassis mounting techniques. If re design is not feasible mechanical stiffener have to be added. This is also one reason why very wide or long boards are not desirable.

c. An incorrect hole-to-wire ratio causes weakening of the joint in two ways. First, it makes the hole difficult to fill if the ratio is large. Second, any stress, like cutting leads after soldering, can cause cracking (Figs. 7-30A and 30B). For a treatment of the hole-to-wire ratio, see Section 1-11.

d. There is poor compatibility between materials when stresses are generated that exceed solder strength. The mismatch of coefficients of expansion is a case in point. A board 0.062 in. (1.6 mm) thick made of polyimide had gold-plated brass pins staked in a hole that was not plated-

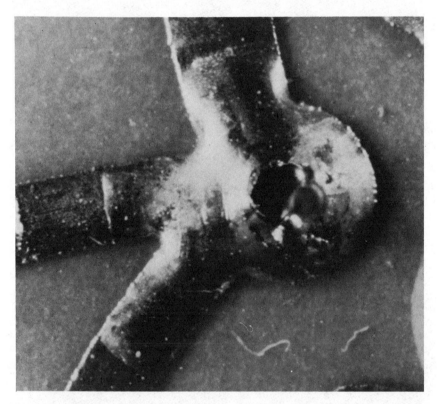

Fig. 7-30A. Crack formed by inline cutting of a joint with a large hole-to-wire ratio. The situation was aggravated by the small diameter transistor lead, which was made of a hard ferrous material. (Copyright Manko Associates)

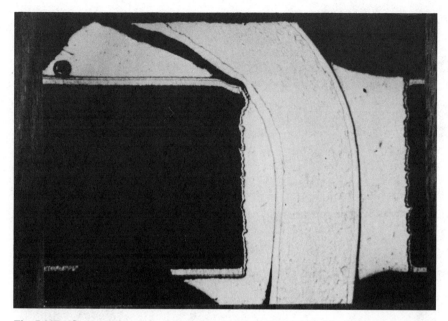

Fig. 7-30B. Cross section through a similar crack as in Fig. 7-30A. The hard transistor lead moved during cutting, causing a crack on the opposite side of the board. (Courtesy Alpha Metals)

through. The device operated at elevated temperatures (350–400°F). This required a high-melting solder, and an alloy of 93.5% lead, 5% tin, 1.5% silver was used. The dramatic mismatch of coefficients of expansion was enough to crack the ductile solder fillet (Fig. 7-31). A similar failure occurred on a thick multilayer card, using eutectic tin-lead solder in the standard temperature range.

In summary, marginal solderability of component leads is the most common cause of cracks in joints. Solderability problems in printed circuit boards are the third most common category. Contaminated solder ranks second, with design problems last.

The touchup and repair of cracks in solder joints is not as simple as it may appear. merely adding flux and remelting the solder may restore the joint only temporarily. Repair depends on the origin of the crack and whether the failure mechanism is still operative. It is imperative therefore, to make repairs only after consulting with the reliability or quality department. Cracks that occur in the field need similar reporting.

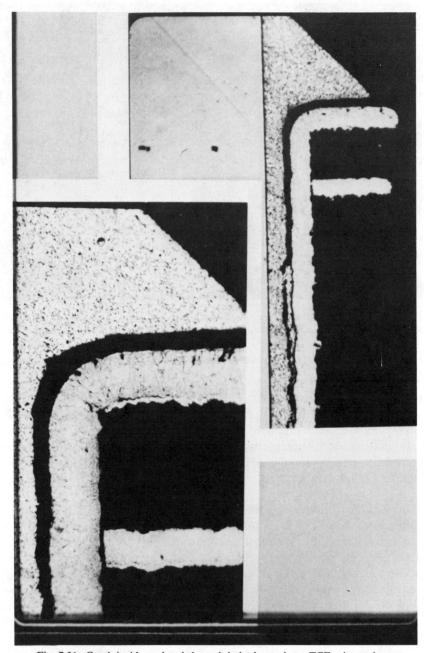

Fig. 7-31. Crack inside a plated through-hole due to large TCE mismatch.

7-24 CRACKED FILLETS IN SURFACE MOUNTING

Solder joints in surface-mounting are much more prone to cracking than through-the-hole fillets because of their smaller size and lesser strength (see Chapter 5). The failure can usually be accelerated by thermal cycling, which is the largest source of stress in this application. Careful thinking is required in planning a practical temperature range for testing. There is no known correlation between large unrealistic temperature excursions and actual joint life expectancy.

In this book, we will assume that the design and materials are correct, and that the assembly has already been successfully soldered. Under these conditions, when a solder joint cracks, check for one of the following reasons:

1. In leadless components, when the crack is in the solder bulk and not along an interface (see item 3 below). The size of the solder joints around the device is probably uneven. This causes the stresses to be concentrated in the failed joint.

Remove excess solder from the device to minimize the *vice blocking effect*. Then resolder the cracked joint by adding flux and heat, making all the joints as even as possible.

If the problem recurs, make sure that future joints will be uniform and of the correct size.

2. In compliant leaded components, when the crack is located in the solder fillet and not along an interface (see item 3 below), it is probably due to too much solder stiffening some of the leads around the device (the solder should not get into the knee). Remember that we depend on the ductility of the lead to absorb any stresses.

Check all leads on the device in question. Remove excess solder from the leads in the knee and above. Then resolder the cracked joint.

If the problem persists, change the soldering procedures to keep the leads compliant.

3. If the crack is located at an interface between the solder and either surface being joined, it is a solderability problem. The solder cannot fully wet the surface, and a crack is formed on stressing.

While this is the easiest failure to diagnose, the solution is complex. There is only a small probability that a single isolated joint surface was nonsolderable, while the rest were good. In such an unlikely event, try the following procedure:

- Remove all solder.
- Clean the joint area with the best solvent.

- Try to wet the offending surface with the strongest flux permitted, and clean flux residues.
- After the surface accepts the solder, remake the joint. Chances are, however, that the poor solderability condition is not localized to one joint. In that case, a similar repair procedure must be used for *all* affected areas. Consult Sections 7-1, 7-4, and Section 2-7 for solderability restoration hints.

4. In leadless devices with fired (silver) electrodes, another type of failure mode is found. The crack is located on the surface of the electrode or between the electrode and the ceramic surface. Here the crack is due to the mechanism of *scavenging* (leaching—see Section 5-12).

If the solder no longer adheres to the termination, there is no way of correcting this fault. Component replacement is required. During repair, scavenging can be retarded by using short soldering times and low temperatures. A solder alloy with 2% silver may also help (Sn62—62% tin, 36% lead, 2% silver). To prevent the problem entirely, a nickel barrier plating should be specified on all future components (see Section 5-12).

If the failure occurred between the termination and the ceramic body, a similar mechanism is at work. The solder migrates through the porous silver fired surface, and the tin combines (alloys) with the silver. The resulting stresses lift the termination. Here, too, a nickel barrier is needed to prevent this failure in future parts. Use the same scavenging retarding precautions during replacement.

5. Excessive stresses can be generated by too much adhesive under the components, since the coefficient of expansion of the adhesive is much greater than that of the materials around it. No generalizations can be made; each case must be studied on its own merit.

7-25 WHITE, GRAY, AND YELLOW RESIDUES

White residues is an all inclusive term used in the industry for any loose or adhering film of material on the boards. Close examination may reveal it to be more gray or yellow than pure white. The residue may be found over the entire board or only on specific parts, like the resist, or the metal.

There are scores of materials and reactions that cause this phenomenon (Figs. 7-32, 7-33, and 7-34). We will cover only the more frequent conditions that are related directly to soldering and/or cleaning. Caution is advised in the study of this defect and the assignment of its cause and cure. Often the condition is lot oriented and may disappear on its own.

Let us classify the colored (white, gray, or yellow) residues by their location where they are found:

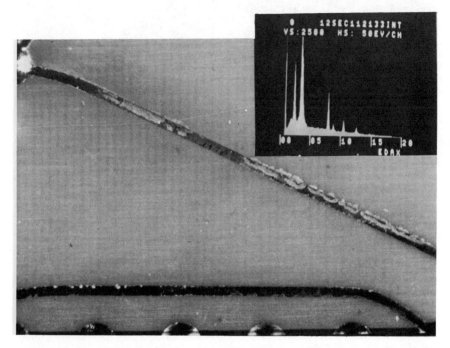

Fig. 7-32. White and gray residues on traces of printed circuit board, as a result of poor etching chemical and flux cleaning. The X-Ray analysis (see insert), showed them to be insoluble salts of tin and lead. (Copyright Manko Associates)

1. Residues all over the assembly
   - Weak or contaminated solvent.
   - Board hotter than the vapor degreaser.
   - Board lowered into dirty boiling sump.
   - Rosin overheated in wave soldering.
   - Delayed cleaning after iron use.
   - Use of nonpolar solvent only.
   - Minute rosin residues that craze.
   - Solvent too acidic in cleaner.
   - Hard water scum.
   - Insufficient water rinsing.
   - Incompatible water system (coating).
   - Inefficient air knife blowing.

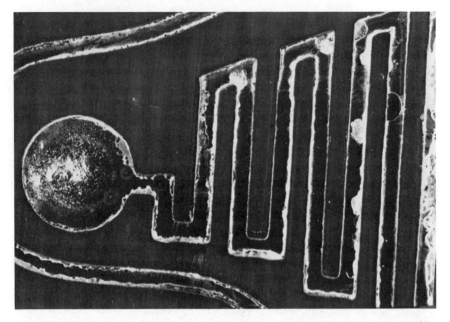

Fig. 7-33. White residues on solder metal only. These resulted form improper fusing-flux removal during board fabrication. The white residues were already visible in the unsoldered PTH. (Copyright Manko Associates)

2. Residues on laminate (insulation) only
   - Improper curing of the laminate.
   - Laminate attacked by a process solution.
   - Halide (bromine) reaction with solder (usually the lead).
   - Solvent attack (especially methylene chloride).
   - Reactions with the saponifier.
   - Protective coating partially removed.
3. Residue on solder mask (resist) only
   - Mask incorrectly formulated.
   - Mask not fully cured.
   - Mask incompatible with flux thinner.
   - Mask attacked by solvent cleaner.
   - Mask reacts with saponifier.
4. Residue on solder only
   - Chloride activator attack on solder.
   - Excessive drying heat.
   - Exposure to humidity and heat in air.

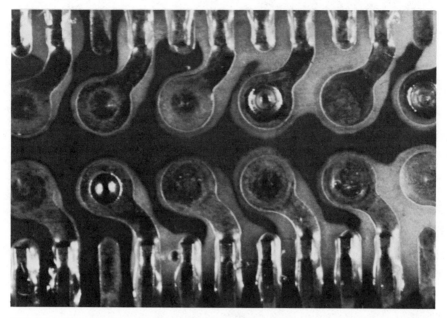

Fig. 7-34. The white halo around these joints is due to the lifting of the conformal coating. No foreign material developed under the coating, but surface contamination prevented proper coating adhesion. (Copyright Manko Associates)

- Residual plating insolubles.
- Traces of etching chemicals.
- High solder temperature.
5. Residues or dull appearance on specific components
  - Leached plasticizer from plastic.
  - Chemical reaction by flux or cleaner.
  - Heat effect on outer coating.
6. Other colors found on boards
  - Brown stains from degreaser sump.
  - Brown rust from cut ferrous lead ends.
  - Green or blue copper corrosion.
  - Stamp or marking ink migration.
  - Adhesive residues or leached labels.
  - Attack by tape adhesive.

The cure for most of these problems is obvious once the source has been identified. The diagnostic process may be a series of trials and errors, which yields the fastest results. Chemical analysis is, unfortunately, more difficult and costly, and the interpretation of the results is cumbersome.

Most residues, once formed, are difficult to remove. Some may be loose and easy to wipe or brush off, but as such are insoluble in commercially safe materials. Others are soluble in water, alcohol, or solvents and pose no problem. The majority, however, must be left on the surfaces, and cause either reliability or cosmetic problems.

The reliability of the printed circuit assembly with residues depends on the answers to two questions:

1. Does the residue interfere with the operation of the board? Residues are normally a current leakage problem, and ionic contamination checks may be helpful. A test with full power on, under humidity and temperature conditions would yield better information. The residue may also interfere with mechanical contacts.
2. Will the residue continue to form, or has the reaction come to completion? This is more difficult to establish, but fortunately, very few reactions continue.

Identification of the problem becomes more difficult if the residues appear with use. While they may have been caused by a material- or process-related problem, they are probably due to external causes. Airborne fumes and particles may settle on the work; some can be reactive under high humidity conditions. In another case, failure was traced to condensation water standing on the surfaces.

7-26 LIFTED PADS, MEASLING, AND BLISTERING

When the metallic conductor (land or trace) separates from the board laminate the problem is called *lifted pads*. *Measling* and *blistering* are faults inside the laminate. Measling first appears as a series of lighter dots in a pattern (Fig. 7-35). Blisters are large, rounded areas where delamination is evident.

All three phenomena are heat and stress oriented. The resin that holds the laminate together, usually epoxy, is weakened by the glass transition phenomena. As the resin goes through the critical transformation temperature T_g, it loses about 90% of its strength. The transition temperature for most laminate resins lies below the melting point of solder. Therefore, every time molten solder is in contact with the board, there is a

Fig. 7-35. Lifted pad and land; also note white measling spots around hole. (Copyright Manko Associates)

simultaneous lowering of strength. In this weakened condition any stress can cause delamination and separation. The stresses may already be in the assembly from such operations as staking, punching, and bending.

In lifted pads (Fig. 7-35), the pressure usually comes from an untrained solder operator with an iron. If pressure is exerted on the pad during soldering (the temperature is always above the T_g) pad lifting will result. It may also result from stresses in component cutting or bending. Repair procedures are based on re-bonding the lands with an adhesive. This is costly, and the reliability of the repair is not as high as that of the original construction.

Measling (Fig. 7-36) and blistering are normally caused by presolder stresses. Punching, shearing, and similar mechanical operations trigger the internal forces that cause separation when the T_g temperature is exceeded. No repair methods are available to restore the original condition. To most none-critical assemblies, measling does not pose a serious reliability problem.

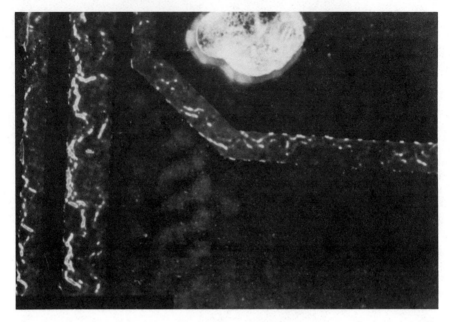

Fig. 7-36. Measling on surface of board, also note crinkling of solder resist. (Copyright Manko Associates)

7-27 COMPONENT TILTING IN SURFACE-MOUNTING (TOMB STONING)

There is a tilting condition unique to leadless chip mounting, called *tomb stoning*. This occurs when one side of the leadless chip solders in place, while the other tilts up from the pad. The result resembles a tilted tombstone, hence the name. This condition results from two separate processes:

1. In wave soldering, it happens only infrequently when several conditions exist:

   - The adhesive has been sparsely applied, and once it was heated above T_g, it can no longer hold the component. Alternatively the adhesive was applied over a dirty surface, with similar results.
   - The component is oriented parallel to the direction of the conveyor, passing one termination through the wave before the second.

- The side that is soldered in place has the metallization wound around the edge.

Under these circumstances, the surface tension of the solder, aided by the motion of the wave, can cause the tilting phenomenon. Logic dictates that the receding edge be anchored, to avoid losing the component in the moving wave. Tomb stone conditions have also been observed on the leading edge.

2. In reflow soldering, tomb stoning is much easier to explain. There is no adhesive to hold the part in place, as in the wave operation. Any uneven melting of the solder will cause one joint to wet before the second, and the surface tension of the solder will do the rest. To avoid this condition:

- Bring up the assembly heat as uniformly as possible to wetting temperatures.
- When using a cream or paste, try to prevent it from climbing up the wrap-around-termination edge.
- Place components as flat as possible.

The last piece of advice reflects the placement problem associated with the pad profile (see Section 5-17). The quantity of reflowed solder on the pad determines the flatness of that surface. A high solder profile, suitable for good component spacing (off the board), is rounded. During component placement, it tends to move and shift the components out of the planar position, helping the tilting during soldering.

7-28 SKIPPING AND SHADING DURING SURFACE-MOUNTING

The phenomena of *skipping* or *shading*, refers to the unsoldered pads and or terminations found on the bottom of a board after wave or dip soldering. For more information see the discussion in Section 5-14. This holds true for the pretinning of board pads by molten solder, including the leveling process, as well as for the soldering of low-profile components to the bottom of the board.

The condition results from gas entrapment on the surface, which must be displaced mechanically. Some additional steps can be taken to improve mechanical devices like double waves. Therefore, if you are periodically plagued by skipping problems, with or without components, look for the following potential causes:

1. The surfaces are dirty and coated with an interference layer that prevents proper flux action. This condition is aggravated by the low surface profile.
2. The surfaces have poor solderability with the flux in use (see Sections 7-1. 2-4, 2-6, & 2-7).
3. The component is spaced too far away from the mating surface.
4. The board is badly warped.

A similar condition can be found in reflow soldering with preplaced flux and solder. Several joints on the board do not seem to solder, while others do. In many respects this is similar to the condition of tomb stoning described in Section 7-27. The remedies are very similar:

- Heat the assembly as uniformly as possible so that all joints reach the wetting temperature together.
- When using a cream or paste, make sure that the quantity is sufficient to bridge the gap between the two surfaces.
- Make sure that the solder cannot run out of the joint by using solder mask as a solder dam.
- Place components as flat as possible and make sure that at least part of the mating surfaces overlap.

The last suggestion refers to self alignment, which will help to prevent skips (see Section 5-7).

7-29 WEAK PLATED-THROUGH HOLE KNEE

The corner that is formed between the pad around the hole and the barrel in the hole is called the *knee* of the plated-through hole. The term *weak* does not denote structural problems, but solderability difficulties. This area is unique when it comes to soldering, and affects the solder's rise to the top of the fillet.

To understand why the knee is vulnerable, let us consider its structure. It starts as a hole edge in the laminate, with copper on the outside surface. It is then covered with a copper plating, at which time the edge is perpendicular, forming a 90° bend. Since bare copper has a relatively short shelf life, it is covered with solder. The tin-lead is applied by electro-plating and reflow or by hot coating. In either case, the solder forms two distinct meniscus-like surfaces that meet at the edge. The molten solder cannot cover the corner, leaving the knee with a very thin protective layer of solder.

The two major mechanisms that cause weakening of this vulnerable area are:

a. Mechanical damage. This is by far the more common cause of weak knees. In its worst form, copper shows through in the knee area. This is caused by:

- Undue scrubbing of the board, after reflow, renders this thin layer unsolderable. Pumice brushing is the most harmful operation, but plain bristle brushing has also caused the problem.
- Rubbing the boards against one another can also cause this condition, although only in isolated areas.
- Rough insertion techniques have resulted in partial weak knees. The damage is found on the part of the hole where the lead wire rubbed the hole.

b. Prolonged aging or adverse storage conditions can cause weak knees. This is obviously a function of the very thin protective solder layer in the bend.

The weak knee forms only after solder reflow or solder leveling. As long as the surface is electroplated, the knee is well protected. In addition, the amount of solder on the knee is a function of the part's geometry. Sharp corners in the knee are better protected than rounded corners. Furthermore, one side of the board is usually more vulnerable than the other. This difference can be traced back to the surface which was the top during reflow. It appears that gravity tends to help protect the knee on the bottom.

The effect of a weak knee on the solder joint is seen only on the top (component side) of the board. The condition prevents the solder from climbing up over the pad surface, while it rises on the lead wire (Fig. 7-37). The appearance is very similar to a crack in the fillet. It is sometimes referred to as a "stress line".

The reliability of the joint top is an interesting subject. It hinges on the acceptance of the plating in the hole as a viable conductor. If the plating is reliable, the weak knee is no problem. In the author's opinion, this is the right approach, Be careful not to allow touchup from the top to correct this condition. The high heat of the soldering iron is too close to the components and can cause damage (see Chapter 9). Here the cure would be worse than the cause.

The weak knee on the bottom (solder side) of the board does not show

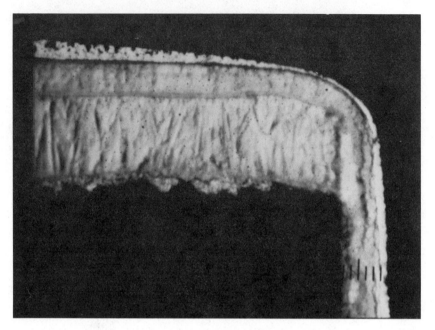

Fig. 7-37. Cross section of plated-through hole knee showing the weak area.

because the solder bridges over the solderability gap created by this defect. There is no reliability problem in this case.

The steps to prevent weak knees from being shipped in, or formed inhouse are simple:

1. Inspect for copper showing in the area.
2. Specify that no brush marks or scratches are allowed on all incoming reflowed solder surfaces.
3. Specify that the boards should be send through reflow upside down. This way the top knee can have more solder protection.
4. Avoid rubbing boards against each other during handling, and assembly.

7-30 SOLDERING CHEMICALS TRAPPED UNDER SOLDER-MASK

The entrapment of soldering chemical under the solder mask (resist) must be avoided for obvious reasons. Fortunately, it is formed only if several conditions exist simultaneously as follows:

- Poor mask adhesion.
- A pin hole in the mask.
- A suction mechanism.

Let us follow the sequence of events in the mechanism of entrapment:

1. When there is poor adhesion between the solder mask and the board, a hidden pocket is formed. This cavity is needed to hold the chemicals or there will be no entrapment. Adhesion problems can result from dirty surfaces, a condition that can be detected during incoming inspection. More frequently the lack of adhesion develops during soldering, when the melting of the solder under the mask shifts (see Section 3-16).

2. If there is a pin-hole or discontinuity in the mask, which provides an opening to admit the soldering fluids. The source for these mask imperfections depends largely on the type of material used. Some typical sources are:

- Brittle resists that flaked off.
- Air bubbles created during screening.
- Solder mask starvation during screening.
- Mechanical damage (gauging).
- Large ground planes without cross-hatching.

3. When there is a negative pressure created during the process, while a liquid chemical is present at the opening. This will provide the force to draw the chemicals into the pocket. The shifting of solder during wave soldering, provides both pressure and suction which makes the solder crinkel (see Section 3-16). A wrong heat profile during washing also creates a negative pressure that can draw dirty cleaning solutions under the mask (see Section 6-18 and Table 6-6).

Trapped soldering chemicals pose a special danger, because their effect is delayed. Their adverse reactions may develop at a later time, usually at the customer's location. If the conditions for the trapping mechanism are unavoidable, judicious selection of soldering chemicals can prevent the problem. By using only rosin based fluxes, of the right strength, trapped flux residues will remain harmless. For those who wish to eliminate any chance of chemical penetration under the mask, two options are open:

1. Use solder mask over bare copper (SMOBC) and prevent the possibility of solder shifting (see Section 3-16).
2. Use a dry-film (sheet) resist with photo-imaging techniques. These

masks are applied as a tough film, which is not likely to develop pin holes.

To summarize, the entrapment of soldering chemicals can be a serious reliability hazard. Fig. 7-38 shows a case history, where the shifting solder created the cavities. In this case a brittle mask provided many openings, where sections flaked off. The water soluble chemistry that was used for this process, aggravated the situation. The organic intermediate flux was trapped under the mask, and was not washed out during the standard inline washing. The units failed after a short time in service. The use of solder mask over bare copper was the solution selected for this problem. This should be considered by all users of water based systems.

Another problem which is generated by the same mechanism is the extrusion of solder balls onto the surface. These are not loose, but adhere to the copper land from which they originate (Fig. 7-39). They cannot be washed off for obvious reasons and must be removed with a hot soldering iron.

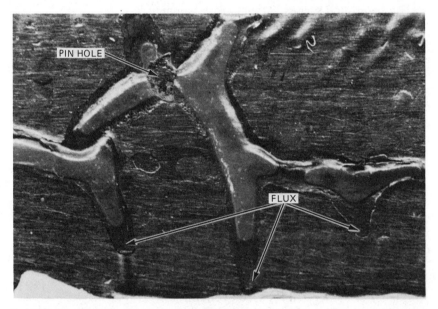

Fig. 7-38. Pin hole in a solder mask has admitted dirty cleaning solution, which is trapped under the mask, and cannot be washed out. (Copyright Manko Associates)

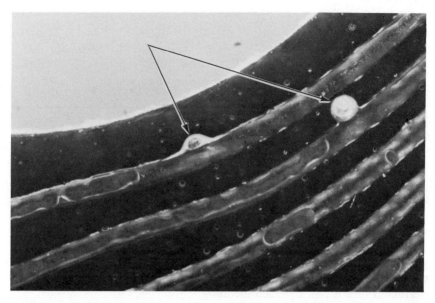

Fig. 7-39. Solder ball squeezed out from under a solder mask on top of board during wave soldering of the bottom. Note adjacent drop which has not formed into a proper ball. (Copyright Manko Associates)

REFERENCES

7-1. Howard H. Manko, "Eliminate Poor Solderability, Don't Bury it under Solder,"*Insulation Circuits*, February 1976.

7-2. Howard H. Manko, "Soldering Defects That Count and Those That Don't," *Circuits Manufacturing*, September 1981.

7-3. J.D. Keller, "Can the U.S. Afford the Cosmetic Look in Soldered Joints?," *Assembly Engineering*, October-November, 1973.

7-4. J.D. Keller and J.L. Waszczak, "The Case for Unfilled PTH's", *Electronic Packaging and Production*, October, 1973, pp. 144-49.

7-5. Howard H. Manko, "Steps to Your Own Solder Training and Quality," *Circuits Manufacturing*, November 1979.

7-6. Howard H. Manko, "Solderability—A prerequisite of Tin Lead Plating," *Plating* July 1967, pp. 826-828.

7-7. IPC—institute for Interconnecting and Packaging Electronic Circuits, 3451 Church Street, Evanston, Ill, 60203.

7-8. Howard H. Manko, *Solders and Soldering*, 2nd ed, McGraw Hill Book Co., New York, 1979.

7-9. Howard H. Manko, "Understanding the Solder Wave", *Insulation Circuits*, January 1978.

7-10. R.A. Bulwith, "Blowholes and Voids", *Electronic Packaging and Production*, November 1976.

8

QUALITY AND INSPECTION

8-0 INTRODUCTION

The feedback that is generated by quality control and inspection is a vital tool in today's manufacturing cycle (statistical quality control). It eliminates the unnecessary cost of repair and guarantees the quality of the end product. In soldering and cleaning technology this area is often badly neglected, yet it has a special meaning.

The *quality solder joint* has been traditionally defined by committees and organizations. Most specifications that prevails today originated in the days of NASA's big push into space. The United States was committed to overtake the Soviet Union in this race, and cost was not important. Under these conditions, the *ideal solder joint* was easy to define:

The solder fillet has to be as close as possible to the ideal, and no fault is tolerated.

If this standard was very stringent, that was considered to be good. Potential damage from unwarranted touchup or repair was completely ignored.

In today's competitive world, this attitude towards quality no longer applies. It is time to think in terms of *reliable solder connections*. These can best be defined as:

A connection that can withstand the expected stresses under operating conditions for the life of the product.

In other words, the solder joint must outlast the equipment under the rigors of use. This definition, in concept, is vastly different from that of the old quality solder joint. The reliability impact of the stresses caused by unnecessary touchup and repair are included in the decision cycle (see Chapter 9). Not all the equipment we build is used in space, nor can the cost factor be ignored. Thus it is important for each organization to have its own quality and workmanship standards rather than to follow the ideal fillet specifications (see Section 8-2).

8-1 VISUAL VERSUS OTHER INSPECTION METHODS

The conventional method for solder inspection is a visual check, which is still the best one available. When comparing the solder joint to given reliability criteria, it sets up a simple yet effective quality control system (Figs. 8-1 and 8-2). The difficulty occurs in selecting the right criteria and training the operator and inspector to abide by them (see Section 10-3).

There are equally good or even better destructive methods of assessing joint quality. However, they have the following inherent drawbacks:

- They are more expensive: in addition, part of the lot must be destroyed to get results.
- They are slow and time-consuming, unlike visual inspection, which is *as quick as a glance.*

Fig. 8-1. "Snifty" the ideal joint inspector. One simply places the solder fillet in front of the schnozzle . . . etc. (Design first proposed by author in *Product Engineering,* June 13, 1960, requesting help with the interior design.)

Fig. 8-2. "Eyeballs" still provides the most practical approach to solder inspection, provided a meaningful quality standard exists.

- They require complex interpretation by off-line technicians, not production line operators.

In summary, destructive testing is slow, costly, and cumbersome. These methods should be used to augment visual examination on a regular basis. They require trained technicians to interpret the results, which is an important equalizing factor. Visual inspectors need just as careful a training program as laboratory technicians. They also need a set of good workmanship standards for proper interpretation (see Section 8-2).

Destructive methods include cross sectioning, tensile and shear testing, vibration analysis, and so on. They are essential in the beginning, during the design stage, and for continuous verification during production.

Some nondestructive automated methods have been suggested and tried. They are standard metallurgical methods like ultrasonics, X-rays, radiation, and the like. Unfortunately, they are not very effective on solder joints containing lead-based solders. The lead absorbs ultrasonics, is not penetrated by X-rays or radiation, and thus remains inscrutable by these means. Lead free solder joints are routinely inspected this way. X-rays can be used to detect the presence of voids and solder balls under components in surface mounting (see Chapter 5).

For very uniform solder fillets, various methods of thermal analysis hold promise. A controlled heat source, like a low-power laser, imparts a known amount of energy to the joint. A sensitive instrument picks up the heat rise in the joint, which is a function of mass. The equipment thus indicates if the solder mass in the joint is within tolerance limits and how well it conducts the heat to the assembly. This test is good for very uniform, repetitive patterns or thermally well-defined assemblies.

8-2 ESTABLISHING YOUR OWN WORKMANSHIP STANDARDS

In the orderly flow of a new product design, the requirements for all solder connections should be calculated. Then good solder joints can be easily defined in specific terms. This, unfortunately, does not happen with printed circuit boards. The designer concentrates on the layout and other aspects of the boards, and the solder joints just happen. As a result, the average organization has hardly any guidelines for inspection.

Fortunately, all is not lost; even the inspector and quality engineer can readily calculate any fillet. A joint design outline is given in details in Chapter 4 of Ref. 8-1. It enables anyone to determine how much of a solder connection is really needed and, in retrospect, to set up quality guidelines. In addition, Sections 1-4 through 1-10, of this book present a detailed design analysis of the plated-through hole connection, and Chapter 5 provides data on all surface mounted joints.

For example, not every plated-through hole has to be completely filled all around the circumference (360°). In many fillets like dual-inline integrated circuits (ICs) or other multi-leaded components, the weight per lead is minute, the current-carrying capacity is limited, and no appreciable heat develops in the device. Under these conditions, if at least half of each hole is filled, there is no need for touchup. Obviously, good solderability and wicking to the top must be within acceptable limits (Ref. 8-2).

The least desirable approach is to copy a set of such specifications from someone else. Each case is unique and requires attention. For example, the author has worked with every major color TV manufacturer in this country and has helped to set up or criticize their standards. None of these companies could justify adopting a competitive document, since each is quite unique. No one quality or workmanship standard could be written for the TV industry. The same holds true for computer, appliances, and other types of equipment.

The government is a unique case that further strengthens this argument. For obvious reasons, their specifications must be uniform throughout. A contractor who wishes to get work must plan and price his product according to these standards. Nevertheless, each program has a large number of variances, which recognize the special needs of each assembly.

The method used to set up your own specifications is given in Section 10-3. It follows the first part of the outline for writing your own training program. See also Section 7-19 in Ref. 8-1.

8-3 INSPECTION OF SOLDER JOINTS

Visual examination alone is also inadequate, if it is not supported by incoming inspection (see Section 8-5) and in-process control (see Section 8-4). With a workmanship standard available, the visual evaluation of solder joints can proceed. The major points of interest are as follows:

- Degree of wetting.
- Joint contours.
- Design requirements.
- Thermal damage.
- Special conditions.

Let us discuss each of these in detail, with inspection in mind. The scientific background for each item is covered elsewhere in this book and Ref. 8-1:

1. *The degree of wetting*: This is our measure of joint perfection. Fig. 8-3 shows a series of visual aids for board laminate. The sample on the right shows good wetting, while the rest show various degrees of imperfection. The nonwetted sample shows an area of exposed copper base metal. In the dewetted sample no copper is visible even though the solder pulled back from the surfaces.

Rejects due to poor wetting should be flagged for special handling and repair. Remember that: *more time or temperature with a soldering iron do not improve poor wetting once it has occurred* (see Section 9-1). The proper repair procedure for inadequate wetting consists of three steps:

1. Remove all solder to expose the offending surface.
2. Restore surface solderability (see Sections 2-5 through 2-7).
3. Resolder the joint with new flux and solder.

For further details on trouble-shooting poor wetting conditions, see Sections 7-2 through 7-4.

SOLDERABILITY TESTS SAMPLES

First Class Wetting   Small amount of Dewetting   Complete Dewetting   Nonwetting

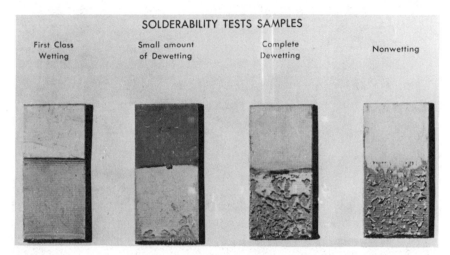

Fig. 8-3. The quality of wetting as demonstrated on four laminate samples. (A color photo was donated by Alpha Metals to the IPC for their use.)

2. *The joint contours*: These are an indication of the volume of solder deposited (Fig. 8-4). A maximum and a minimum amount of solder should be called out in the workmanship standard for each type of joint. The solder quantity per fillet is dictated by electrical conductivity, mechanical integrity, and heat dissipation. In addition, the contours should be concave and the outline of both surfaces visible. A smooth surface is required to exclude disturbed joints (see Section 7-5), and the presence of intermetallic compound contamination (see 7-21).

Note that excess solder hides true wetting conditions without adding to strength or conductivity. This condition is labeled *Lack of Inspectability* and is reason for rejection.

3. *Design requirements*: These are needed to fulfill special joint requirements. This includes quality compromises on such things as degree of hole filling, height of solder rise, and other factors. Again, these requirements must be spelled out in the standards and, if possible, backed up by color photos or sketches.

4. *Thermal damage*: Thermal damage to the fillet is rare, but to the plastic materials surrounding the joint it is more common. The results may be lifted pads, measling, and blistering (see Section 7-26). Other conditions include solder webbing (see Section 7-13) and various residues (see 7-25).

Fig. 8-4. A good solder joint with the right contour. (Copyright Manko Associates)

Some types of thermal damage, like board warpage, require process adjustments. Most of the conditions described above are material oriented and require a different approach.

5. *Special conditions*: These include peripheral details like length of lead protrusion (see 2-9), stress relief on component leads, and so on.

In general, final inspection is only a small part of the program that guarantees reliability. Remember that the inspector does not change the quality of the lot he views. He merely ensures compliance with the standard. Final inspection alone is not efficient and is very costly because it feeds back information after the fact. The rejects found are expensive to repair. Corrective measures to eliminate the fault can only be incorporated in a later manufacturing cycle.

8-4 IN PROCESS CONTROL

A very beneficial effort to ensure the reliability of solder joints is control during the process itself. At this stage, the feedback cycle is much shorter. The production, can take corrective action before a large quantity of rejects is produced. Production is also the location where a computer exercises most of the control over the process in any computer aided manufacturing CAM system (see Section 4-19).

Many of the defects that can be avoided by this step may not be visible or detected by final inspection. For example, overheating may shorten joint life with no visual damage, an electrostatic discharge or excessive component stress, will damage joints internally (see 2-10).

In-process control should be a quality assurance function, or at least audited by the quality assurance department even if it is assigned to production. In either case, high reliability and low cost must be the two ultimate goals. Let us review the major points of interest:

- First item audit (FIA).
- Soldering and cleaning equipment performance.
- Soldering and cleaning materials maintenance.
- Special conditions surveillance.
- End product inspection.

Most of these items are covered in other parts of the book. Let us summarize them briefly:

1. *First item audit*: The FIA is an effort to forestall trouble by a preproduction run. Every company must have an efficient incoming inspection function (see Sections 2-2 and 8-5). After receipt, there is a period of long or short storage when quality may deteriorate. This refers not only to solderability of the surfaces but also to moisture absorption of the boards and other factors.

It makes good sense, therefore, to preassemble a small sample of boards and solder them on the production equipment. These boards are then subjected to careful scrutiny for any signs of trouble. Any corrective measures at this stage are easy to make and lowest in cost. The FIA must therefore be executed after kitting or staging, and before any assembly work is done. This is true especially before sequencing and automatic insertion or surface mounting with a permanent adhesive. At this stage, a single component or the bare board is easy to rework. Once components have been mounted on the board, corrective steps are very costly and cumbersome. Such remedies include the restoration of solderability (see Sections 2-5 through 2-7) and prebaking of boards (see Section 2-12).

2. *Soldering and cleaning equipment performance*: Control of the soldering and cleaning equipment parameters must be left to the operator, but the monitoring is a quality control (or the computer's) function. Making sure that temperatures are correct and calibrating the equipment on a schedule are useful illustrations of the concept. In addition, records must be kept for future reference and equipment maintenance schedules monitored.

3. *Soldering and cleaning materials maintenance*: The soldering and cleaning chemicals require control just as the equipment does. Here, too, hour-by-hour control must be placed in the hands of the operator. The inspector, however, is responsible for double checking the results less frequently. Such items as flux density and solder purity should be part of quality control.

The control of the major materials in this category is described in Section 8-5.

4. *Special conditions surveillance*: Many assemblies require special process conditions or a unique sequence of operations. This may include the use of heat-sensitive components, the handling of static sensitive devices, or the mounting of nonwettable components during cleaning. The role of the inspectors here is clear. It is their responsibility to see that the procedures are followed. The damage that results from lack of compliance may well be hidden until the work reaches the field.

5. *End-product inspection*: The quality of the final product should also be continuously audited by production and QC. This is needed for rapid adjustment of the process before a substantial number of rejects are made.

This is very important for record keeping and the establishment of a production data base. In most cases, soldered boards go directly from the wave or reflow to touchup. Their quality is thus altered before they reach final inspection, and no true statistical picture can be obtained. This yields a false solder process profile and eliminates the data needed to fine tune the process.

A computer tied into the process can perform many of these functions automatically (see Section 4-19). This does not relieve the quality assurance department of the responsibility for auditing the process; it merely simplifies the task. It is important to view the in-process function as a quality assurance step, since it is related directly to product reliability.

8-5 INCOMING INSPECTION

This is the first inspection step. At the receiving end, there must be close cooperation between the engineering, production, purchasing, and quality control department.

The sequence must start with good engineering specifications that clearly define, without any ambiguity, what is needed. This includes functional requirements (electronic, mechanical, etc.), physical size measurements, and material composition including solderability. Such requirements should be based not only on design but also on production requirements for a full value analysis. Here total cost and quality are important, not just the initial purchase price. On the basis of these detailed documents, the purchasing and quality assurance departments can identify a number of approved vendors.

Let us review some of the standard items that must be included for soldering and cleaning:

- Solderability.
- A termination finish to withstand aging.
- Stability at soldering conditions (temperature and time)
- Resistance to cleaning agents.
- Safe inner packaging.
- Smooth, clean plated-through holes.
- Properly cured plastic coatings.
- Correct hole to wire ratio (lead versus plated-through hole diameter).
- Product oriented parameters gathered by experience.

Let us cover each one in more detail:

1. *Solderability*: There are specific tests available for qualifying solderability before and after aging (see Sections 8-6 and 8-7). This is one area where a substantial amount of attention is mandatory, but where the results are very rewarding.

2. *A termination finish to withstand aging*: The termination finish is crucial not only for proper soldering and storage but also for joint reliability. Gold-plated lead wires are a good example, since they cause a number of well documented problems. In a well-run government or commercial organization, all gold plated leads are pretinned on receipt, using a double dip process. Rather than go through this added expense, it would be wise to prohibit the purchase of gold finishes. If this is ignored, and gold surfaces are used, there is a danger of added costs due to production problems or field failures.

Similar termination problem exists in surface-mounted devices, where silver fired films are used. Unless a nickel barrier plating with a solder coating is specified, silver scavenging can become a serious production problem (see Section 5-12).

3. *Stability at soldering conditions (temperature and time)*: All parts of the assembly must be able to resist the soldering heat. Sometimes low cost parts are heat sensitive, which creates a costly production situation, because these devices must be hand soldered at a later stage. A cost analysis will reveal that in addition to the disruption of the manufacturing flow, such components also increase costs. The added expense is usually much higher than any reduction in the purchase price.

4. *Resistance to cleaning agents*: The resistance to cleaning agents is a function of the solvents used. The cleaning system should have been carefully selected for quality and cost considerations and should not be compromised by the quality of the purchased parts. Too often a more expensive and less efficient cleaning solvent is used, because of poor purchasing control. Swelling of components in solvents, loss of marking, and attack on plastics make components incompatible with the cleaner. Remember to test solvents at the actual cleaning temperature (i.e., soak components in boiling solvent if vapor degreasing is used).

5. *Safe inner packaging*: The inner packaging material (usually a plastic film) should be differentiated from the outside shipper (wooden crate or cardboard box). Usually the environment and the outer container are detrimental to the solderability of the surfaces. Plastic bags 4 mils thick (minimum), are considered good inner packaging materials. They must be free of silicone mold releases and ionizable contaminants. Electrostatic sensitive components require additional protection, which should not violate good solderability protection. For further details, see Sections 2-2 and 2-3.

6. *Smooth, clean plated-through holes*: The quality of the plated-through hole is critical for good solder rise (see Section 7-6), and the prevention of blow holes (see Sections 7-16 and 7-17). Inspecting the hole with back-lighting yields the required information (see Section 8-9).

7. *Properly cured plastic coatings*: The curing of plastic coatings, like the solder mask, is entirely in the hands of the vendor. This incoming check is needed to identify formulation difficulties in two component systems or improper curing cycles. It prevents such problems as solder webbing (see Section 7-13), white residues (see Section 7-25), and embrittlement. The best method used to check these parameters is the *float test*. The sample is fluxed with the production flux and floated on top of a solder pot for 8-10 seconds. It is then cleaned in the actual cleaning system and carefully inspected.

8. *Correct hole to wire ratio*: Dimensional checks like the lead wire and plated-through hole diameter for the hole-to-wire ratio, or board dimensions for the conveyor opening, are obvious. They do save the company a great deal of unnecessary expense.

9. *Product oriented parameters gathered by experience*: This list is only partial and will grow with experience. Each manufacturing facility encounters a number of problems. It is wise to add only repeated defects to the list, or it will become too long and cumbersome.

One beneficial result of record keeping during incoming inspection is the ability to create a *vendor rating grid*. If the acceptance and rejection data on each part are tabulated, a ranking of vendor quality results. It can be the basis for the removal of frequent offenders or the reward of a reliable source. This ties in with the value analysis that an organization must run on each purchase. A rejected part requires action which increases the unit cost. It is short-sighted to procure the least expensive part without considering the reliability impact and the added cost in production, repair, and so on.

8-6 SOLDERABILITY DEFINED

Solderability is a specific property of a surface, yet the term is often misused. It has become a catch-all phrase for soldering problems of all kinds. Let us define it in its true sense:

Solderability is the degree of wettability of a surface with molten solder under defined conditions of time, temperature, and environment (Flux).

This definition emphasizes that we are dealing with the surface of the base metal, not the bulk. Furthermore, we are concerned with a surface condition that is affected by the history (aging) of the metal and the environment in contact with it. While the intrinsic properties of a metal are important, surface treatment by the flux is needed to overcome the aging effects.

The above definition is appropriate for solid, uncoated metal surfaces, usually referred to as the *base metal*. In the case of a part coated by electroplating or hot dipping, solderability remains a surface property. The quality or integrity of the solder joint, however, depends on a different critical area, namely, the interface between the base metal and the surface coating.

This shift from the surface to the interface requires additional analysis (Ref. 8-3). We will divide the coatings according to their behaviour at soldering conditions (see Table 8-1).

Table 8-1. Critical Surface Versus Solderability.

METALLIC STRUCTURE	REACTION TO MOLTEN SOLDER	CRITICAL LAYER
Solid	Standard	Outer surface
Coated	Soluble in solder	Interface
Coated	Fusible at temperature	Interface
Coated	Not soluble or fusible	Outer surface

Let us cover these coating categories in the order in which they appear in the table:

1. The first coatings to consider are the *solder soluble* ones like gold, silver, cadmium, and zinc. Not all of these coatings are now used in printed circuit assembly. The last two are detrimental to the behavior of the bulk molten solder. Gold in large quantity also reduces joint quality. Obviously, we are talking about relatively thin coatings, as in electroplating or hot dipping. Heavier layers may take too long to dissolve down to the interface.

When a part like a component lead wire or a printed circuit board is coated with such a metal, wetting starts on the surface. With time the thin coating is dissolved in the molten solder, and wetting to the interface must take place. Thus the quality of the joint depends on the thickness of the plating and the solderability of the interface. Unfortunately, it is possible that the electroplating was deposited over a slightly contaminated surface, which will be coated by electroplating but will not wet with solder. This contamination residing at the plating-base metal interface may thus prevent good solder wetting. This problem can be prevented by close quality control before plating. A simple solderability test of the base metal prior to plating is required (Ref. 8-3).

2. In the case of a fusible electroplating, like tin or tin-lead coating, the same holds true. Wetting starts at the surface, but when the coating is heated to soldering temperatures, it also reaches the melting point and fuses, mixing with the solder. Wetting is again required at the interface with the base metal. Here too, solderability of the base metal prior to electroplating needs control.

3. A number of metallic coatings are neither soluble in solder nor fusible at the wetting temperature. These include nickel, iron, and the like. Here solderability is merely a surface wetting condition, as with the uncoated base metals. Joint strength is obviously also a function of coating adherence to the base metal.

8-7 SOLDERABILITY TESTING AND ACCELERATED AGING

A number of tests are used to assess solderability. The best and most universal one is based on a controlled dip. To understand the principles involved, let us outline the manual version:

1. Flux the test specimen (in the as-received condition).
2. Allow excess flux to drain and volatiles to evaporate.
3. Skim molten solder in the pot to remove dross.

4. Dip the specimen vertically in solder. Let it dwell long enough for wetting.
5. Withdraw the specimen and allow the solder to freeze without vibrations.
6. Clean off flux residues.
7. Inspect the surfaces for wetting and compare the results to visual standards (Fig. 8-3).

The concept here is to simulate defined soldering parameters and inspect the results. Dip soldering was selected because here the wetting forces are countered by gravity. Solder will not adhere falsely to a poor-quality surface. In addition, dipping provides easy heat transfer when the rate of insertion and the dwell time are correct. The visual wetting standards are for the use of novices only; an experienced technician soon does not require them.

Table 8-2 lists some of the published specifications used in the industry.

Table 8-2. Solderability Test Methods.

TEST NAME	SUITABLE PCB	LEAD	SOME PUBLISHED REFERENCES[a]	COMMENTS AND RATING
Dip test	Yes	Yes	MIL-STD 202 Method 208, MIL-STD 883 Method 2003, IPC-S-801, IPC-S-803, EIA-RS-178B, EIA-RS-319A, IEC-68-2-20, Test T	Visual rating; most common in the industry; requires the least amount of equipment
Wetting balance	b	Yes	MIL-STD 883 Method 2022, IEC-68-2-20, Test T	Numerical & graphic ratings; high cost
Globule test	No	Yes	IEC-68-2-20, Test T	Numerical rating; popularity declining
Spread test	No	No	Federal QQ-S-571, MIL-F-14256	Numerical rating for flux, solder, etc.
Twisted pair test	No	No	None known	Metallurgical interest only

[a] IPC—The Institute for Interconnecting and Packaging Electronic Circuits. 3451 Church Street, Evanston IL. 60203.
EIA—Electronic Industries Association, 2001 Eye Street N.W., Washington, D.C. 20006.
IEC—International Electrotechnical Commission, U.S. National Committee, c/o ANSI, 1430 Broadway, New York, NY. 10018.
[b] Suitable for raw stock (double-sided clad laminate) only.

The specific parameters used in each test determine its suitability for a particular application. The following general concepts, however, are also important:

1. The majority of tests use a weak flux in order to detect marginal surfaces. The assumption is that the good parts will always pass, while the bad ones will always fail. The marginal parts are thus weeded out by the flux used. This makes sense for incoming inspection, but not for pre-production analysis (FIA).
2. To simulate possible surface deterioration in storage, accelerated aging is included in the test. This, again, is invaluable for incoming material evaluation and should always be done. The *steam aging* procedure is considered the most appropriate. It combines heat, air, and humidity to accelerate any deterioration on the surface. Dry heat was found to give only partial information, since chemical processes often depend on humidity to develop a surface condition.
3. Thought must be given to the interpretation of the test results when these are used to arbitrate between the vendor and the user. In this respect, numerical results are easier to use but more costly.

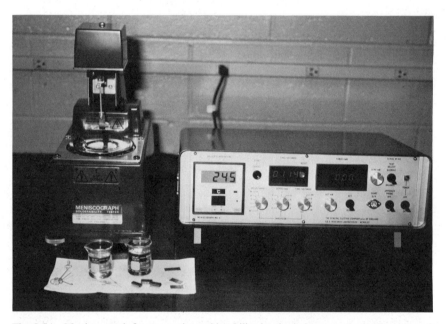

Fig. 8-5A. Meniscograph for measuring solderability by the balance method. (Courtesy of Hollis Automation)

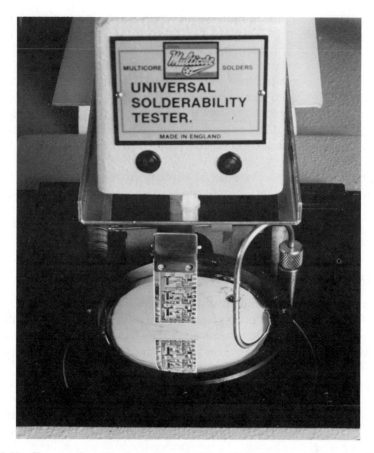

Fig. 8-5B. Closeup of wetting balance head. This unit can be adapted to a number of different solderability tests. (Courtesy of Multicore Solders)

4. When evaluating soldering materials (fluxes, coatings, solder alloys, etc.), surfaces are often treated chemically and/or heated. These methods are used rather than steam because a known coat of metal reaction products is desired. The results of steam aging depend on the history of the part, not its inherent properties.

The most accurate and acceptable numerical evaluation is the wetting balance test (Figs. 8-5A and 8-5B). The equipment consists of a controlled dip process, in which the specimen is suspended from a sensitive electronic balance. The output is either graphic or numerical and records the forces exerted on the sample during dipping. At the beginning of the immersion cycle, negative buoyancy or expulsion energy exists. The unwetted specimen is lighter than the molten solder, and thus will float. Once the solder wets the surfaces, however, the force reverses direction and turns into a positive pull into the pot. The time, rate, and extent of this force balance are used to assess solderability.

While the wetting balance test is not a full duplication of production conditions, it bears a close correlation to most soldering processes. For example, the wetting time must be shorter than the actual solder contact in the wave.

Unfortunately, the test cannot be used on etched printed circuit boards because of the variations in pattern. Uneven metallic coverage causes a tilting force to be exerted on the immersed specimen. This is the reason why single-sided laminate (unetched) also cannot be checked. The method is acceptable only for double-sided clad raw materials.

8-8 PRE-PRODUCTION SOLDERABILITY AUDIT (FIA)

In addition to solderability testing during incoming inspection, there is a need for a check just prior to actual use. The first test is intended mainly to check on the supplier and to take action, when warranted, at the time of receipt. But good parts may also have been contaminated in storage and handling, and their solderability may have deteriorated.

It is good practice, therefore, to check the condition of all surfaces just prior to assembly and soldering. This is normally termed a *first item audit* (FIA). For this check, the actual production materials and equipment are used. Several boards are hand assembled after kitting or staging and before automatic assembly. These hand assembled boards are then wave soldered with the usual flux under normal production conditions and closely inspected. Any trouble with solderability is easy to correct at this stage (see Sections 2-5 and 2-7). Only then are the components sequenced for automatic assembly or prepared for manual insertion. Remember that once a board is assembled, restoring the quality of any single part is cumbersome and very costly.

Figure 8-6 shows a typical flow chart for solderability monitoring, as well as actual evaluation. This is the best way to eliminate unnecessary rejects.

The first item audit test obviously also helps to reveal other board problems, like blow holes (see Sections 7-16 to 7-19), webbing (see Section 7-13), measling and blistering (see Section 7-26), and the like. Where cleaning is part of the wave process, cleaning problems like white and grey residues (see Section 7-25) are also predictable.

MANAGING SOLDERABILITY

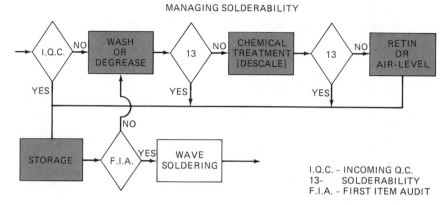

Fig. 8-6. Flow chart of solderability management.

I.Q.C. - INCOMING Q.C.
13-   SOLDERABILITY
F.I.A. - FIRST ITEM AUDIT

8-9 INSPECTING THE PLATED-THROUGH HOLE

The holes and their plated layers are important to the overall quality of the solder joint. There is no good solderability check for the inside of the hole. Some companies assess the number and fill of the holes when the board is floated on a solder bath. This was never accepted by any specifying authority, although the IPC float test is often used as a basis. There is also a modification of the globule test that measures the time needed for solder to rise through a plated hole. Both of these methods are destructive.

The Author has used a nondestructive visual test that can provide much information. The board is mounted at an acute angle under a 10-30X power microscope (stereo preferred). Light is focused from the back at a right angle with the optical system. This back light enables the inspector to view fully the inside finish of the hole. This test resembles the old fashioned inspection of the bore of a rifle. It is suitable mainly for reflowed or air leveled boards. The critical conditions inspected are as follows:

1. The wall must be smooth and uniform, with only slight ridges representing the inner glass layers. No nodules, loose jutting-out fibers, or cracks are allowed. (Note that some of the specifications allow up to 20% circumferential cracks; this does not bring into consideration the quality of soldering, only current conductivity.)
2. The walls of the hole must be shiny, with no foreign materials visible. This is a guarantee that the plating was properly reflowed all the

way to the cooler center. Foreign materials include solder resist, dirt, and particulate matter. One should also look for any slight haze, normally white, that indicates poor reflow flux removal techniques. This last material is very detrimental to soldering.

This inspection procedure, when used with surface inspection for *crowning* of the reflowed solder, is sufficient for all solder coated boards. The appearance of uncoated (with solder) copper additive boards cannot be judged this way. No good nondestructive method is known at present. The measurement of conductivity alone has no correlation with solderability.

The thickness of the copper in the hole is important for joint quality, since it relates to the generation of blow holes (see Section 7-17). It has no bearing on solderability or solder rise in the hole.

The quality of the plated-through hole knee should also be checked at this point. Look for abrasion marks on the surfaces, which could give rise to these defect of *weak knee*, and should not be allowed. In the worst case bare copper can be seen in the knee area. For further details on this problem, see Section 7-30.

8-10 SOLDERING MATERIAL CONTROL

The control of such soldering chemicals as flux, solder, oil, and cleaners, is an important part of the quality assurance program. It is not, however, used to compare competitive materials and select the best for the application (see Chapter 4 in Ref. 8-1). Since each material is also discussed in other parts of this book, some sections will be quite brief. We will discuss the materials in no order of importance:

1. *Flux and thinner*: These are used in liquid form for wave, drag, and most other mass and manual soldering techniques. For a full discussion of fluxes, see Sections 3-1 to 3-9. Methods of flux application are discussed in Sections 4-2 to 4-6.

Fluxes are controlled by the following parameters:

a. *Specific gravity or density*. This is a measure of the solid content in the flux. While automatic equipment to monitor this parameter is available, its present cost may be prohibitive. A manual check using a hydrometer is adequate. The frequency of testing depends on the application (2-4 times a day). The test is temperature sensitive, and a calculation or temperature adjustment is needed. The appropriate thinners are added for adjustment of density, since they are the most volatile part of the system.

Some automatic equipment measures viscosity and correlates the results with the solid content. Each flux requires calibration for this test, and few data are available from the vendors on viscosity changes with aging or contamination.

b. *Color and clarity*: These are indications of contamination buildup in flux. A color change normally signals flux breakdown, while turbidity indicates moisture, oil, or particle contamination.

c. *Ionic content*: This can be used to monitor flux strength even when the composition is unknown. For ionic content, inject a controlled amount of flux into an Ionograph or Omegameter (see fig 8-7 and 8-9). The water extract resistivity should always be within a narrow range, as agreed on by your supplier.

d. *Chemical analysis* This is normally for chloride or organic acid oriented fluxes since these are the common activator ingredients. Consult your vendor before establishing such controls.

2. *Dross inhibitor and reducer or solder blanket*: These materials are designed to inhibit or prevent dross formation. Because of the nature of the wave process, the best we can expect is a substantial reduction in dross losses.

Such materials can be controlled only with the vendor's help and information. They vary in chemical origin, ranging from heat-stabilized resins to organic acids and salts. The success of the dross reduction program serves to assess their relative quality. The soldering oils have additional functions, which are discussed in item 4 below.

3. *Mask (temporary)*: Masks are generally organic coatings that are screen printed or squeezed onto areas to prevent solder contact. The applied coating must be dried or cured before it can withstand the rigors of soldering. Some masks are designed to wash off readily in the subsequent cleaning operations, while others must be peeled off the board manually. Near board edges or on flat surfaces, tape or mechanical clamps fill the same purpose. Wooden pegs or toothpicks are also helpful in keeping a hole free of solder. The usefulness of these materials is a function of their behavior on the production line. Unique conditions often determine whether they can be used. Make sure that they do not clog up cleaning equipment, or drains.

Material control relies on the manufacturer's information. In-use tests may suffice, but are difficult to transmit to the supplier in case of trouble.

4. *Oil—soldering*: *Oil* is an archaic term dating from the time when real peanut oil and similar oils, were used in the wave. Since then, a number of

non-oil-based materials have been used for the same purpose. Some of them are water soluble. A better term might be "soldering fluids" or "tinning liquids," but soldering "oil" has become the accepted norm.

Heat stability, the effect on surface tension, and the density or viscosity of the product can be used as control parameters. There are a variety of tests for these properties, but there is no uniformity in the soldering industry. In practice few users actually check any of these properties, much to the supplier's delight.

One condition which can cause processing problems is the amount of low-boiling fractions (volatiles) in the oil. They cause foaming and spitting when placed on the hot solder and may continue to give trouble even after prolonged heat exposure. Vendors will usually replace such defective materials.

5. *Saponifiers:* Also referred to as *detergents*, these are water-based cleaning additives, to remove nonsoluble materials like rosin, oil, and dirt (see Section 6-10). Small concentrations of saponifiers are used even in all water systems to remove accidental deposits of unknown source. They react with the contamination to form a soap-like compound that is washable. Some organic alkaline compounds are suitable for the electronics industry because they are mild.

The pH of the saponifier and its solution have been erroneously used as a control. This parameter has very limited value because the solutions are highly buffered, making this test insensitive. Many vendors supply titration information with their product, which gives a much more accurate result.

A general word of warning: if your system tends to foam due to water agitation and aeration, use only silicone-free antifoaming agents. While silicone contamination introduced during washing takes place after the wave, soldering is still needed in touchup and repair. Such residues also interfere with the adhesion of conformal coating materials.

6. *Solder alloy:* Specific information on solder metal is given in sections 3-10 to 3-14. In house analytical facilities are seldom available for tin-lead alloys. There are, however, a number of reputable solder companies that provide this service. Because of their high-volume business their prices are moderate and their accuracy is high. Metal analysis is good for incoming inspection as well as in-process control. Plot the contamination buildup in your equipment as a function of time for your records. It will facilitate trouble-shooting and failure analysis.

Certain precautions should be used when taking a sample from the wave or molten reservoir. The contamination content is affected by fresh solder additions, which dilute it. In addition, segregation in a stagnant container may affect the results. Therefore:

1. Always sample the pot at the same part of the cycle. It is best to do so after it has been brought up to the correct fill level.
2. Make sure the pot is well agitated. In a wave, make sure that it has been pumping for over an hour, while a solder pot should be stirred for 1-2 minutes.
3. The ladle and mold used must be at their correct temperature. Heat the ladle in the solder until it reaches the same temperature, and fill it with molten solder. Cast into the mold, which should be at room temperature. This will ensure a homogeneous sample.

7. *Solvent cleaners*: These are discussed in detail in Section 6-13. They are well described in the literature, and technical data are available from the manufacturers. The incoming tests as well as the in-process control procedures recomended by each vendor should be followed.

An important incoming check is the quantity of residue left on evaporation. Good solvents leave quantities in the part per million range.

The in-process control tests include the pH of water extract, acid acceptor content, and stabilizer content. These guarantee that the product will remain stable in use. It is also important to monitor the increase in boiling temperature with solid content buildup to prevent thermal decomposition.

8. *Water*: Water for cleaning is discussed in Chapter 6; the grades of water are covered in Section 6-9.

The monitoring of water depends on the grade used. There is little a manufacturer can do to change the quality of public water supplies, which can vary with the season. Well water does not seem to change much seasonally. While soft and deionized water becomes a check of the equipment and its efficiency.

8-11 CLEANLINESS MONITORING METHODS

The type of dirt present dictates the method of testing. For all types of dirt (particles, wet or dry films), visual observation is good only for large quantities. Real testing requires specific methods aimed at the type of contamination involved. These methods focus on selective removal coupled with a chemical or gravimetric measurement. Much must be known about the nature of the dirt, or generalized results are obtained in units such as *total particle count, insoluble weight per surface area*, or *total soluble and insoluble contamination*. To date there is no known universal dirt meter or method of testing.

One known category of contaminants are the ionic materials, which promote electrical leakage and corrosion. Fortunately, ionic contamination is amenable to electrical measurements—generally the change in conductivity (or resistivity) of a polar solvent after extraction of the dirt from the surface. Here a variety of tests exist. The methods discussed below have been approved by MIL-P-28809. They rely on electrical changes caused by ionic material dissolved in a polar solvent consisting of 50–75% isopropanol and the balance is deionized water. The alcohol is only a weak polar material and has limited nonpolar solvency power. It thus can remove such nonpolar materials as oil, wax, and rosin, revealing ionic material trapped underneath. The test methods fall into the following two major groups:

1. *Static limited time extract resistivity*: In this method, we test for a limited fixed extraction time. The rationale is that a 5- or 10-minute contact between an agitated solution and the parts to be measured gives a good indication of cleanliness.

If used for quality control on a continuous basis, such tests will monitor cleaning equipment performance so long as all conditions stay the same. Since the tests use a fixed volume of solution to extract the ionic materials, the inherent properties of the contaminant interfere with the interpretation.

2. *Dynamic absolute extract resistivity*: This method, while slower, gives more reliable results than the static limited time extract resistivity method. It recognizes that dirt may be trapped in hard-to-reach places, such as underneath flat components or in various cavities. Furthermore it acknowledges that the solubility of materials differs, as does their ionization potential. This measurement thus eliminates time dependency, and the results are more accurate. This is particularly so since, in actual usage, the time under humid conditions may be very long.

Highly automated equipment for ionic measurements is now available. Figures 8-7 and 8-9 show two popular units in use by industry. Note that the computer in the unit can calculate actual contamination in weight per unit area or give a visual record of the dirt extraction rate versus time (Fig. 8-8).

Surface insulation tests on the board are often included in cleanliness analysis, although they cannot be considered a measure of contamination alone. Many parameters affect insulation resistance that are not caused by dirt. These include:

- Underetched specs of copper.
- Metallic smears—copper, tin lead, etc.
- Surface conditions—roughness, composition, etc.
- Adsorbed high boiling organic materials.

Fig. 8-7. Ionograph® by Alpha Metals. This unit can do a static limited time change extract resistivity test by using a fixed extract time. It can also do a dynamic absolute extract resistivity test. (Courtesy of Alpha Metals)

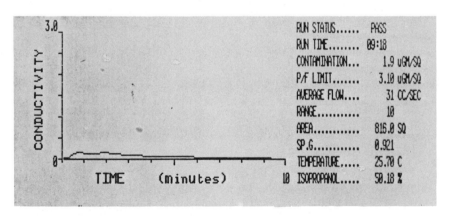

Fig. 8-8. A typical readout from the Ionograph®, which displays actual test data and the extraction rate vs. time.

Fig. 8-9. The Omegameter® by Kenco. This unit uses a static limited time change extract resistivity, using a fixed extraction time. (Courtesy of Kenco Alloy & Chemical)

Insulation resistance is usually measured on a special pattern, but may also use any two parallel lines. The specimen is conditioned under high humidity at an elevated temperature. A direct current (DC) potential is applied, and the leakage over the pattern is used to assess the relative value of the surface.

For specific applications of the insulation resistance test, see the case histories in Section 8-12.

8-12 CLEANLINESS TEST LIMITATIONS

The limitations of cleanliness tests in general, and the resistivity of extracts in particular are:

1. They are not universal dirt measurements, but rather specific to certain types of dirt (ionic, particular, etc.)
2. They do not indicate the exact location of the dirt (i.e., trapped under a specific component or located in a particular area of the board).

3. By themselves, they do not indicate the source of the contamination (flux residue, fingerprint, etc.).
4. There are bothersome contaminants that will not register since they are insoluble or nonreactive (silicone, oil, and compounds).

In addition, a clean board does not guarantee electrical integrity. Even if we discount such problems as shorts (since they are seldom related to cleanliness) other problems may exist. The following case histories illustrate this situation:

Case History I. A high-impedance board showed small amounts of ionic contamination when measured on an Ionograph. Thirty day humidity testing (95% RH and 100° F) showed a steady deterioration in insulation resistance. Investigation revealed that the board fabricator had not cleaned the board surface properly before the application of solder resist. It took approximately 20 hours under humid conditions to permeate the epoxy coating and activate the ionic material underneath.

Remedy: Have the board manufacturer check for ionic contamination prior to solder mask application. An alternative solution, and a monitor on the vendor, is to print an insulation resistance test pattern on the board (see Case History II below).

Conclusion: Surface cleanliness of assemblies is not the only factor in cleanliness.

Case History II. A clean board, without solder resist, was tested for insulation resistance (using an IPC test pattern). After a 24-hour humidity soak, erratic failures were noted. These could not be correlated with ionic measurements before or after the test. Nor could any correlation be found within the same lot received from the vendor. Microscopic examination revealed that the failed boards were underetched. They had specks of laminated copper adhering to the printed circuit board surface.

Remedy: Print an etch-target pattern on the printed circuit board for an easy visual check on etching during incoming inspection. Also make sure that the vendor has adequate lot control.

Conclusion: Metallic contamination also lowers surface resistance. Be especially careful to avoid solder smears that result from rubbling soldered boards against each other during handling, storage, and shipping (see Section 7-12)

Case History III. A specific type of board continued to register ionic contamination even after prolonged cleaning cycles. Exhaustive studies of the water washer revealed nothing unusual. Other types of boards did not show a similar failure. When the components were cut off and studied, a specific lot of electrolytic capacitors was found to leak. Since there were four capacitors per board, the reason for the failure became obvious.

Remedy: Study components separately from the board. One additional component was identified as a source of continuous ionic contamination. The problem stemmed from an impurity in the resin used to encapsulate the device. The amount of contamination per device was relatively small.

Conclusion: Cleaning process efficiency is not the only reason for poor cleanliness test results.

8-13 HOW CLEAN IS CLEAN?

There are literally millions of dirty printed circuit assemblies in use. The question is, how much contamination is tolerable? The answer, unfortunately, is complex and unique to every assembly. In general, however, it can be said that a board will function as long as dirt does not interfere with its operation. In the case of ionic or hygroscopic contamination, this depends on board design (spacing, voltage gradients, the presence of coatings like masks or conformal coatings, etc.). In addition, the in-use ambient conditions of humidity and temperature play a major role.

Let us not lose sight of the fact that board cleanliness steadily deteriorates after processing. Unless an assembly is sealed, contamination buildup continues even at the final user's location, especially if the equipment is air cooled by fans.

It appears that the majority of boards are immune to small dirt levels. Gross contamination is obviously a hazard. However, a small yet ever increasing percentage of high density boards is sensitive to the presence of dirt. It is the manufacturer's responsibility to know or study his own assemblies.

With the advent of surface mounting and the fine line boards associated with this technology, cleanliness will become more of a standard. The use of better solder masks and the exclusion of dirt will help the industry to maintain cleanliness. However, it is doubtful whether universal cleanliness standards will ever be applicable to the entire industry.

Several references have been made throughout this book to the MIL-P-28809 cleanliness standard. This document cites values considered acceptable after processing and cleaning. A part cleaned to this level will pick up contamination in a normal room environment and fail the test several days later. There are no data in this specification to help resolve problems of cleanliness levels needed for specific operational conditions of the end product.

8-14 FINAL INSPECTION VERSUS IN-PROCESS CONTROL

In this chapter, we have covered the various aspects of inspection in relation to soldering and cleaning. Each of these subjects is important in its own right. If we want to relate this information to the final product, an important concept develops: *it is possible to identify a problem at an early stage, thus preventing costly repair and touchup later.*

Let us take this concept one step further. In the economic environment in which we operate, we need to produce at the lowest cost and the highest reliability. This excludes the option of passing marginal quality products and letting the "buyer beware". Our industry must learn to deliver products that do not fail in service. At the same time, we must keep costs down, so that excessive touchup is not acceptable.

This brings us back to an old industrial engineering principle: *use inspection to detect problems at the lowest material and labor content level.* By the judicious application of quality resources, we can prevent costly production problems. In general, it is advisable to spend about 40% of the quality dollars for incoming inspection. Another 40% are needed for in-process control. The remaining 20% can be applied to final inspection.

It must also be kept in mind that no part of the soldering joint benefits from the additional heat of repair and touchup. The dangers of reheating the connection are:

- Additional metallurgical reactions.
- Heat damage to the plated-through hole or pad.
- Heat shock to the component.
- Complex flux removal.

In reality, the joint may be less reliable after an additional second soldering operation than a connection that was made correctly the first time.

The cost effect of touchup is also dramatic. Table 10-1 shows that the cost of reworking a joint at the end of the wave is six times (6X) greater than that of the original connection. It was also calculated that, on the average line, five touchup operators double the cost of wave soldering and cleaning.

8-15 INSPECTION OF SURFACE-MOUNTED SOLDER JOINTS

The inspection of surface-mounted joints involves additional quality issues. The basic problem stems from the inability to inspect most joints by visual examination. See Chapter 5 for more details on surface mounting and the types of solder fillets in each type of component. Because of the different nature of each components, we will treat the inspection in two groups, according to the method of soldering used:

1. Components glued to the bottom of the board and wave soldered in the upside-down position.
2. Components soldered in the upright position.

We must further divide these groups according to the type of joint formed in relation to the size of the device:

a. Lap joints with small devices (leadless Chips of 0.125 in. (3.2 mm) in length, like capacitors and resistors).
b. Lap joints with large devices (large leadless chips, leadless chip carriers, etc.).
c. Lap Joints with small outline devices with compliant leads (SOTs, SOICs, etc.).
d. Butt joints with leaded components (chip carriers with "J" leads, etc.).
e. Lap joints with large chip carriers.

Let us look at each group by itself and analyze the inspection criteria needed:

1a. Small leadless chips soldered upside down by wave, dip, or drag. Here the critical plane is the lap joint, which is hidden. In addition, gravity has a tendency to pull the solder down to the top of the device on the outside fillet. This visible fillet serves as our criterion for inspection. The small size of the chips eliminates stress problems due to the mismatch of thermal coefficients of expansion.

1b. Is very similar to 1a, but has larger and/or more numerous fillets. The same considerations, however, apply.

1c. Small outline devices soldered upside down by wave, dip, or drag. Here the critical stress absorption is supposed to be in the compliant lead. Gravity, however, will fill the joint, preventing the lead from flexing. This is acceptable only for limited component width and a small thermal cycle.

2a, 2b, and 2e. Lap joints formed in the upright position can have a controlled fillet. This lap joint, however, is hidden and cannot be inspected by visual means. Two approaches are possible:

- Use a controlled amount of solder (preform, paste, etc.), and check on the solder penetration by the amount left outside the joint. The solder will wick into the joint only under good wetting conditions.
- Use x-rays to view the fillet under the device. This, however, tells us little about the state of wetting (adhesion).

The reliability of the joints depends on fillet uniformity. A big or beefed-up connection does not support its full share of stress but rather directs it to another, weaker joint. Solder cracking can result from this problem, referred to as the *vise-blocking effect.*

2c. Lap joints on *gull-shaped* compliant leads. Here the lead is meant to absorb the stresses, and the solder should not be allowed more than halfway up to the knee. This is a visible joint, and the easiest to inspect by eye.

2d. Butt joints formed with "J" leads. Here we can inspect the joint. We should look for small, uniform fillets no more than one-fourth the lead height. This type of connection can absorb the required stresses.

In all of these cases, good wetting must be observed on both surfaces. If a pad or termination cannot be checked directly, look for a similar surface to indicate quality. Also, remember that the surface of an unprotected silver fired termination may have a rough appearance when wetted. If the termination has a nickel barrier plating, it should be smooth.

For fillet appearance, remember that some alloys are dull by nature, and that the fillet must be only as shiny as the parent alloy. Some solders freeze in a rough pattern, and the surface appearance of such alloys in the fillet is also rough.

REFERENCES

8-1. Howard H. Manko, *Solders and Soldering*, 2nd ed, McGraw-Hill Book Co., New York, 1979.
8-2. Howard H. Manko, "Must I Fill That Plated-through Hole?", *NEP/CON Proceedings*, 1978, pp. 293–295.
8-3. Howard H. Manko, "Solderability a Prerequisite to Tin-lead Plating," *Plating*, June, 1967, pp. 826–828.

9

TOUCHUP AND REPAIR

9-0 INTRODUCTION

The words *touchup* and *repair* express a different concepts for different people. We associate them with good quality as a result of a deliberate action. Further they refer to the correction of faults that do not meet our standards of workmanship. How does this relate to product reliability?

For our purpose, reliability can be defined as: *the ability of the solder joint to outlast the rest of the assembly under the stresses of anticipated use.* In other words, the solder joint must not fail within the life expectancy of the product in spite of the abuse it is exposed to in use (see Section 8-0).

Thus, meaningful repair involves more than applying a hot iron to the bottom of a board after wave soldering. We must learn to distinguish between functional repair and useless cosmetic overlays. The use of additional heat reduces long-range reliability and must be curtailed. The best principle would be:

When in doubt, don't touchup: leave it alone.

This is quite different from the common misconception that each joint must meet universal quality criteria. Such requirements have developed historically and were specified by committees. They refer to the ideal solder fillet, without regard to commercial practicality. There is no doubt that these joints would be as close to perfect as possible, but in most cases such quality is excessive. In our competitive world, we must learn to define the solder fillet in terms of reliability versus cost. This dictates a compromise between the ideal joint and a commercial solder fillet that is inferior, yet reliable. The degree of solder fill in a joint is a good example. Classical quality criteria permit only complete solder joints, 360° around and all the way to the top. In commercial soldering, unfilled holes with 180° well-wetted fillets may be acceptable (Refs. 9-1 and 9-2). This issue would involve component weight per lead, number of leads per device, double- versus single-sided boards, and so on. In addition, solder sag up to 25% of board thickness may be perfectly acceptable. For further details on setting up apropriate solder requirements, see Section 8-2.

We must understand the true purpose of each touchup. This operation must be limited strictly to meeting well-defined workmanship standards. It must go beyond satisfying the quality department per se, or reworking everything that is doubtful just to ensure that it is not rejected by the inspector. Untrained operators often want to look busy in order to justify their job. We must train operators and inspectors alike to guarantee a common understanding.

When concentrating on the quality of the solder fillet in a printed circuit board, it is obvious that it must remain a good *stress coupler*. The metallurgical properties of the solder are best preserved if we limit the amount of metallic contamination picked up from the basic metals surrounding the fillet (copper, gold, etc.). This contamination is picked up in appreciable amounts only during the heat cycle while the solder is molten. Thus we must minimize both the time and the temperature of contact with the liquid solder during the added touchup cycle. Diffusion after solidification has a minimal effect over the life of the joint. Of course, the ideal process would eliminate the need for touchup entirely. *Doing it right the first time* is better in terms of reliability and, obviously, cost.

The extra heat of touchup is also detrimental to the plastic resins in the laminate. These materials degrade from the heat of the iron, which is more severe and localized than the wave. The components may also be damaged from repair heat, especially if the heat is applied near the cold device and causes localized heat shock. Touchup from the solder side of the board, away from the components, should be the rule. Applying a hot iron on the device side can cause component failure or reduced life.

The sources of printed circuit faults and corrective procedures were discussed in Chapter 7. The folly of adding solder to a dewetted joint have also been stressed (Refs. 9-3, 9-4, and 9-5).

In general, quality problems can be divided into three categories:

• Functional material-oriented problems.
• The effect of poor design on the final result (repeat offender faults).
• Process-oriented problems.

9-1 PLANNING FOR SCIENTIFIC TOUCHUP

Soldering is a reversible process, and remelting the solder makes it possible to correct industrial defects. In addition, solder removal makes component replacement relatively easy. Before embarking on a touchup operation, however, we must have certain prerequisites:

- Definitive workmanship standards.
- Qualified touchup operators.
- Trained quality inspectors.
- Correct tools (soldering irons).
- Appropriate flux and solder.
- Peripheral equipment (sponges, holders, etc.).
- Cleaning procedures and materials (optional).
- Specific repair techniques.
- Logical flow of operations.

Armed with these tools, we can initiate a touchup procedure which will enhance the quality of the product at a reasonable cost. This list, however, requires more qualification.

Section 8-2 describes a time proven method of setting up your own training and workmanship standards. It is based on the realization that each branch of the industry has its own reliability requirements. Manufacturers often have different requirements due to design and end use. Therefore, no universal quality or workmanship standard can exist. The reader is urged to reevaluate any standards now in use against product reliability and cost.

Training is the most efficient way to get optimum output from a work force. Training is not limited to operators; it also includes the inspectors. Higher level supervisors, engineers and managers must also be aware of the process details in order to help enforce them. No deviation from the standard or training program should be made without proper documentation, or it will undermine the quality of the product (see Section 10-3).

9-2 THE SOLDERING IRON

The heart of the touchup operation is the soldering iron and similar hand tools. While there are a few very sophisticated automatic machines for rework, this is generally a manual operation. It is likely that in the near future more robotic operations will take over some repair and replacement functions. However, touchup as we know it will remain a human manual operation because of the very nature of the *hunting and seeking* required. This will not change until we discover a scientific principle for joint quality inspection to replace the human eye.

The soldering iron has not received much attention from the engineering community, probably because of the low status of hand soldering. Yet

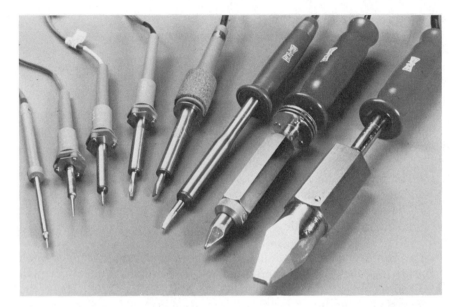

Fig. 9-1. Selection of steady output irons. (Courtesy of Hexacon Electric)

the wrong tool can be very costly in terms of reliability and scrap generation. This applies not only to large, discrete components but also to surface mounting. For a detailed analysis, see Chapter 7 in Ref. 9-1, as well as Refs. 9-5, and 9-6. Let us briefly review and classify the irons. They fall into three major categories:

- Steady-heat output irons
- Temperature-limiting irons.
- Temperature-controlled irons.

Since selection of the iron is critical, further discussion is in order.

1. *Steady-heat output irons*: These are by far the simplest and most common type of iron (see Fig. 9-1). Their low cost no doubt adds to their popularity. These tools are "on" all the time and supply a steady flow of heat. While they are rated by wattage, this can be a misleading yardstick. This property does not define the thermal character of the tool. One should consider:

- Tip temperature.
- Heat capacity.

- Temperature recovery rate.
- Temperature variations during idling and under load.

With judicious selection and planning, this iron can easily become the work-horse of the production line. It is very suitable for continuous soldering.

To match the iron to the task, it is easy to measure tip temperatures under production conditions. A thermocouple is located in the tip, and a thermal trace is recorded. Adjustments are made in two ways:

- The input voltage (an electrical adjustment) or the wattage of the element (a mechanical interchange).
- The Heat content of the tip, is adjusted by changing dimensions like tip diameter or length.

The iron's temperature at idling is the most significant parameter when considering touchup. The intermittent use of the tool and its low thermal demand require a working zone of 800° ± 20°F (455° ± 10°C). This is the highest tip temperature range for all irons used in resoldering, but is more restricted than that of original iron applications. For surface-mounted touchup and device replacement, the working zone is 650° ± 20°F (345° ± 14°C).

2. *Temperature-limiting irons*: These are more specifically designed for rework operations but cost 4–5 times more than the previous category (Fig 9-2). In an effort to maintain tip temperatures within a preset range, some irons are equipped with a temperature-limiting device. While these devices are effective in preventing overheating, the iron is slowed down. It becomes a partially "on" tool, with slow heat recovery under load.

The iron's ability to maintain a heat range is a function of the switching mechanism. These mechanisms range from magnetic efficiency to bimetal switches and have relatively low sensitivity. Even though the working heat range is relatively large, the temperature is much lower than that of steady-heat output irons. A 700°F (370°C) temperature-limiting iron is best for standard applications. For surface-mounted devices, a 600°F (315°C) limit is more suitable.

3. *Temperature-controlled irons*: These are by far the most sophisticated tools and are suitable for sensitive work (Fig. 9-3). They cost at least twice as much as the temperature-limiting irons, although the prices are steadily decreasing. Basically, they use a high-output element (as high as 100 W) which is switched "on" only on demand. Their switching system is more sophisticated than that of the limiting irons, hence their increased price. They can hold a set temperature within a narrow tolerance range of less than 10°F even under load. These units have what

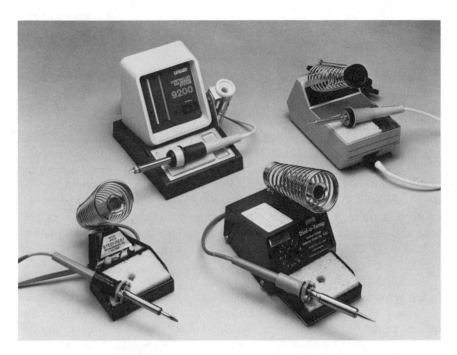

Fig. 9-2. A number of temperature limiting irons.

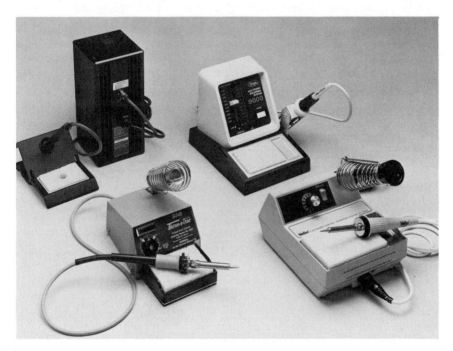

Fig. 9-3. A selection of temperature-controlled irons.

amounts to an instantaneous recovery rate for most touchup and repair applications.

This type of iron is highly suitable for intermittent usage and eliminates the need for more than one soldering tool per work station. With conventional irons, more than one iron is needed if loads are vastly different. In the case of touchup, for instance, removal of solder with a braided wick would require more heat than the physical repair of a joint.

The cost of the soldering iron increases from type 1 to type 3 according to the complexity of the equipment. For touchup the author would recommend either a well-matched steady-output iron or a fully temperature-controlled unit. The specific choice depends on the sensitivity of the assembly and the frequency of repair. Temperature-limiting equipment has been widely used in this type of work but does not justify the added cost.

Tip temperature and control, however, are not the only considerations in the selection of an iron. The physical configuration of the iron is also important, since it is related to handling. A comfortable tool improves the quality of the repair. You must have a well-balanced unit with a tip that makes easy contact with the work (Chapter 7 of Ref. 9-1). Avoiding damage like burning adjacent areas should require little effort. Irons must also be easily repaired and maintained by the user. Where needed, temperature recalibration of the equipment must be simple in order to meet MIL specifications.

We cannot end this discussion of soldering irons without touching upon the electrical damage which may be done to voltage-sensitive circuitry (Ref. 9-6). Such components as MOS-FET and CMOS devices are recognized as sensitive to voltage gradients. Special precautions are taken to prevent electrostatic shock.

Temperature-limiting soldering stations often use switches which produce high-voltage discharges for brief periods. Typically, 150 V_{pf} spikes have been measured in the nanosecond range and occur on a completely random frequency. Voltage spikes of this intensity can cause component failure on unprotected thin film devices.

Component damage from static electrical charges has been well documented and recognized. Static damage from soldering tools is rare because a continuous metallic conductive path carries off any charges generated in the handling of the iron. Ungrounded two-wire irons with all-ceramic elements should not be used on static-sensitive work.

Steady-state voltage leakage through element insulation exists in any ungrounded soldering iron; therefore, ungrounded (two-wire) tools have no place in soldering sensitive components. Even grounded irons should be monitored. Aging of ground path connections may cause damaging levels of voltage to appear at the tip. The use of positive certified grounding methods as prescribed by MIL-S-45743E, can avoid this element leakage. Typical levels in the microvolt range (10^{-6}V) can be achieved by well designed positive grounding.

Most of these detrimental electrical energy effects can be eliminated by specifying soldering equipment that conforms to MIL-S-45743E standards.

9-3 SOLDERING AIDS

Several standard pieces of equipment and supplies are necessary to make the use of the soldering iron efficient and easy. They are referred to as *soldering aids* (Fig. 9-4). Let us review the more common ones:

1. The stand or cage in which the soldering iron is kept. It should have the following characteristics:

• It must be cool on the outside for safety so that a worker will not get burned.

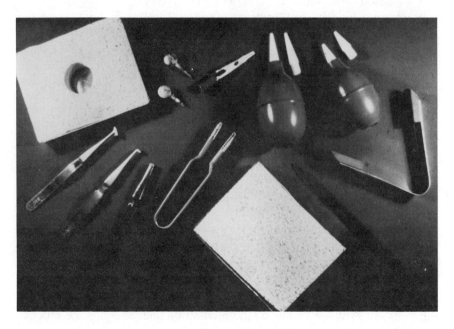

Fig. 9-4. Soldering aids for improving hand soldering. (Courtesy of Hexacon Electric)

- It must be at an angle which will make insertion easy, while preventing hot air convection from heating the handle.
- Its design must avoid heat sinking through contact with the tip so that its temperature can be maintained.
- Its design should allow for easy removal of solder drippings and dross.

2. A wiping sponge. This sponge, soaking wet, is used to *steam clean* the tip in order to remove burned flux and dross from the tinned area. The sponge wipe is not abrasive to the thin iron cladding over the copper tip. It thus prolongs its life, and yet is very efficient in cleaning action. The sponge should:

- Have a horizontal, not a vertical, configuration, since the former retains more water. It is also much easier to rinse clean (this should be done daily).
- Have a collection cavity in the center for trapping excess solder and dross, eliminating the danger of reintroducing contamination into the work. Always start the wipe from the hole.
- Be sulfur free; sulfur contributes to dewetting of the iron tip.
- Have no binder or other organic material that will contaminate the work.

3. Various supplies which are needed for hand soldering:

- Spool of solder.
- Flux container.
- Desoldering wick, syringe, or tool.
- Printed circuit holder.
- Heat sinks and antiwicking tools.
- Stripping, cutting, and shaping tools.
- Static protection equipment.

4. The solder iron analyzer. We refer to this tool repeatedly in this chapter. Its proper use is described in the literature (Refs. 9-1 and 9-6). It is essential to understand that the analyzer is not only an important laboratory tool but is used under actual manufacturing conditions. The soldering irons are evaluated while on the production line, enabling the process engineer to make vital corrections.

9-4 THE IRON USAGE SEQUENCE

The right sequence in which the iron is removed and returned to the cage requires amplification. While this seems an unimportant detail, it has a direct bearing on the tip temperature, and its ability to hold solder. It is also related to the life of the tip and iron. Therefore, it affects the uniformity of soldering results. The best sequence is as follows:

1. Remove the iron from the holder.
2. Sponge-wipe the tip, starting at the collection cavity, to remove dross and burned flux. Use a light wipe; do not scrub.
3. Solder or do the touchup.
4. Return the iron to the holder without wiping.
   (Note: If the tip is to idle in the cage for more than 5 or 10 minutes, add more solder to the tip.)

The rationale behind these steps is simple. The sponge wipe is meant to steam clean the tip. The collection cavity is meant to hold all solid debris. After these cleaning steps, there will be only a thin layer of tin on the tip. This is ideal for the soldering operation, so that the joint will not be contaminated. During soldering, the tip will pick up some flux and solder. These protect the working surface during short idling periods. Longer intervals demand the deliberate addition of more flux and solder for tip protection. However, if we clean the tip and then let it idle in the cage, the small amount of tin will burn off and the tip will become *dry*. In this condition, the working surface cannot hold any more solder and is therefore ruined. See Section 9-5 for details on tip restoration and maintenance.

The making of the solder bond itself follows precise rules. They are detailed in Chapter 7 of Ref. 9-1, where the reasons for this procedures are also discussed. Let us summarize the most common steps for conventional work:

1. Apply the tip to the lead wire away from the printed circuit pad, if possible.
2. Simultaneously, touch the core solder to the junction between the tip and the lead. This will create a heat bridge (Fig. 9-5).
3. Draw core solder around the joint and remove the solder first.
4. Watch the fillet form and remove the iron last.

Remember, the critical time for hand soldering boards is 1 ± 0.25 seconds. The *critical time* is defined as the time the solder is molten. This

Fig. 9-5. Solder bridge formed. Note the flux location near the core solder. (Copyright Manko Associates)

is also the time the board is kept above its glass transition temperature.

For surface mounting, the procedure is quite different because there is no lead wire to contact. The board pad, however, is still prone to lifting, and no pressure on the board is allowed. To limit laminate temperature damage do the following:

1. Use a lower tip temperature with the same time limitations.
2. Flood the pad and its immediate surroundings with liquid flux.

The flux acts as a *heat cushion*, keeping the joint area much cooler. The flux must also be present on the joint to provide the right soldering sequence for surface-mounting. Only when solder paste is applied does the role of the liquid flux change. The steps for soldering are as follows:

1. Flood all pads and terminations with liquid flux.
2. Place additional solder, if required, in the joint area (wire, preform, paste, etc.).
3. Touch the hot working tip to the solder, since it is the least heat-sensitive element in the joint.
4. The melting solder will act as the "heat bridge" to the component and the pad.

5. Remove the tip only after the fillet meets your workmanship standards. (Remember the 1 ± 0.25 second critical time limitation, which also applies here.

The quantity of solder per joint depends on the workmanship standards set for each organization (see Section 8-2). It also depends on good quality control and inspection procedures.

9-5 IRON START-UP, SHUT-DOWN, MAINTENANCE, AND TIP REPAIR

In order to get the best results from the soldering iron, we must consider it a precision tool. With proper care, it will reproduce the same thermal conditions every time. We must also remember that many devices and circuits are voltage sensitive. Good maintenance procedures will minimize temperature variations and current leakage problems. The irons should, however, be tested periodically.

The start-up procedure is the operator's responsibility. In the morning, the cold iron should undergo the following procedure:

1. Remove the tip from the cold iron daily (Fig. 9-6).
2. Wipe the tip shank to remove loose heat scale products (Fig. 9-7).

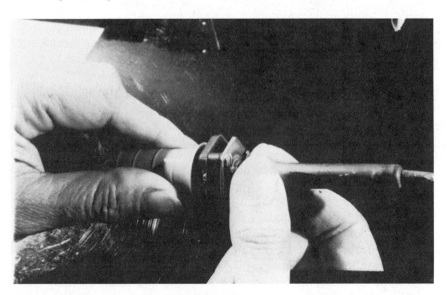

Fig. 9-6. Cold tip being removed for maintenance. This daily routine prevents tips from freezing in heating element. (Copyright Manko Associates)

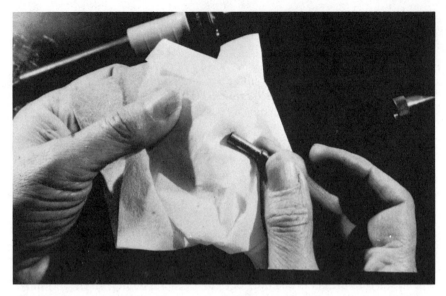

Fig. 9-7. Wiping tip to remove loose oxides that prevent uniform heat transfer. (Copyright Manko Associates)

3. Occasionally, clean the inside of the iron shaft with a soft material like a pipe cleaner (1-2 times a month).
4. Inspect the tip for cracks or erosion on the working surface.
5. Reinsert the good tip in the iron and tighten the holding mechanism lightly. (Forcing screws or outside threads may cause them to *heat-size*, follow the manufacturer's instructions.)
5. Shake the iron slightly, listening for any rattles or indications of loose parts.
6. Inspect the cord for damage.
7. Clean out solder splashes from the cage.
8. Wet the sponge.
9. Switch on or plug in the iron.

This procedure may vary slightly from model to model. Damaged irons or tips should be replaced and fixed by trained personnel.

The shut-down procedures are much simpler. Make sure that the hot tip is protected from dewetting (drying up) between shut-down and start-up. Follow these instructions:

1. Sponge-wipe the tip to remove all burned flux, dross, and excess solder.

2. Melt a generous amount of fresh solder on the tip.
3. Return the iron to the cage without wiping.
4. Shut off the current to the iron.
5. Clean the sponge and remove solder splashes from the sponge and tray. (The cage is cleaned out in the start-up procedure.)

Regular iron maintenance should follow the manufacturer's recommendations. These should include:

• A recalibration of temperature-controlling devices or a check of tip temperature.
• A check for voltage leakage or static damage potential.
• An inspection for physical damage (to the cord, plug, handle, etc.).

Replacement of the defective parts is easy and is the responsibility of a trained technician. Several spare irons should be kept on hand for replacement. It is customary to heat spare tools for routine rotational maintenance. Thus the operator receives a replacement iron that is hot and ready for use.

Tip problems fall into two major categories:

1. Eroded or cracked tips, which must be discarded.
2. Dewetted or dry tips, which will not hold molten solder. These can be repaired.

The repair of dewetted or dry tips can be done in several ways. It is preferable to assign this job to the maintenance technician and to collect a number of tips before restoration. These tips are expensive and should not be discarded.

The process focuses on the removal of heat scale from the thin protective layer. This coating is usually an iron plate to protect the copper based tip from dissolving (eroding) in the molten solder. The plating is too thin to withstand much mechanical abrasion, although gentle tumbling or light buffing may be used. Most restoration procedures use a combination of a very aggressive (corrosive) flux and molten solder to retin the surface. This can be done by slowly dipping (at a rate of 1 in./sec—2.5 cm/sec) the loose tips in the flux and a solder pot. Several repeat dippings in both the flux and the solder may be needed. Thorough flux residue removal must follow. None of these flux residues can be left on the tip. Experience has shown that a technician can rework up to 100 tips an hour, dipping them in small groups at the same time. These retinned, cleaned, and dry tips are returned to stock for future reuse.

Retinning can also be achieved by the operator while the tip is still in the iron. This is less desirable for several reasons. The flux fumes can damage the iron, the heating elements, and the temperature control mechanism. It is also poor practice to use aggressive fluxes at an electronic work station. These fluxes and their fumes may cause damage unless properly cleaned off the assemblies.

9-6 TOUCHUP TECHNIQUES

Touchup is only slightly different from primary hand soldering. See Chapter 7 of Ref. 9-1 for a detailed discussion of hand soldering. The major differences are as follows:

- Solder must often be removed in this operation.
- Core solder is not always added in touchup.

Therefore, external liquid fluxes must be applied in many situations in order achieve the short contact time and efficient heat transfer required. The fluxes are normally dispensed from a needle point attached to a plastic bottle (Fig. 9-8). This enables the operator to squeeze minute quantities into the exact areas to be repaired. The use of a flux dispenser requires an additional hand motion (i.e., the operator must put down the core solder, pick up the flux bottle, and than pick up the solder again). A simpler and very convenient technique requires only a small cup to hold the flux and the standard solder wire (cored or solid). The cup is filled with the touchup flux and discarded when the flux becomes dirty or too thick. The operator dips the solder wire in the flux and transfers the liquid to the work even when no additional solder is required. This way, the solder is always held and the work is simplified. Remember that small diameter solder [less than 0.025 in. (0.6 mm)] does not hold enough flux for most touchup applications even with a flux core.

Even when printed circuit boards are not cleaned and rosin residues are left behind, external flux is still required. However, it is recommended that a combination of cooling and cleaning be used in all situations where a second soldering operation on the same joint is anticipated. *Cooling and cleaning* is basically the application of cleaning solvent with a brush. It serves the following purpose:

- Cools the laminate and strengthens it by ensuring that it goes back below the T_g.
- It loosens the heated, spent rosin so that the next soldering operation with fresh flux is easier.

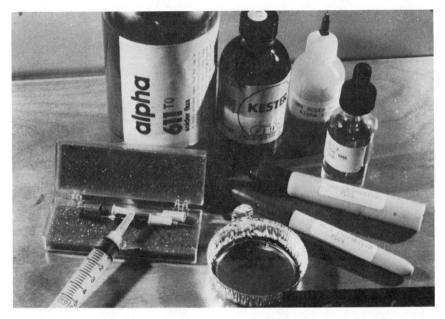

Fig. 9-8. Several methods of applying flux. Note the small open dish, which is used with wire solder to transfer small amounts of flux. (Copyright Manko Associates)

- To facilitate inspection and automatic testing, and enhance the general appearance of the product.
- It is an essential operation when total removal later is part of the process.

Let us now discuss the three major forms of touchup and how they differ from hand soldering. The add-on operation, in which sensitive or nonwettable components are soldered on after wave soldering and cleaning, should not be included here. The three touchup categories are:

- Adding solder.
- Reflowing solder.
- Removing solder.

1. *Adding solder to an existing joint.* This is one of the applications in which liquid flux is mandatory. The amount of flux in core solder, even of a large diameter, is not sufficient because only a partial amount of solder

382 SOLDERING PRINTED CIRCUITS AND SURFACE MOUNTING

is added to the fillet. The sequence used to apply the iron to the printed circuit joint is very similar to that of the original soldering:

1. Add a small amount of flux to the joint.
2. Place the iron next to the component lead, using the broad side of the tip. Touch any solder fillet on the printed circuit board with the edge of the tip only.
3. Bring the core of the solder immediately to the junction between the work and the hot tip to create a molten heat bridge.
4. Draw the core solder around the joint to form the desired fillet.
5. Remove the core solder first, but only after a correct amount of solder is deposited.
6. Remove the iron last, after the fillet has reached perfect wetting and meets workmanship standards.
7. Cool-and-clean the joint.

2. *Reflowing solder*: This is done when no additional solder is required by the existing fillet. The repair is intended to reshape the deposited solder. This is another application in which additional flux is mandatory. The sequence is as follows:

1. Coat the solder fillet with a thin layer of external flux.
2. Place the broad side of the tip near the component on top of the solder joint. At the same time, try to maintain contact with the solder, using the edge of the tip. Do not contact the board itself.
3. Heat the joint until all the solder is molten.
4. Remove the iron only after the fillet has reached the configuration dictated by workmanship standards.
5. Cool-and-clean the joint.

3. *Solder removal*: The methods used to remove solder are described in Section 9-7. Once the solder is gone, clean the joint with a liquid. This also cools the board material. The steps are as follows:

1. Remove the solder.
2. Cool and clean the joint with liquid.
3. Reflux the joint with external liquid.
4. Resolder the joint.
5. Cool-and-clean the joint.

It would be advantageous to delay resoldering for a while. After solder removal, the resin under the board circuitry is weak because it has exceeded the glass transition temperature. This delay would help the resin regain full strength, which takes time after reaching lower temperatures.

9-7 SOLDER REMOVAL

Removal of solder from a flat surface is relatively easy. The real problem is to evacuate solder from a hole or recess. In most cases, solder removal is associated with the replacement of a component. Some methods are aimed at such replacement without necessarily removing all the solder. These are covered in Section 9-10. Let us look at the common solder removal methods:

- Wicking.
- Vacuum.

1. *Wicking*: Solder removal is easy to achieve with the use of a fluxed wire braid. The braid is made of fine copper wire that acts like a wick. The wire can be used bare and is protected from oxidation by the flux coat. In another construction, the copper is tinned but the wick is still fluxed. The tinned braid seems to act faster but does not show the extent of wicking. The copper wire has a different color, making it easy to know how much was used up.

In this process, the solder is absorbed by the capillaries in the braid. When the fillet is remelted, the solder is removed from the joint area (Fig. 9-9). The wick is preferred for solder removal from flat surfaces, because it is difficult to empty a plated-through hole with this procedure.

An iron with a larger heat capacity is required because of the heat sinking effect of the copper braid. Unless a temperature-controlled tool is used, a second iron is recommended for this purpose. This larger iron is too hot for the normal touchup operations.

One must be careful to remove the braid while the solder is still molten. Otherwise the wick will be bonded to the circuitry and will stay there. Pulling on it would cause lifted pads.

In cases where solder is removed from a fillet with a wick, the procedure is as follows:

1. Make sure that the wick is fluxed adequately or add external flux to facilitate heat transfer.
2. Interpose the wick between the hot soldering iron and the fillet, and maintain contact until all the solder is drawn into the wick.
3. Remove the wick with the iron, while the solder is molten, avoiding accidental bonding to the copper land.

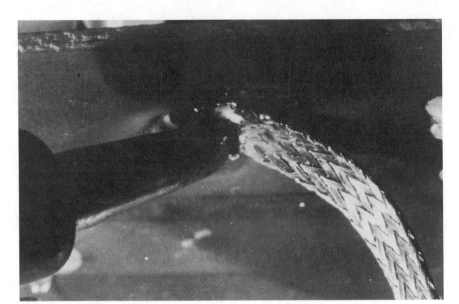

Fig. 9-9. Wick being used to remove solder. (Copyright Manko Associates)

4. If the component wire is still held by a thin layer of solder, reheat the wire and remove it from the board. (Do not heat the printed circuit board or use a wick.)
5. Clean-and cool the area with liquid; this also cools the board.

2. *Vacuum*: This can also be used to remove the solder. The suction may be applied through the center of a heated iron tip, or it can be created with a rubber bulb and a Teflon tip (Fig. 9-10). Hot solder removal using blowing air rather than suction is not recommended. The liquid solder is not easy to collect and may lodge in other parts of the assembly or adjacent areas.

In cases where suction is used to remove the solder from a printed circuit board, follow these steps:

1. Apply a thin coat of external flux to facilitate heat transfer. This is mandatory for product reliability, even though the equipment may require more frequent maintenance.
2. Position the heat source (tip) directly on the lead wire and/or the solder; avoid touching the board.
3. Apply the vacuum only after the entire fillet is molten. Try to move

Fig. 9-10. A selection of desoldering devices. Note that liquid flux is used with most of them for faster heat transfer. (Copyright Manko Associates)

the wire gently to see if it is free to move; this is an indication that all the solder is liquid.

4. If possible, continue moving the tip-and-wire (*walking* it around the hole) while prolonged suction is applied. The long flow of cool air will move through the joint and prevent resoldering of the lead to the wall of the hole.

5. Remove the tip without touching the board; then remove the lead wire.

6. Clean-and cool the joint area with liquid cleaner.

7. Inspect the hole for damage before resoldering.

In cases when walking the lead with continued suction for cooling is not possible (like using a one shot suction tool or a syringe). Be very careful to clean, cool, and inspect the evacuated joint before the wire is removed. A trace of molten solder left behind can solidify and re-attach the lead to the hole wall. While it will not offer much physical resistance during wire removal, it can damage the heat-weakened plated barrel. If during inspection you find that solder was left in the hole, follow these instructions:

1. Gently blow air through the joint to make sure it is dry (the previous clean-and cool may have left some liquid cleaner behind.)
2. Resolder the joint with fresh flux and solder.
3. Clean-and-cool it with liquid.
4. Repeat the removal steps above.

Remember that the board exceeds the glass transition temperature during each soldering step. In this condition, copper peel strength is very low and lifted pads or damaged walls can result.

9-8 LOCALIZED CLEANING AND FLUX REMOVAL

Final cleaning and flux removal are desirable for all high-quality, reliable assemblies. For a discussion of the need for cleaning, see Chapter 6. The major reasons, again, are:

- Removal of potentially corrosive materials.
- Elimination of potential current leakage sources.
- Cleaning of the fillets for automatic test probes.
- Ensuring good contact on mating connector surfaces.
- Preparing the surfaces for proper conformal coating adhesion.

Cleaning after wave soldering is easy because the entire assembly can normally be immersed in liquids. After touchup, and repair, many so-called nonwettable components are added. Now total immersion is no longer possible. Thus localized cleaning takes on special significance.

The immediate use of a liquid cleaner after hand soldering is always recommended. It serves several functions:

- It loosens the flux residue for eventual total removal.
- It cools the board material, which has been above the T_g.
- It exposes the solder fillet for easy automatic testing.
- It makes clear visual inspection possible.

The liquid cleaner used may be a nonflammable solvent, which may pose a health problem. It may also be a cheaper but flammable isopropyl alcohol, which is medically safe. In either case, it should be applied from a closed container with a pump on top (Fig. 9-11). A cleaner in an open container will evaporate and become contaminated from the dirty brush.

Obviously, the dirt is not removed by such an operation. A final cleaning is still needed. Where possible, a vapor degreaser or water washer will remove the final dirt. However, when nonwettable components are in-

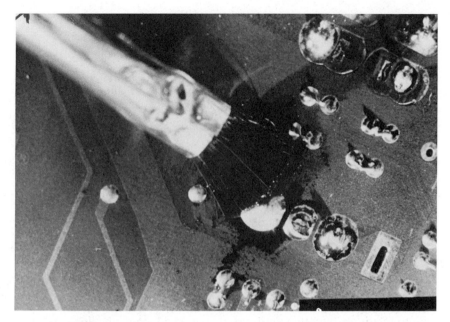

Fig. 9-11. Brush used for cooling and cleaning during touch up. The solvent cools the assembly and loosens the flux for easy total removal at a later stage. (Copyright Manko Associates)

volved, this is not feasible. Nor is it possible to clean assemblies which are already installed in a chassis. A special device is available for localized cleaning (see Section 6-17).

9-9 TOUCHUP OF SURFACE-MOUNTED DEVICES

Touching up surface-mounted joints involves some additional considerations. See Chapter 5 for more details on surface-mounting and soldering. As in Section 9-6, we will divided touchup into two categories:

* Removal of surface-mounted components.
* Replacement with a new surface-mounted device.

The repair and touchup of conventional faults like bridges, poor wetting conditions, and solder balls remains the same. However, each category of surface-mounted devices requires some special techniques.

1. *Removal of surface-mounted devices*: These devices require additional classification into three categories:

- Devices that are not to be saved.
- Components that are needed for testing only (leaded devices may have distorted contacts, not suitable for reuse).
- Devices that must be saved for resoldering.

There are no easy procedures for saving leaded surface-mounted devices; each case must be reviewed in detail. It is customary to discard removed devices because of heat damage. This is especially true of leadless chips, small outline components (SOT, SOIC, etc.), and low cost chip carriers. When testing is desired for failure analysis, the leads may be mutilated during removal.

The simplest method for component removal involves melting all the solder joints simultaneously and removing the component while the solder is molten. This method tends to heat the device more uniformly, causing less heat shock than other procedures. It does bring the component to relatively high temperatures. This procedure can be done with the following tools:

- A Soldering iron with a special tip.
- Hot air from the top and/or bottom.
- A molten solder reservoir (pot, wave, etc.).
- Any other convenient heat source.

We will cover only the instructions for soldering iron applications. All other equipment processes follow the same rationale. The reader or equipment manufacturer can modify these procedures.

A more complex procedure involves remelting each joint separately, removing the solder, and freeing the embedded connection. The last step requires flexible lead wires and is not suitable for leadless devices. If the heat is applied properly, it remains localized and the component stays cool. The procedure is lengthy, expensive, and not always successful.

Let us discuss these methods using the soldering iron. Figure 9-12 shows a variety of iron tips used for this application. The removal is classified by component type:

a. *Leadless chips (capacitors, resistors, etc.)*: Figure 9-13 shows a chip resistor being removed by a special tip. The procedure is simple:

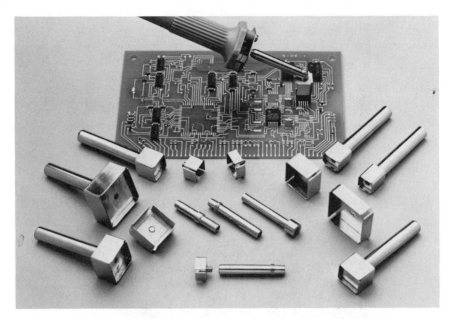

Fig. 9-12. A variety of surface-mounted component removal tips. (Courtesy of Hexacon Electric)

1. Identify components by size. Note that each outside dimension requires a different soldering iron tip. (A decision must be made on whether to equip one operator with several irons or have an operator dedicated to one size. It is not feasible to change the hot tips during the operation.)
2. Flux all solder fillets on the device.
3. Clean the working surfaces of the tip (inside the fork) with a wet sponge wipe.
4. Fit the tool over the joints and contact solder.
5. After the solder is molten, twist or push lightly to dislodge the adhesive spot. (The cured adhesive will be weak because it too is above its T_g when the solder is molten. It can thus be disrupted by a small mechanical movement.) Try to hold the critical time (when the solder is molten) to 1 ± 0.25 sec. as in standard hand soldering.
6. Lift the iron from the work. The component will be retained between the wetted surfaces of the tip. (In stubborn cases, use tweezers or an orange stick to pull the component off while the solder is molten.)
7. Deposit the faulty component in the collection cavity of a wet sponge. Do not hit the iron or your hand against a hard surface to

Fig. 9-13. A chip resistor being removed with a special tip. Note that the tip can also be turned 90° for tight spaces. If there is enough room, a twist is sufficient to break the weak (hot) adhesive. In tight spots, move the component in any possible direction to achieve the same results. (Courtesy of Hexacon Electric)

dislodge the component. This damages the iron and may cause personal injury.
8. Retin the inside of the working tip if the iron is to idle for more than 5 minutes.
9. Use liquid cleaner to cool-and-clean the pads.

b. *Small outline compliant leaded components (SOTs, SOICs, quads, flat packs, etc.).* Many of these can also be removed by special soldering iron tips. Figure 9-14 shows an SOIC being removed by a hot tip. The

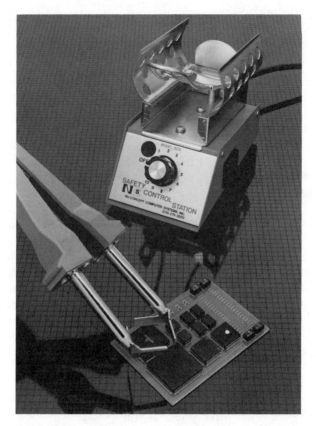

Fig. 9-14. Removal of an SOIC with a special tip. (Courtesy of Nu-Concepts Computer Systems)

process is nearly identical to leadless chip removal when the devices were wave soldered to the bottom of the board. Device removal is obviously much simpler when no glue was used, as in top-side soldering on the board.

If the tip is well designed, it will fit around the device. Many of the heavier components are also retained by the solder's surface tension. However, for those components that are not retained, vacuum pick-up through the tip is also available (Fig. 9-12).

c. *Leaded chip carriers (with "J" leads, etc.)*: These can also be removed with special tips and an iron. As with the small outline devices discussed in item b above, this is a question of fit (to the leads) and weight (for pickup).

d. *Leadless chip carriers (with castellations or bottom terminations only)*: This group is the most complex to remove because there are three possible types of construction that affect heat transfer:

- Short bottom pads with castellations. These have good heat transfer from special tips.
- Short bottom pads only. These have a difficult path for heat transfer but behave uniformly.
- Irregular bottom pads, some recessed beneath the device. These have a very poor path for heat transfer. While some of these devices can be removed with a special tip, most require more sophisticated equipment.

After any surface-mounted device is removed, the pad must be inspected for damage (lifting, measling, etc.). It must also be covered with a known amount of solder to make replacement easy. Remember that the quantity of solder on the pad determines the surface curvature (see Section 5-7). It has been established that a minimum amount of solder at this stage provides the best rework conditions. It is therefore recomended that all excess solder be removed from the pad at this stage. *Excess* is defined here as any visible curvature of the solder surface. The process, using a simple wick, is as follows:

1. Flux the pad or section of the wick.
2. Place the wick over the pad and melt the solder.
3. Remove the wick while solder is still molten.
4. Cool-and-clean the pad.

2. *Replacement with a new surface-mounted device*: This process too depends on the component's configuration. However, there are some unifying characteristics for hand soldering (with an iron) which will be described below. Many more complex procedures are possible, depending on the equipment in house (cream application and hot air or IR reflow, etc.).

The steps for hand soldering are as follows:

1. Start with a low solder profile pad from which excess solder was removed with a wick or similar device. Uneven solder pad height can result in tilted or mechanically stressed components.
2. For uniform stress distribution, all solder fillets located around a surface-mounted device should be uniform in size. This is difficult to achieve by free-hand feeding of the solder wire into a joint. It is even

more complex when part of the connection is hidden underneath the device. It is thus necessary to use a pre-metered amount of solder, such as a preform (wire segment, sphere, or stamping—in order of decreasing cost). Solder paste can also be used if it is certain that there will be no solder ball problem during heating.

3. The device should then be anchored to the board on one lead first. This can be achieved by pre-melting the preform on the pad and forming a meniscus. The lead or termination is then soldered to the deposited solder (remember to keep the critical time within the 1 ± 0.25 sec. time frame.) It is also possible to hold the device in place and use direct soldering, similar to the soldering instructions given in item 4. below. A chisel tip end on the iron, as shown in Fig. 9-15, is very suitable, and flux should always be used where needed.

4. It is simple to solder the rest of the leads or terminations. The anchored device from step 3 is firmly held by one connection. After flux is applied a preform is fed to each connection and melted in place. The iron is removed only after the predetermined quantity of solder has wetted and assumed the correct shape (always within the

Fig. 9-15. Chisel tip being applied to solder preform in the resoldering process. (Copyright Manko Associates)

1±0.25 sec. critical time frame). The same chisel tip soldering iron used in step 3 above can be applied here.

5. After the device is soldered, the flux is cleaned and the area cooled.

During the removal and replacement process, a tool is required to hold the device. Tweezers are also needed to handle the preforms and hold them during soldering. It is best to use non-heat-conducting equipment for these applications. Glass filled teldrin and similar nonmetallic materials are used.

Let us discuss the resoldering (iron) methods used with each type of surface-mounted device:

a. *Leadless chips (capacitors, resistors, etc.)*: The process can include or exclude the anchoring step 3 described earlier. Instructions should be issued to the operator, describing the exact shape of the joint desired. Remember that some chip devices, like capacitors, have the termination wrapped around the top and bottom. Other components, like resistors, may have metallization only at the vertical edge.

b. *Small outline compliant leaded components (SOT, SOIC, Quads, Flat Packs, etc.)*: The process follows the full procedure outlined before. Keep in mind that the stresses on the assembly are relieved by the compliant lead. Therefore, make sure that the solder does not rise to more than half the height of the knee. Too much solder in these joints will stiffen the lead and be subject to stress cracking. The small outline transistors is excluded from this statement because of its small width.

c. *Leaded chip carriers (with "J" leads, etc.)*: These can also be soldered according to the instructions. As with the small outline devices in item b above, the stresses are absorbed by the "J" lead. Too much solder or solder that climbs too high will create a joint cracking hazard under stress.

d. *Leadless chip carriers (with castellations, or bottom terminations only)*: Here there is very little space to locate the preform and melt it in. Pad design and joint height are critical and must be known before specific instructions can be developed. Consider placing a spacer under the chip carrier; it may be temporary or permanent. (Use a lower TCE material than the solder for permanent spacers.)

Solder joint integrity is strongly dependent on fillet uniformity. Great care must be taken to ensure that all joints are equal after melting. This is especially important when there are castellations. If varying amounts of solder wet up on the side of the metallization, the critical amount of solder on the bottom of the joints will change.

9-10 REPLACING MULTI-LEADED COMPONENTS IN PLATED-THROUGH HOLES

Up to now, we have concentrated on hand soldering techniques suitable for touchup and repair. In this context, component replacement is just another operation, consisting of solder removal (Section 9-7) and resoldering. This is suitable for the average two-leaded device, both axial and radial. But when we consider the removal of multi-leaded devices, these techniques become slow and expensive.

There are several additional methods available for single inline packages (SIPs), dual inline packages (DIPs), and other multi-leaded components. For details on surface-mounting replacement, see Section 9-9). These include:

- Conduction, using a molten solder reservoir.
- Convection, using hot air or gas.
- Radiation, using infra red focused and non-focused sources.

Let us discuss these methods, independently of specific equipment, which will vary from case to case.

1. *Conduction from molten solder in a wave or a dip pot*: In this method, the solder itself serves as the heat source. Since it is applied at a lower temperature than the soldering iron, heat damage is reduced. It simulates the original soldering operation. Figure 9-16 shows such a unit.

Not all the holes are cleared from solder during this process. A method to be used in resoldering with this equipment is described later in this section.

2. *Convection from hot air, or an inert gas*: This method allows localized heating in the component vicinity without delineated temperature boundaries (Fig. 9-17), making it relatively safe from heat shock. To speed up the process, heat can be applied from both sides of the board. For efficient heat transfer, the air must be above the remelt temperature, and care must be taken not to overheat the assembly.

Here too, the holes are not always cleared. It is dangerous to blow the air hard enough to clear the holes, because of the poisonous lead in the solder. In addition we do not want to have solder particles lodged all over the assembly.

3. *Radiation using infrared focused and unfocused sources*: This method has an inherent appeal because it can penetrate confined spaces, and produce a fast temperature response. Laser radiation is also a possi-

Fig. 9-16. Desoldering with the use of hot solder. (Courtesy of Air-Vac Engineering Co.)

bility. Temperature control, however, is difficult, and diaphragms or apertures are needed to protect the board material itself.

Now that we have covered some additional heating methods, let us
discuss component removal. Check to see if any of the leads are bend
over (clinched), which makes removal more cumbersome. This removal
can be achieved in several ways:

1. Cut all leads from the top side of the board and remove the solder
   and lead from the bottom. This method destroys the device but
   causes little damage to the plated-through holes. It is suitable for
   field repair but depends on access space on the top.
2. Remove the solder from each plated-through hole from the bottom
   side of the board. Use method 2 described in Section 9-7. Cool-and-
   clean the work and inspect the hole for damage to the barrel.
3. Melt all the joints simultaneously by contact with molten solder. If
   possible, replace them immediately with a prefluxed component.
   This avoids two hazards. First, it is not necessary to clean out any
   empty but plugged holes, avoiding additional heat shock. Second,
   the replacement part is soldered in place immediately, with no addi-

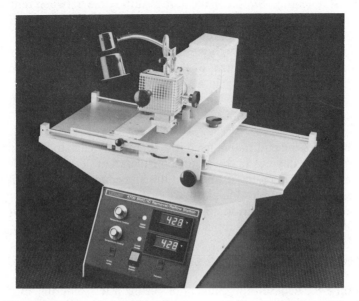

Fig. 9-17A. Hot air removal of multileaded components. (Courtesy of Ungar Div. of Eldon Industries)

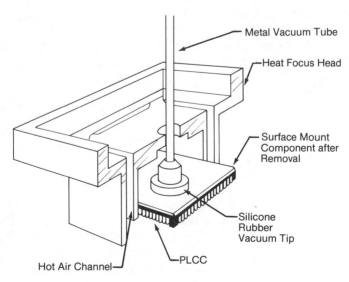

Fig. 9-17B. Diagram showing hot air distribution and PLCC removal with a vacuum probe on unit in Fig. 9-17A.

tional reheating. Simply make sure that the time between removal of the old component and resoldering of the new device is within the thermal limits of the board.

4. Melt all solder fillets without contact by using convection or radiation. Temperature control of the equipment and exposure time determine the safety of this equipment. It usually heats up a larger portion of the assembly than needed. It is very difficult to localize the heat without automation.

Regardless of the equipment used, the thermal damage to the board and the surrounding devices is the major consideration. The application of a flux often helps reduce this damage in the immediate area of application.

The use of low-temperature alloys for original soldering and for replacement deserves consideration. In critical situations, it will minimize the heat needed for repair and replacement.

9-11 CUTTING LEADS AFTER SOLDERING

Section 2-8 discussed the dangers of cutting after soldering. As a brief review, the hazards are as follows:

- Cracking the solder joint.
- Exposing ferrous lead wire ends.
- Damaging the component itself.

There are always repair or touchup situations that do require lead trimming after soldering. In these cases, we must prevent loss of reliability by:

- Resoldering all cut connections. This corrects any cracks that may have formed in the solder fillet. It also coats the exposed ends of ferrous lead materials to prevent rust.
- Using only a sharp cutter with a minimum of twisting during cutting. Air-actuated cutters are available; they cause minimum component damage.

For further details on cracked joints due to cutting, see Section 7-23. Figure 2-2 shows a board where rust formed on steel lead ends. The rust migrated from one joint to another, causing an electrical fault. The unit did not operate in a very humid environment. Rusting is also found at the ends of steel leads, exposed by cutting during lead forming (see Fig. 9-18). Note that the rust migrated to adjacent component bodies.

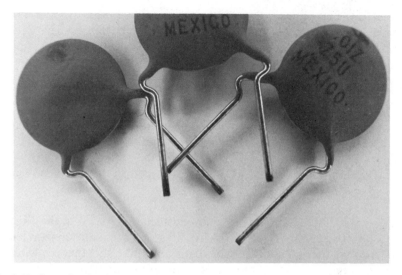

Fig. 9-18. Rusted ends of steel leads exposed by cutting before soldering. Note that the rust migrated up on the lead away from the cut and this effects solderability. (Most rosin fluxes cannot remove rust)

9-12 ENGINEERING CHANGES—POINT-TO-POINT WIRING

Engineering changes in existing printed circuit designs are part of repair and touchup. Circuitry changes that are perceived after the board has been etched require the addition to or interruption of existing conductors. They may also call for the addition of components to the present assembly.

There is seldom room for all such changes on the top of the board. Therefore, components and wires may be added to the bottom (solder side) of the printed circuit.

Some of these engineering changes, which have not yet been covered, are as follows:

- Cutting of traces to disrupt a circuit.
- Soldering of wires to add circuitry.
- Addition of components to existing circuitry.

These tasks should be divided further into those that must be preformed on old assembled boards and those that must be done on new unsoldered assemblies. In this case, the procedures described below should be employed. However, they are slow, expensive, and often associated with reliability hazards. Therefore, whenever engineering changes are planned on new boards, much of the work should be done before wave soldering. Some suggestions on this are also included in the instructions.

1. Cutting of traces requires caution to avoid penetrating the laminate structure. The cut should be treated immediately with a sealer or conformal coating. This prevents the penetration and retention of any soldering chemicals.

2. The addition of wires to extend and modify the circuitry can be divided into two categories:

a. On soldered assemblies, a wire is usually lap joined to the existing land pattern on either side of the board. Make sure that the lap joint length is at least two times that of the wire diameter. Make sure that the wire has between a one-half and a three-quarter turn covered with solder.

Soldering of wires to component leads on the top of the board is not recommended. It causes a great localized heat shock to the component and thus can cause reliability problems.

b. On new and unsoldered boards, try to use the smallest-diameter wire that can carry the current. It may well fit next to a component wire in the unsoldered through-hole. In this case, push the additional wire from the top into the hole and join by wave soldering.

3. The addition of components to existing circuitry, is similar to the addition of wire.

a. On boards that have already been soldered, all additional components should be added to the bottom of the board. This is more difficult because it requires tracing the correct joints on the solder side. However, the heat shock danger described above makes this the more reliable approach.

b. On new unsoldered assemblies, try to use the existing holes to hold the added component wire. If this is not feasible because of the tight fit, add the components after soldering.

When bending and fitting wires and components for addition to a circuit, strain relief practices must be used (see Section 2-10). When the length of the wire exceeds 2 in. (5cm), it must be anchored to prevent vibrations from stressing the solder joints. Hot glue or other adhesives are used.

REFERENCES

9-1. Howard H. Manko, *Solders and Soldering*, 2nd ed, McGraw-Hill Book Co., New York, 1979.

9-2. Howard H. Manko, "Must I Fill That Plated-Through Hole?" *NEP/CON Proceedings*, 1978. pp. 293-295.

9-3. Howard H. Manko, "Eliminate Poor Solderability: Don't Bury It Under Solder," *Insulation/Circuits*, February, 1976

9-4. J.D. Keller and J.L. Wyszczak, "The Case for Unfilled PTH's," *Electronic Packaging and Production*, October 1973, pp. 144-149.

9-5. Howard H. Manko, "Tune Up Your Hand Soldering," *Assembly Engineering*, July 1976.

9-6. Nicholas Rusignuolo, "Nondestructive Hand Soldering of Microelectronics," *SME Technical Paper EE77-117*, Dearborn MI., 1977.

10
PROCESS ECONOMY AND MANAGING
THE LINE

10-0 INTRODUCTION

The U.S. electronics industry has been under global competitive pressure for the last two decades. In the 1980s, however, it has lost considerable ground because of high labor costs, and unfavorable exchange rates. At the same time, the electronics industry has overtaken the automotive market in dollar terms. As a result, a major economic factor may be lost if this trend is not reversed.

The solutions to this problem are very complex and go beyond the realm of this book, but there are actions that can be taken by every branch of the industry. Lowering costs is one undisputed contribution action we must make if we want to maintain our market share. We can reduce costs greatly by the judicious application of the soldering and cleaning process. This reduction in cost does not require large investments. It is based on a rational management approach and the implementation of a manufacturing system.

In soldering and cleaning, this reduction in cost is associated with a dramatic improvement in reliability. In this book, we have separated the classic concept of *quality* from the more modern one of *reliability*. Reliable solder joints have been defined as connections that outlast the assembly under their unique *in-use* conditions. This approach is important because we can no longer afford to pay unnecessary production costs (see Section 8-0).

A combination of improved reliability and reduced costs is difficult to beat. One wonders why managements have not rushed to adopt these methods and systems at a time when they are trying desperately to cut costs. The answer is embarrassing: it is a combination of lack of respect for soldering and cleaning as a technology and plain ignorance. It is hoped that this chapter will help to destroy management apathy.

10-1 MANAGEMENT'S ROLE AND APATHY

Perhaps this chapter should have been the first in the book, since it is aimed at the people who can make a difference. For years the author has been seeking a way to reach the decision makers in a company. They are the people who seldom pick up a technical book or attend seminars. They relegate these professional tasks to engineers but hardly ever listen to their suggestions.

Part of this problem is due to the different "jargons" that these two groups use in their everyday discussions. Management is heavily involved in economic decisions, and their concern is with the bottom line. The engineering and manufacturing communities are normally charged with getting the product out the door, and they are interested in quantity. While managers are not expected to study technology, engineers are well advised to learn the principles of cost accounting and economics. A well-prepared proposal with clearly defined cost consequences, will always be honored by management. The purpose of this chapter, is to present the economic impact of soldering and cleaning processes. It is hoped that this will help you initiate worthwhile changes, regardless of whether you are a financial manager or an engineer.

This chapter and this book are not restricted to the manufacturing technocracy. For that matter, the reader does not need any special background. The material is based on economic studies performed in conjunction with specific companies. The data are dated and are applicable to a particular case history, but they are general enough to indicate the trends and to help you draw some conclusions. Obviously, each organization must do its own analysis under its own unique conditions. Pay scales, overhead ratios, productivity, degree of automation, and similar factors differ from case to case.

The lack of respect for soldering and cleaning stems from a long held misunderstanding. Both of these processes have definite technologies with their own set of rules and regulations. What is more, they have a profound effect on two very important factors—reliability and cost. It is hoped that management apathy can be overcome and progress is made.

One should never generalize and talk about management as a cohesive entity. Let us look at the specific types of corporate functions that must be convinced if we are to succeed. We can refer to those who are unfortunately, not yet involved in soldering and cleaning as the *missing VIPs*. There are three types of characters who require mention here.

Using some levity, let us call them:

- Mr. TightFist—representing financial management.
- Mr. FalseSave—representing purchasing.
- Mr. BlueSky—representing design and product planning.

You will note that the last group is not isolated from us like management, but actually is part of the engineering world. Let us consider these persons, one by one, and see what changes are needed:

1. Mr. TightFist is part of top echelon management, which sets fiscal priorities. To him, soldering and cleaning are such a mundane subject that allocating any funds for sophisticated equipment is low on the priority list. In an imaginary confrontation, one must convince him that there is a direct correlation between the following:

Cost of Solder joint   = product life expectancy.
                         = performance in service (humidity, temperature, vibrations, etc.).
                         = low-warranty cost and a good Reputation.

There is no top executive who does not want to gain such advantages for his organization. But most managers do not realize that this requires action on their part, and not always money. As we will see, most solder improvement programs pay for themselves and are profitable in the long run.

Section 10-2 gives a detailed economic analysis of the soldering process. For a more universal discussion, see Table 10-1, which shows the cost progression of a single solder joint as it moves through the various steps of production. These calculations are based on a medium sized production line making about 750 Ft^2 (6.4 m^2) of boards per day. The cost of routine touchup after the wave was based on 1984 figures, when activated rosin flux was still being used. At that time, the company had three touchup operators.

The cost progression in Table 10-1 is very illuminating. The cost of *doing it right the first time* is less then a penny per connection. The reliability of that connection is also of the highest order. Any additional touchup or repair increases the cost dramatically, while creating a reliability hazard. Figure 10-1 shows that each touchup operator on the line accounts for roughly one-sixth of the total cost.

The lesson to be learned from this trend is simple. An investment in the right material and production system is paid back in a very short time.

Table 10-1. Cost Progression of a Solder Joint.
(Jan. 1985)[a]

COST PER JOINT	JOINT TYPE	CALCULATION DETAILS
$ 0.0083	Wave and wash only	Includes loading and unloading of wave, direct supervision, and maintenance.
$ 0.0520	Routine touchup	After wave, without inspection[a] (No add-on of components).
$ 0.110	Inspected touchup	Repair only what inspector found.
$ 3.880	ATE-found repair	Correct solder-oriented problems after automatic test equipment identification.
$65.000	Field service	National average for service call, (No data available for "Hunt & Seek" fault).

[a] Costs include equipment, energy, materials, and Labor (no overhead). The calculations are based on 750 Ft²/day (6.4 m²/day) double-sided FR4 boards with one million plated-through holes, using Sn60, rosin flux, oil, and water wash.
[b] Based on 1984 figures with RA flux and three touchup operators.

Management's role here is to allocate the resources of talent and people to study the situation. If your touchup rate is higher than 1 joint in 5,000, you can make dramatic savings—and, above all, improve the quality of your product.

2. Mr. FalseSave is our well-meaning purchasing agent, who is looking after the company's interests. He is usually not aware of the added manufacturing cost of improper purchases. But here the blame is to be shared by engineering. As we indicated before (see Sections 2-1 and 8-5), an organization needs suitable incoming material specifications. These tell purchasing and the vendor exactly what is needed. Without this documentation, the purchasing department should seek the lowest-prices items and save the corporation as much money as possible.

Two points must be made to illustrate the most common problems in this area—the printed circuit board and the components. Let us discuss them separately.

a. The normal lead time for printed circuit fabrication runs in month. No large, high-quality fabricator would interrupt his schedule for rush orders that become routine. There are, however, many smaller shops where a fast turn-around is feasible. The quality of these less highly engineered companies is variable at best. As a result,many rush and priority orders have technical defects. Let us amplify this statement further by considering the cause of these troublesome rush orders.

Most production facilities have trouble scheduling the purchase of printed circuit boards months in advance. This is a situation that is aggravated by constant design changes. As a result, pressure is brought on the purchasing agent to get boards in a hurry. Therefore, quality and price must often be sacrificed to get fast service.

But sacrificing quality means that a great deal of extra money is spent by production. Poor yields, increased scrap, and lost time are just part of the story. In the end, the customer, too, does not get the product reliability he wants.

There are no easy solutions to the problem of printed circuit board buying, but top management must get involved. By sorting out the real rush orders from those that can be scheduled in advance, progress can be made. Only top management can help solve this kind of problem.

b. The solderability of components under aging conditions (see Sections 8-6 and 8-7) is another good illustration. Unless we specify exactly what we need, a good purchasing agent will buy the lowest-cost item, even if it has unsolderable leads. In addition, he will try to find a large list of suppliers so that he can always get devices quickly (to minimize the cost of inventory).

An unsolderable lead is, by definition, not useable in making a quality soldered product. Hopefully, our industry has passed the stage where we just hide our mistakes under a pile of solder (see Ref. 10-2). This refers to the practice of adding solder over an unsoldered lead so that inspectors cannot find the problem. The unsolderable devices thus need special handling (see Sections 2-5 to 2-7), or they will become points of expensive touchup without the benefit of reliability.

The role of management in this situation is also clear. Even without extensive cost studies, vendor rating programs can be easily initiated. The issue here is one of recognizing the problem, rather than implementing a solution.

Keeping a record of vendors' performance will result in a helpful profile. Known offenders can be eliminated, while good performers are retained. By implementing this approach, one manufacturer has been able to eliminate 87% of his touchup.

3. Mr. BlueSky is responsible for all new design and product planning. Unfortunately mundane items such as soldering and cleaning are farthest from his mind. They certainly do not need his attention, but decisions made at this design stage do influence production profoundly. For example, in Chapter 4 of Ref. 10-1, the flow chart for solder joint design recommends a *manufacture-ability* check. Let us use this procedure to illustrate how it works in an actual case history.

The designer needed a number of switches in his circuit, which would be user operated. In order to keep the product price down, he decided to use the lowest cost unit in the catalog. This happened to be an open (not sealed) switch which could not be wave soldered because the rosin fumes would foul the contacts. Nor could it be washed, because the cleaning solutions would remove all lubrication. This switch was therefore classified as a nonwettable device, which had to be hand soldered. Now let us see the impact of this switch on manufacturing cost, and reliability.

In order to solder the switch in place after wave soldering, the production department had to:

a. Place a temporary solder mask on the board to keep the holes open during wave soldering.
b. Remove the temporary mask by peeling it off or washing it away.
c. Insert the switches into the board and solder them by hand.

The total cost for this manual operation was $0.84 for each switch, which was more than twice the cost difference between the cheap switch and a sealed unit.

Now let us follow the finished product to the user. The open switch was also subject to contamination while in use, and eventually this led to "sticky" intermittent contacts. The reliability of the unit was compromised and could not be compared to that of a unit with a more economical sealed switch.

To summarize, the false saving gained by selecting a nonwettable switch resulted in higher soldering costs and a less reliable end product. Here, too, management's role is clear: new designs must be subjected to production engineering scrutiny. They should never be imposed on the manufacturing division without the right to make changes.

Awareness of these roles, can usually bring top executives into action. With their involvement, costs can be reduced, while reliability is increased.

10-2 AN ECONOMIC ANALYSIS OF SOLDERING AND CLEANING

There are many variables in each process and operation which affect an economic analysis. Therefore, the data presented in this section should be considered only for the trends they show. Specific calculations based on your own conditions may be quite different in magnitude but would show very similar trends.

In Table 10-1 dramatized the effect of economy of scale on the cost progression of a solder joint. The cost of the mass-produced wave solder joint is only 1/6 that of routine touchup ($0.0083 to 0.052). The routine touchup costs 1/75th of the price of repair after an automatic test ($0.052 to 3.88). This, in turn, is only 1/17th of the cost of a field repair ($3.88 to 65.00).

Let us look at another cost progression in soldering, that of the material and labor investment in the board. Good manufacturing principles dictate that failures must be identified at the lowest cost of labor and materials. Let us take an actual example, involving a batch of unsolderable components, and follow the cost progression:

- If found at the vendor's location, this would save the cost of shipping, returns, and customer service.
  (Scrap—none, because the units can probably be reworked on production equipment.)
- If detected during incoming, even when returns are not feasible, this would require only batch treatment (see Section 2-5).
  (Scrap—Very little, if any, provided proper restoration equipment is available.)
- If identified after assembly and before soldering, the entire unassembled board would have to be treated or the bad component disassembled and restored.
  (Scrap—medium to large, depending on the assembly process.)
- If found only after soldering, real repair would be expensive, since it requires desoldering the bad components, correcting the problem, and resoldering (Ref. 10-2).
  (Scrap—very high; may include fully populated boards which are declared unusable because of lifted pads, measling, delamination, etc.).
- If not detected by final inspection and allowed through without being corrected, may cause field failures and possible law suits.
  (Scrap value may include entire systems.)

It is obvious that early remedial action will result in the lowest cost. It is also linked to a definite increase in reliability and long term consumer confidence. We have also shown the potential increase in costs due to scrap, a factor that increases with the complexity of the assembly.

Let us now concentrate on some specific cost comparisons between soldering systems. This will include:

1. The cost makeup of a soldering and cleaning operation.
2. The impact of touchup.
3. The impact of fluxes and cleaning.

1. *The cost makeup of a soldering and cleaning operation.* The cost of an inline soldering and cleaning operation depends on what parameters are included. We will use the following factors, listed in order of increasing impact:

1. The cost of equipment over a five year tax write-off period. This formula includes the following items:
   a. The purchase price of the units.
   b. Installation and start-up costs.
   c. The cost of money.
2. The cost of energy to run the equipment. This can usually be estimated from the equipment specifications and adjusted later on the basis of actual consumption.
3. The expense of soldering and cleaning chemicals. This is based on the following items:
   a. The flux and thinner.
   b. The oil, if used.
   c. The dross blanket or dross reducer, if used.
   d. The cost of cleaning chemicals:
      I. Water (with treatment when used).
      II. Saponifier.
      III. Solvents, their recovery and disposal.
4. The price of the solder metal, including:
   a. The metal for the solder joints.
   b. The loss of dross, if any.
   c. The replacement of contaminated solder.
5. The cost of labor. This is the most difficult part, since it may include actual or artificial overhead factors. It may also include quality, maintenance, and indirect supervision. We have included only:
   a. Direct labor (loading, operating, and unloading).

b. Direct supervision.
c. Routine maintenance. We specifically excluded the cost of quality, and the cost of touchup and repair. These will be treated separately. in making the calculations, we used a medium-sized inline wave and water cleaner for rosin, making 750 ft² of boards per shift with 1 million plated-through holes. The number of operators needed for this line is 3.67, which is a practical figure based on industry norms.

The breakdown of costs is given in Table 10-2. The calculations were compared to actual figures from several diversified plants of similar size. They are based on an average wage for the industry, with a 35% overhead. The 35% is the minimum needed by law to strictly cover social security and workmen's compensation. Each company must modify these figures by adding its own overhead.

A review of these figures establishes some basic facts. The yearly equipment cost is minimal, and the purchase of the most automatic equipment makes economic sense. The table is based on an inline process, which saves the costs of an additional operator. Batch equipment is very costly in terms of labor and cannot be justified. A batch vapor degreaser, for example, costs more than an inline unit on an annual basis, even though the original investment is much smaller.

Table 10-2. Breakdown of Inline Soldering and Cleaning Costs. (In increasing order)[a]

ITEM NAME	PERCENT OF TOTAL	COMMENTS
Equipment	9	Over a five year tax write-off period (includes installation + cost of money)
Energy	11	For wave, cleaner, and hot water
Soldering chemicals	16	For rosin flux, oil, saponifier, and soft water
Solder	24	For 60% tin, 40% lead, no dross
Labor	40	Without inspection or touchup (no overhead)

[a] Calculations based on 750 Ft²/day (6.4 m²/day) double-sided FR4. boards with 1 million plated-through holes, using Sn60, rosin flux, oil, saponifier, and soft-water wash.

Savings on energy can be substantial in all stages of soldering and cleaning. The yearly cost of a polypropylene washer is dramatically lower than that of a well-insulated stainless steel unit. Using infrared preheat banks only when needed, lowering the temperature of the cleaning water as much as possible, running water only when work is in the chamber, and optimizing drying are some of the ways of reducing the energy costs.

Little can be done to cut the cost of the soldering chemicals and metal. Subsection 3, of this section gives some hints on material system changes. Obviously, the "cheapest" is not always the best. The true impact on labor costs must be weighed against the per gallon or per pound savings. If low-cost materials increase touchup rates by even 1%, they become too expensive to buy. Experience has shown the best materials are also the most economical. Manufacturing and engineering should make the decision on material quality on a *per unit produced* basis, not purchasing on a per unit material price.

Finally, we come to the cost of labor, which has a major impact on the process economy. Any steps taken to lower these costs have beneficial consequences and must be seriously considered. If we could fit the wave and cleaning modules into an inline conveyorized process, for example, we would save the loading and unloading costs. This would cut labor costs by 54.5% and our overall annual costs by 18.2%- an important saving. Even automatic loading and unloading equipment would pay for itself in a very short time. The dramatic impact of touchup costs on this operation will be discussed next.

Remember that U.S. labor costs are a major factor in our economic disadvantage. By lowering the labor content, we become much more competitive in international markets and against imports.

2. *The impact of touchup.* Unfortunately, many companies still have an unnecessary large number of touchup operators behind the wave. This situation is usually the result of poor process and system control, and, as shown in Chapter 8, does not translate into high reliability joints. The cost is prohibitive and worth detailed analysis.

Using the labor rates presented in Table 10-1, we can calculate the impact of each touchup operator on annual costs. Figure 10-1 shows a step cost analysis of the operation. Note that the annual costs are fixed, and are more than doubled by seven touchup operators.

Reducing or eliminating the touchup operations is the most effective way of reducing total costs. Remember that the touchup operator does not improve product reliability; (see Section 8-14). This should be the first order of business if there is more than one inspector/repair operator at the end of the line. Management should strive to eliminate any manual *add on* operation for the same reasons. We have already analyzed the cost of

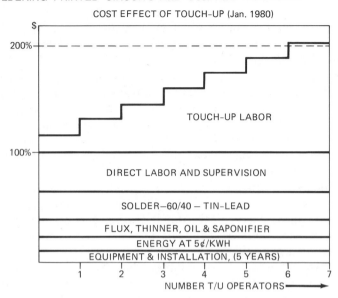

Fig. 10-1. The large burden that each touchup operator places on the cost of soldering and cleaning.

nonwettable components (see the discussion of designers in Section 10-1). Any manual routine operation after wave soldering is a tremendous cost burden.

3. *The impact of fluxes and cleaning.* A company is not always locked into the material system being used. Often the selection of a flux and a cleaning system is historical and can be improved in respect to quality and costs. Government contracts and some industrial agreements may not give the user this same flexibility.

Table 10-2 was based on a system using rosin flux, an oil, saponifiers, and soft water. The costs of this system were taken as 100% for comparison purposes in Table 10-3.

The calculations are based on efficient equipment with minimum solvent evaporation losses. The estimated costs of solvent reclamation is included. None of the water systems need effluent treatment. Note that the first line represents a poor practice that, unfortunately, is still in use. Chloroethane only does not remove any ionizables, thus providing poor but cheap cleaning.

While the switch from rosin flux to organic water washable flux reduces overall costs by 14%, the material savings are closer to 40%. In addition, all the calculations were based on hot water (ca. 150°F or 66°C); a shift to

Table 10-3. Cost Comparison of Flux and
Cleaning Systems.
(Figures include equipment, material, and energy)

| MATERIAL | | EQUIPMENT | ANNUAL |
FLUX	CLEANER	TYPE	COST (%)
Rosin	Chloroethane only[a]	Degreaser	73
Rosin	Chloroethane blend	Degreaser	80
Rosin	Freon blend	Degreaser	107
Rosin	Saponifier, soft W.	Washer	100
Rosin	Saponifier, D.I. W.	Washer	115
Organic	Rinse aid, D.I. W.	Washer	98
Organic	D.I. Water	Washer	96
Organic	Soft Water	Washer	81

[a] Does not remove ionizable contamination.
Note: The calculated 100% figure is based on 750 Ft$_2^2$/day (6.4 m^2/day),
double-sided FR4 boards with 1 million plated-through holes, using
Sn60, rosin flux, oil, saponifier, and soft water.

cooler or even cold water can also result in substantial savings. Cold water systems require special process and material modifications.

While the material presented in this section gives an idea of cost trends, you must do your own analysis. If you have no background in this area, seek the help of a good cost accountant in your company. Understanding the economic makeup of your operation is an excellent way of improving efficiency. This always leads to beneficial improvements in the solder system. Beware of making cost decisions without the facts; you are liable to reduce reliability and raise your total costs. Each step must be justified by an overall value analysis.

10-3 ESTABLISHING YOUR OWN TRAINING PROGRAM

Throughout the book, we have stated that you need your own training program. This is similar to the need for a unique workmanship standard for your company (see Section 8-2). The fact that good, complete solder training programs are not available needs careful analysis. Some commercial programs have been marketed. They are usually targeted to a specific industry; they are definitely *not* universal. Material suppliers, equipment manufacturers, and government agencies promote such programs, but with limited success.

The problem is best understood when we consider the basic need for either training or quality standards. Simply stated it is: *Make reliable and economical solder joints.* But what is reliable for one organization may be inadequate for another or too good for a third. Let us take a close look at the following requirements imposed on some solder joints:

- Vibrations of varying frequencies.
- Shock and stress.
- Temperature cycling.
- Humidity variations.
- External contamination.
- Corrosive environments.
- Handling, storage, and transportation.

As we review this list, one thing becomes obvious: not every assembly undergoes the same hazards. In other words, each case is unique and any attempt to generalize quality is dangerous, resulting in unjustified costs or lack of reliability.

The method described in this section is a time-proven procedure that can be used to set up your own program. Numerous benefits accrue from this program, including a reduction in overall costs, and an increase in reliability. In addition, it develops a more knowledgeable, less confused workforce. The steps are as follows:

Step 1. A management definition of the *cost/quality compromise.* As stated in Section 10-1, in many organizations part of the problem is top management's apathy. This usually results from lack of attention to, or respect for, the needs of satisfactory soldering.

To reverse this trend, top management must identify the factors related to soldering. Armed with this information, they can easily define the cost/quality compromise. The parameters involved are:

- Product market penetration.
- Market segment desired.
- Price strategy and competition.
- Life expectancy of equipment.
- Warranty cost and service philosophy.
- Special considerations (UL, OSHA, etc.).

Once these items are identified and analyzed, management will know the price associated with their desired quality. Now the solder engineer can proceed to design a meaningful program. A hidden advantage of this procedure is that top management comes to recognize that *soldering is definitely a technology.*

Step 2. Product review. Armed with the cost/quality compromise, we are in a position to develop quality specifications and translate them into workable standards and a training program. This is a multidisciplinary function which requires a committee made up of decision makers from the following fields;

- Electrical and circuit design.
- Materials and process technology.
- Manufacturing engineering.
- Quality assurance.
- The trainer.

The first task ia a total product review. A qualified expert or the trainer collects samples of all commonly used assemblies and printed circuit boards. They will serve to analyze the different types of solder connections and their use. This may seem to be a difficult and complex task. However, the types of joints in the average assembly can be reduced to common denominators, as shown in the following representative list:

- Round wire into round hole.
- Square peg into round hole.
- Stranded wire into round hole.
- Wire to cup terminal.
- Wire to tab terminal.
- Surface mounted but joint.
- Surface mounted lap joint.

A display board, often called a *story board*, is prepared, with samples of each joint attached to the display. Extreme examples of each configuration may require to be subdivided into several categories (e.g., a small transistor lead to a round hole vs. an oversized jumper wire to the same type of hole). The number of different joints even in complex equipment is not large. In setting up training programs for several major computer companies, the author has identified only 15 to 20 different connections in each case.

Next, the committee must evaluate each type of joint for its unique needs within the equipment. If clear design criteria were available, this would be simple. But information such as current-carrying capacity per joint, stresses anticipated versus solder strength, and thermal cycling data are usually missing or sketchy. The committee thus must do a miniature *design review*, comparing each connection to the six basic design requirements (see Section 1-4). To do this, the sample are evaluated on the basis of the following six solder joint criteria:

Material oriented parameters:

1. Electrical conductivity
2. Mechanical durability
3. Heat dissipation

Process oriented parameters:

4. Ease of manufacturing
5. Simplicity of repair
6. Visual inspectability

A difficult and time-consuming task for the first two or three solder joints, the evaluation becomes easier as the task force gains experience. While the material-oriented parameters are technologically easy to control, the process-oriented ones depend more on production experience, common sense, and specific plant conditions.

After evaluation of each configuration, an appropriate safety factor must be included. The information is then translated into the quality or training outline and is ready for implementation.

The following guidelines are then developed, showing:

- A scale for maximum, preferred, and minimum solder quantity.
- Geometry of the fillet (length, radius, diameter).
- Degree of perfection (solder rise, percent of fill).
- Specific requirements (insulation stripping, direction of the assembly).

A sketch, diagram, or photograph is helpful in setting these simple guidelines (Fig. 10-2). In addition, the trainer can use these guidelines to prepare samples of each configuration. Often the samples will show not only the preferred solder joints, but the extreme maximum, and minimum, and unacceptable cases as well. Once the samples are ready, the committee must review them carefully. If there is full agreement, each sample should have a tag attached to it, and everyone should sign it. This avoids conflict later.

Now we are ready to write the actual workmanship standards and quality control documents. In addition, the training program can be prepared, with the photographs and sketches turned into slides. A formal

Fig. 10-2. A good solder joint that shows excellent wetting (Copyright Manko Associates Inc.)

training syllabus is recommended, with specific questionnaires, hands on exercises, and a certification procedure (see Step 5 below).

Step 3. Process review to establish the state of the art. (runs concurrently with step 2). Many corporations do not review their soldering processes often enough to ensure that they are following the state of the art. Any solder improvement program with new training and workmanship standards must take a hard look at the materials, equipment, and processes used.

Consider for example, the profound effect of a hand soldering iron, the cage, the sponge and other soldering aids on the solder joint. Before training starts, specific tools need to be evaluated and compared with those that are now available. The same holds true for automatic soldering, since equipment and materials change continuously and new and better methods are constantly developed. We must place strong emphasis on training the operator of automatic equipment, as well as those who do final touchup and repair.

A *cleanliness philosophy* is another important aspect that needs close scrutiny (see Section 6-20). Cleanliness in many cases is a question of attitude, as is materials and process selection.

Personnel safely, Occupational Safety and Health Administration (OSHA) requirements, and Environmental Protection Agency (EPA) regulations must also be reviewed. A relevant training program cannot be successful without them.

Step 4. Preparing the training program. This can become a lonely task for the solder engineer aided by the trainer. They must now generates the actual training material, including the syllabus, slides, hands-on work stations, and so on. The building blocks for a typical successful program may consist of the following 10 items:

1. Motivational material to set management's tone for the program.
2. Awareness of the pertinent cost/quality compromise rationale.
3. The basic principles of the soldering process.
4. A review of the tools and materials required for the specific operation.
5. A soldering workshop with hands-on demonstration.
6. Procedures for associated disciplines (e.g., wire stripping, tinning, wiring, and cleaning).
7. Unique workmanship standards applicable to the company.
8. On-the-job working experience under supervision on actual assemblies.
9. Work practices such as safety and cleanliness.
10. A final quiz and intermediate question-and-answer sessions.

Similar programs should be developed for diversified soldering operations such as wave soldering, furnace and preform soldering, paste and reflow surface mounting, tinning, and the like. Additional training courses are also needed for touchup and repair operators, and inspectors.

Step 5. The training sessions. To be successful, the program must start from the top and permeate to the bottom of the organization. A training program for the operator alone does not guarantee success. Remember that soldering personnel at all levels are subjected to pressures from supervisors, engineers, quality inspectors, and management. Since the bench operators interest is stimulated to a great extent by personal economic incentives, they want to satisfy their supervisors. The trainer and the training program are not related to future raises and promotions. Therefore, training only the operators usually backfires. An operator who returns to the work station and tries to use new techniques may find a co-worker, supervisor, or inspector inquiring, "Why did you do it this way?". The reason was clear to the operator during training, but will it be so now?. In addition, there is a constant fear of antagonizing the wrong person.

To prevent this from happening, training should proceed from the top to the bottom of the company. Realistically, it should start with management, including all the department managers who will have a future impact on the soldering operation. This will also acquaint them with the result of the task force's effort. The next training session should involve the engineers, supervisors, quality inspectors, and other nonproduction personnel. Reinforcement of the training program will depend on these individuals, and their understanding and cooperation are essential. This meeting should focus on workmanship standards, quality requirements, and specific work practices, while omitting the hands-on portion.

Step 6. Operator training and certification. This phase should become an ongoing task. The success of a soldering program requires the use of only certified operators and inspectors for production, touchup, and repair. Excluding additional training subjects for new employees, like orientation, benefits, and component recognition, solder training takes three days with hands-on work. Some on-the-job training under the lead operator or trainer will also be needed before the students are left on their own.

Once certified, however, operators should also be retrained periodically. This retraining is intended to reinforce the original training program and review concepts missed in the first session. It also provides a good opportunity to introduce controlled changes in the process, materials, or techniques. Recertification also creates an attitude of respect for the task. Retraining schedules can be periodic or the supervisor's responsibility. Often trimmed down to 6 or 12 hours, retraining is not as extensive as the original program. With heavy emphasis placed on the hands-on sessions, the instructor can spot deviations from good practices.

Step 7. The annual review. Good management calls for periodically reconvening the task group which originated the program. This group is often also termed the "soldering committee." In the fast-moving electronics industry, changes occur continually and updating is always required. The training program, workmanship standards, materials, and processes must be kept abreast of the state of the art.

Part of this effort involves an orderly, controlled change procedure which prevents unauthorized modifications. With time and experience, new materials, equipment, or procedures may prove beneficial. However, an uncontrolled introduction of such items will undermine the entire program. The soldering committee is the only authority for making changes, since they have the methodology to evaluate the organization's needs. While a change should be easy to request, it must be evaluated before it is implemented. Thus the company gains the benefits of the best available materials, equipment, and procedures, but in a well-controlled, documented procedure.

The soldering committee also reviews new designs and products to cut down on the lead-time from prototype to full production. Finally, the committee provides backup for the solder engineer on trouble-shooting cases. There are many situations in which the program breaks down unexpectedly. Soldering problems often originate in diversified areas not directly related to the process and not apparent to the solder engineer.

Useful material to augment your own workmanship and quality standards is available. This is generated by private noncommercial organizations, and by the Institute for Interconnecting and Packaging Electronic Circuits (IPC), American National Standards Institute (ANSI), International Society for Hybrid Microelectronics (ISHM), and the Tin Research Institute.

Having discussed the steps involved in setting up a company program, let us now review the type of person who would make an ideal trainer. Usually the training department is staffed with a number of communications specialists. These do not make good soldering instructors because they lack the ability to demonstrate the hands-on portion of the program. The ideal trainer is a lead operator who has been involved in soldering on the line. If this person lacks communications skills, these can be acquired. Operators will absorb more from one of their peers and accept changes in their way of soldering if such new practices can be demonstrated skillfully.

10-4 FUTURE TRENDS

Like any industry in a free market, printed circuit manufacturing and assembly is subject to many outside pressures. There are two opposing trends that should be considered when contemplating the future: the cost of labor and the degree of automation.

Both parameters are economically driven and address the same problem. Their impact on any national economy is profound. One cannot, however, forget the marketplace, which demands reliability (quality) and not just price in the product.

The inherent high labor cost in the United States makes offshore facilities attractive. In general, labor rates are a function of the national economy and the stage of industrial development. There are many underdeveloped countries with vast pools of low-cost labor. If the political climate permits, they are open to foreign investment. Thus offshore facilities can greatly reduce final costs. Their use is one way to make a U.S. company very competitive in free world markets.

This picture is not entirely rosy, however, and is wrought with a series of pitfalls. The difficulties of doing business in underdeveloped countries, can add substantially to the unit price through unexpected costs. Let us review the major obstacles:

1. Long supply lines mean not just shipping costs but an increase in inventory. With the value of money at today's interest rates, this factor has a major impact.
2. Limited local resources make reactions to technical problems very slow and cumbersome. Often there are no nearby sources for technical materials, and the line may have to shut down until support shipments arrive. Airlifting of chemicals or components to prevent slipped schedules is a typical unforeseen cost.
3. Expensive management and engineering support from the home office or plant must be included in the calculations. While these personnel may be charmed by the opportunity to travel, the cost is prohibitive.
4. Loss of manufacturing know-how and control is an important consideration. By closing local facilities and moving manufacturing offshore, a company loses basic know-how. There are many little manufacturing details that cannot be easily relegated to paper. Once a product is no longer in production, these skills dissipate. Pulling a line back to home base involves an expensive relearning problem. The loss of know-how, also involve a loss of control over the product. The company will have to yield control to a foreign crew even if U.S. management is resident.
5. It is hard to maintain industrial security at these distances. Many U.S. companies have found cheap copies of their own designs competing with their products in the home markets. While Apple Computer, for example, was able to stop such imitations from selling in the American market, the copies are still available in the Far East.
6. Finally, we must consider the political dangers, which may shorten the terms of the investment. These factors are not predictable, although there have been no major surprises in the recent past.

To make a valid economical decision on a move to offshore facilities is not easy. A company cannot allow itself the luxury of letting personal emotions affect such decisions.

2. Automation is the force opposing any move into cheap labor markets. Automation involves lowering the labor content of a product. The development of computer-aided manufacturing equipment (see Section 4-19), and the use of robotic systems are technically feasible today. In most cases, they are economically beneficial and circumvent all of the problems listed above.

There are two stumbling blocks to the use of automation in the printed circuit industry:

1. The need for an initial capital investment.
2. The lack of middle and upper managers who understand the use of the computer.

1. U.S. companies, unfortunately, still look more closely at current bottom line profitability than at the long-range plan. Thus we are slow in converting to a technology that originally made the United States into an industrial giant.

Fortunately, automation can be justified even for the short term and smaller quantity production if the cost are analyzed correctly. Foreign price and quality pressures are driving the industry in this direction.

2. While the minicomputer has made sizable inroads into many office functions, it is still a stranger to most industrial applications. The new generations of engineers have basically replaced the slide rule with the computer keyboard. However, it will take some time before this young generation can rise high enough in the industrial hierarchy to make decisions. Meanwhile, the older generations feel threatened by this new tool and are trying to slow down this inevitable trend.

There is no doubt that automation and low-cost labor markets offer two different solutions to the same problem. In large scale production with an unavoidable high labor cost, offshore facilities may offer the better solution. Where full automation is feasible in large and medium sized production, it offers a simpler solution. Each case must be judged by it's own unique and inherent problems.

REFERENCES

10-1. Howard H. Manko, *Solders and Soldering*, 2nd Ed. McGraw-Hill Book Co., New York, 1979.
10-2. Howard H. Manko, "Eliminate Poor Solderability; Don't Bury It Under Solder," *Insulation/Circuits*, February 1976.

INDEX